Electronic Technology and Civil Procedure

IUS GENTIUM

COMPARATIVE PERSPECTIVES ON LAW AND JUSTICE

VOLUME 15

For further volumes:
http://www.springer.com/series/7888

Miklós Kengyel • Zoltán Nemessányi
Editors

Electronic Technology and Civil Procedure

New Paths to Justice from Around the World

Springer

Editors
Miklós Kengyel
Department of Civil Procedure
and Law Sociologie
University of Pécs
Pécs, Baranya, Hungary

Zoltán Nemessányi
Department of Civil Procedure
and Law Sociologie
University of Pécs
Pécs, Baranya, Hungary

ISBN 978-94-007-4071-6 ISBN 978-94-007-4072-3 (eBook)
DOI 10.1007/978-94-007-4072-3
Springer Dordrecht Heidelberg New York London

Library of Congress Control Number: 2012942329

Printed on acid-free paper

Springer is part of Springer Science+Business Media (www.springer.com)

This volume is based on the lectures of the 2010 IAPL conference

Contents

Introduction

Peter Gottwald

I

In the past decades, modern means of electronic communication, data processing and data transmission have had a considerable impact both on the business world and on private life. These technological developments now are increasingly affecting courts and civil proceedings. E-commerce inevitably entails e-justice.

1. The IAPL realised the consequences of electronic communication for court proceedings early on. In 1999, at the Vienna World Congress on procedural law *Helmut Rüßmann* and *Wouter de Vos* reported on the "Challenge of Information Society" to civil litigation. In Salvador/Bahia, Brazil 2007, *Janet Walker, Garry Watson, Emmanuel Jeuland* and *Angel Landoni Sosa* reported on new information technologies in civil procedure. These were excellent reports that have outlined all basic problems. Yet there is a need for more detailed discussion, not least because of the continuing technological development and the practical experience already gained in some countries.
2. For politicians, electronic justice relates primarily to the creation of information systems for the organisation of enormous amounts of legal questions and data. Such systems are considered a competitive advantage within the global economy.

   The German website www.justiz.de, for instance, provides online information services ranging from alternative dispute resolution to dates of compulsory auctions of real estate. It offers information on federal law and the law of the *Länder*, and access to the commercial register and the insolvency register. More than 1.4 million inquiries into the commercial register are handled online every day. Simply by using a personal computer, notary publics can browse the land register online and can obtain copies of entries with full probative value. Regarding the European Union, in March 2009 the European Council accepted an e-justice action plan for the period from 2009 to 2013. Part of this plan is the creation of a uniform "European e-Justice Portal" to simplify the use of information and communication technologies in the field of justice at European level. On 17th July 2010, the European portal's website was launched. The first version already contains more than 12,000 pages with information and links relating of the law and legal practice of EU Member States.

3. As an example from outside of Europe the civil branch of the New York State Supreme Court has introduced mandatory e-filing in 24 May 2010 for certain commercial cases. At the same time, very detailed state wide rules for filing were enacted.

   In Germany, the only widely used e-filing system to date is that for online applications for national payment orders. Lawyers have been required to use this electronic device since December 2008. In addition, briefs may be filed electronically with all Federal Supreme Courts.

   With regard to ordinary claims only Bremen, Brandenburg, Baden-Württemberg and Hesse have introduced an e-filing with all or at least some civil courts, all other German *Länder* are hesitant to introduce a e-filing system. They are not content to introduce merely a closed system between courts and lawyers by way of a so-called *Elektronisches Gericht und Verwaltungspostfach* ("electronic court and administration mailbox"). Instead, they want to prescribe the use of claim forms in order to drastically reduce the work load of court registries. The publicly available papers on the preparations do not openly state this, but the officials in charge speak of the introduction of "*xJustiz-Datensätze*" (xJustice data sets). These would be used to automatically fill in the first page of court files, the addresses of information letters to the opposing party and their lawyers and so on. But it is obvious that this could happen only if claims, defences and further replies were to be filed in a strictly formalized manner, at least on the front page.

4. On a global level, UNCITRAL is working to create an online dispute resolution system relating to electronic cross-border commercial transactions. It is designed for both business-to-business as well as for business-to-consumer transactions.

   All these developments of electronic communication will impact on classic civil proceedings and could be a challenge for the overall quality of justice.

II

This volume contains the submissions for the IAPL colloquium on "Electronic Justice – Present and Future", which took place during September 23rd–25th 2010 in Pécs, Hungary. For this publication, the submissions have been arranged differently and more systematically.

1. The first part of this book deals with general questions: does the increased or even general use of electronic communication have an impact on the traditional procedural principles? Is it, without more, reconcilable with the right to a fair trial? Will IT help the judge in finding a just decision or is there the danger that judges become just subsumtion machines by applying preformulated text components?

(a) While *Masanori Kawano* worries about drawbacks if the immediacy of a court proceeding with the personal presence of the participants were to be abandoned, *Paul Carrington* believes rather in the emergence of a new and better procedural culture; he suggests the use of electronic communication allows for a "cooperative and permanent dialogue". Carrington views the comprehensively electronically captured "truth" as a dream goal, though he does not address the limits posed by legitimate data protection measures. *Nikolaj Fischer* believes that comprehensive e-justice must primarily lead towards a modification of court organisation (presumably the concentration of court locations; dissolution of smaller courts). He criticizes excessive formal requirements for electronic briefs. Fischer, too, considers the oral proceeding with personally present participants old-fashioned and outdated, though on the other hand for him a public proceeding is absolutely required. Fischer speaks out against the duty of electronic submission of briefs and for the development of "social electronic justice". The oral proceeding, he says, ought to be intensified on the basis of electronically exchanged information.

(b) *J.E. de Resende Chaves Jr.* describes the vision of a future court proceeding "on the web". This process, he suggests, is no longer just a mere paperless image of the classic file proceeding; as a result of the integration into the global information networks it would have a higher quality than hitherto. The author develops a vision of a strongly inquisitive proceeding because the court seems to have access to the entire knowledge stored on the web and because the parties and the court are permanently connected by way of interactive memoranda which allow them to mutually influence the proceeding. These "virtual proceedings – online and on network", the author suggests, would lead to a "hyper-reality" and thus in his view to a higher quality of legal protection.

(c) Concluding the first part, *Viktória Harsági* examines whether the traditional procedural principles also retain their validity under the circumstances of electronic communication or whether they must by modified. In her view, the duty to submit briefs electronically violates the right to free access to justice because not every natural person has a computer and internet access. Therefore, she argues, free choice between electronic and traditional methods must remain available at least to natural persons without a lawyer. To Harsági, the possibility of video conferencing is merely a replacement of personal and direct contact so that witnesses should only exceptionally be questioned in this way. Harsági sees more room for this in regard to party hearings. In regard to online service of documents and electronic access to records, too, Harsági attempts to highlight practicable boundaries for the use of electronic communication technology.

2. The submissions in the second part explore more specific questions which arise in connection with electronic service of documents, electronic exchange of information and proof by electronic documents.

(a) *Fernando Gascón Inhausti* highlights that the submission of briefs, the transmission of documents and the communication of decisions ultimately lead to questions of procedural security. Every electronic transmission of documents requires a reliable, forgery-proof system of transmission with which delivery and time of delivery can be proven. Further problems follow from the legal requirement of electronic transmission and from cross-border transmission. Gascón presents exhaustively the current state of admissibility of electronic transmission of data in the civil process for most European states, the USA and for the international courts of justice. He then concerns himself with the technical requirements for an electronic communication system in the judiciary. At least regarding businesspeople and lawyers, Gascón has no concerns regarding a duty of the defendant to maintain an electronic address already for the service of the claim itself. In regard to cross-border proceedings, too, he advocates the general introduction and promotion of electronic transmission. This, he argues, improves legal protection rather than deteriorating it. Where however mistakes occur in the electronic transmission system, the option of resorting to the traditional ways of transmitting real documents must be made available to the parties.

(b) Whether and how far electronic information technology can simplify the compulsory execution of titles for payment is examined by *Michele Lupoi*. With the help of electronic delivery of titles, he believes the procedure can be sped up. Internet auctions would make the realisation more transparent, would increase the proceeds of the realisation and would prevent corruption. Cross-border distraints would be possible, too. Finally, he argues, the creditor (as proposed in the Commission's Green Paper) could more easily obtain information on the debtor's financial situation electronically.

(c) *Robert Kulski* presents the electronic procedure for the acquisition of a polish order for payment.

(d) This first part is concluded by *Pantilimon's* thoughts on the enforcement of claims with the help of new information technology in the European Union. The author refers to the current state of the European e-Justice Portal and reports on which systems of communication by now have been installed in Romania, on case management modules, on e-filing and on a insolvency bulletin.

3. The final part of the reports is dedicated to the taking of evidence with the help of electronic documents and other applications of information technology in the proceedings.

(a) In a first report, *Helmut Rüßmann* explains the requirements of security and authenticity of electronic documents under German law. He explains the ongoing efforts to introduce electronic files and make them manageable. His report ends with a review of electronic documents as pieces of evidence (evidentiary value; prima facie evidence when using qualified electronic signatures and duty to submit electronic documents).

In a second survey, *Georg Kodek* explores questions of the electronic demonstration of proof and the possibilities of evaluating such evidence.

(b) Both papers are complemented by a submission on e-discovery by *Zoltán Ambrus*. First, Ambrus points out that today a major share of the information within a company is stored only electronically and, of course, must be made available for purposes of evidence. While observing data protection requirements, a forgery-proof copy must be produced and made available to the participants. In detail, Ambrus explains how in practice the specific data sought after can be filtered from large amounts of data.
(c) While Rüßmann assumes the highest standard of the qualified electronic signature in regard to German law, *Maniotis* explains the EU directives on the probative value of electronic documents and focuses on the probative value of an "advanced electronic signature" and the position of the "trusted third parties".
(d) An entirely different area of application is explored by *Maria José Cabezudo*. He concerns himself with the use of DNA data by Interpol for the identification of offenders and for combating serious national and international crimes. Specifically, he presents the standards that must be met for legal data to be obtained and which security requirements must be observed for data protection and the personality rights of affected and unaffected persons to be respected. He sees particular danger in the increasing involvement of relatives ("family researches") and in the varying data protection requirements under EU law and the respective national law. For the improvement of data protection as well as for pragmatic reasons, Cabeduzo advocates a harmonization of these rules on the EU level.
(e) Finally, *Baosheng Zhang* and *Huangxun Chen* report on the application of electronic evidence in Chinese proceedings. Specifically, they deal with the fact that digitally stored data is not as such examined by the court or an expert, but that copies are made in the process of transmission. The authors deal in depth with the dangers of forgeries and explain the existing guidelines for the prevention of such forgeries. Further, the authors explain how the traditional rules of evidence, such as the best evidence rule, are to be applied to computer stored data and how cross examination is to be conducted in cases of digital evidence. Finally, the authors deal with how digital evidence can be suitably evaluated (ratification rules). Regarding China, the authors hope for uniform legal regulations on the digital taking of evidence, which would replace the present uncoordinated coexistence of laws, judicial interpretations, international rules and other standards. In conclusion, the authors emphasise their view that in relation to digital evidence, too, the traditional rules of fair procedure have to be observed both in civil and in criminal matters. This means that strict technical standards must be followed and that at the end the judge retains the authority of free evaluation of evidence.

III

At the end of this introduction my thanks and appreciation go to all contributors for their efforts. Their papers reflect the pros and cons of e-justice as it stands today, from the welcoming vision of a future virtual trial to the fear of a brave new world. Seeing that insurance companies, social insurance, tax administrations, banks and investment companies are primarily using electronic processes and files, I am optimistic that court administrations and practising lawyers will overcome the problems of e-justice. Very soon, we shall use e-filing, e-service, e-discovery, e-proof taking, e-attachment of bank accounts and so on. The legal profession will profit from all these technical improvements.

Our good friend *Miklós Kengyel* invited us to come to Pécs where we could meet within a charming *Art Nouveau* auditory. We are much indebted to him and his staff for leading us through the conference so perfectly, and to the University of Pécs for its generous financial support.

We also thank our colleague *James Maxeiner* for having spontaneously agreed to publish the conference papers within his publication series.

Intervention of Honorary President Marcel Storme at the Colloquium on E-Justice in Pécs (September 2010)

The report of our good colleague and eminent Brazilian proceduralist Petronio Calmon was a brilliant and, for an Association as ours, masterly provocation. We need indeed firework. But I do not agree. How can we accept that "the soul of the process will be transformed with the adoption of the electronic process".

I am not a laudator temporis acti, but let me remind you the final report of the late French professor André Tunc in 1978 at the Florence colloquium, closing the evaluation of the magnum opus of our great master Mauro Cappelletti: the title was a quest for Justice.

And exactly 30 years later the new president of the supreme Court in Holland, Mr. Geert Corstens, declared: "The only task of the judge is the search of Justice."

Do we change with E-Justice the basic principles of our civil procedural law? (cfr. Art. 6 E.C.H.R): The independence and impartiality of the judge; the fair trial; the Waffengleichheit der Parteien;...

But I can, my dear Petronio, follow one of your conclusions: "The electronic procedure places the litigants in a position of collaboration and change the formal adversary system into a cooperative and permanent dialogue".

So doing you join Rosenberg who wrote already in the 1950s: "So liegt das Wesen des modernen Zivilprozesses in einer Arbeitsgemeinschaft von Richtern und Parteien, die zusammen dafür zu sorgen haben, dass dem Richter die sichere Findung der Wahrheit ermöglicht und in einem lebendigen Verfahren der Rechtsfriede unter den streitenden Parteien wiederhergestellt und damit der Frieden der Allgemeinheit gesichert werde" (Leo Rosenberg, Lehrbuch des Deutschen Zivilprozessrechts, 1949, p. 6).

We do not have to look for new codes of procedural law. In a cybersociety we have to draft a new code for the actors of Justice (Attorneys and Judges): a code of conduct.

If we can, thanks to the techniques of E-Justice, solve the everlasting tragedy, qualified by Sir Jack Jacob as a three-headed hydra (costs, delays, frustration), then we will enter into a new procedural Era.

Pécs, 24 September 2010
Marcel Storme
Honorary President of IAPL

Part I
Adapting Systems of Civil Procedure to Electronic Technology

Chapter 1
Electronic Technology and Civil Procedure – Applicability of Electronic Technology in the Course of Civil Procedure*

Masanori Kawano

1.1 Introduction

Over nearly the last two decades, the rapid development of electronic technology (ET) for communication, and its prevalence in our contemporary society has been startling[1]; nowadays highly developed, sophisticated and compactly constructed electronic communication equipment is regarded as a necessity of our daily life. ET has dramatically altered our methods of daily communication, both in private and in public. Such remarkable changes caused by the rapid development of ET for communication have also had an impact upon the civil procedures of many countries. Because of their extreme convenience and effectiveness, the application of newly developed ET for communication to the civil justice process is now regarded as an inevitable and appropriate improvement and already now in many countries its introduction into civil procedure is accelerating, with future developments planned. The legal implications of applying ET for communication to the civil process have been discussed on many occasions and now these plans are being realized, by successive amendments of the law of civil procedure

*This is a general report for the International Conference of Procedural Law in Pécs, Hungary, on September 23–24, 2010.

[1] Since the beginning of the 1990s, there has been innovation and development regarding the personal computer (PC) and related software. Because of their reasonable price and usefulness for the most people, the PC has become popular all over the world.

M. Kawano (✉)
Nagoya University Graduate School of Law, Nagoya, Japan
e-mail: mkawano@fukuoka-u.ac.jp

M. Kengyel and Z. Nemessányi (eds.), *Electronic Technology and Civil Procedure*,
Ius Gentium: Comparative Perspectives on Law and Justice 15,
DOI 10.1007/978-94-007-4072-3_1, © Springer Science+Business Media B.V. 2012

in many countries.[2] Such tendencies have been extended not only in domestic, but also in international civil procedure, especially in the member states of the European Union.

Civil procedure is a social system which is composed of a collection of mutual communications exchanged between the court, on one hand, and the parties, other interested parties to the litigation and their representatives[3] on the other hand. Such activities should be authorized and regulated by procedural provisions based on fundamental principles or requirements of civil procedure, either explicitly or implicitly. Such principles have been developed by long tradition and are based on the fundamental importance of keeping and guaranteeing various procedural and constitutional rights of citizens. Well-established communication must also be an indispensable part of civil procedure, by which mutual understanding of the pending case between the court and litigants can be promoted in the course of proceedings. Such communication should be effective and convenient. Its improvement might be one of the most important steps to bringing about effective justice.[4]

The newly developed ETs for communication have such a radical and profound influence on our daily life that their application to civil procedure might bring about some significant changes or influences upon the fundamentals of our traditional civil process. Any significant changes brought to the fundamentals of civil procedure by the introduction of a new system of procedural communication would have to be examined with reference both to principle and to practice.

Civil procedure is a social institution to obtain a final adjudication by the judgment of a national court to resolve private disputes by a series of mutual activities on the part of the court, both parties and any other persons affected. The court's judgments are preceded by the course of proceedings which is of fundamental importance for the legitimacy of the final decision by the court which could authorize the coercive resolution of private disputes. Because of the nature of civil procedure

---

[2] To give only a few examples: in the *United States of America*, in 2001 the Federal Rules of Civil Procedure were amended and some provisions relating to electronic service and other matters have been introduced; in *Germany,* in 2001 by "Gesetz zur Anpassung der Formvorschriften des Privatrechts und anderer Vorschriften an den modernen Rechtsgeschäftsverkehr vom 13. 7. 2001, BGBl. I 1542", some provisions in the code of civil procedure were amended and the application of ET between parties was introduced; in *England,* in 2003 the Pilot Scheme for Communication and Filing of Documents by email (PD5B) was introduced and in 2010 the 51st Update of the Civil Procedure Rules introduced PD5C Electronic Working Scheme, which allows for the electronic submission of claims and subsequent steps in the following jurisdictions: Admiralty, Commercial and London Mercantile Courts; in *Japan,* in 2004 general provisions relating to the filing by electronic system was introduced, but even before that amendment by the new legislation of Code of Civil Procedure in 1996, some applications of new electronic technologies, *facsimile*, *TV conference system* and *trio phone*, had been introduced into the civil procedure and are used generally in practice.

[3] Henkel, Die mündliche Verhandlung im Zivilprozess aus kommunikations-psychologischer Sicht, in *ZZP*. Vol. 110, 1997, 91.

[4] The English Civil Procedure Rules of 1998 in Part 1 provides as one of the overriding objectives the court's duty to manage cases; active case management includes 'making use of technology', CPR 1, 1.4.(2)(b).

it is necessary that the proceedings and related activities of the court and parties should be regulated, based on some fundamental principles of civil procedure, in turn based on the prevailing social values in our society; disregarding such fundamental values on grounds of expediency could give rise to violations of procedural rights of citizens.[5] It would cause new disputes and undermine reliable fair justice in our society. It is therefore necessary to investigate and to consider carefully some special problems relating to the applicability of ET for communication to the civil justice process.

In civil litigation, a pending case will proceed through different stages with their own procedural purposes and requirements; different fundamental principles should be considered for each stage of proceedings. Civil procedure is commonly commenced by a claimant filing a case with the court, and ends with the court giving final judgment. But the procedural structure is not identical in every jurisdiction; the detailed technicalities of litigation differ from country to country due to their long-standing legal traditions and practices. In investigating the applicability of ET for communication to civil procedure, we have to consider some different procedural structures and some sensitive reactions in each jurisdiction, as well as different stages of procedure.

In our following investigation into the applicability of newly developed ET in civil litigation, it seems to be indispensable to take into account such differences in procedural structure and procedural values prevailing in each legal system. Our main purpose in the following discussion is to investigate and to analyse some newly arising problems brought about by the rapid development and increasing use of ET, and its impact on the further possible application to civil litigation. In this discussion our investigation should be mainly focused on ordinal civil procedure,[6] and will be traced according to its proceedings from beginning to its end. In addition to this, however, we have to briefly mention its applicability to special simplified procedures. Indeed, in such simplified procedures, the greater use of new ET for communication may and can be encouraged more extensively.

1.2 Functions of Newly Developed ET and Civil Procedure

1.2.1 General Remarks

For our following discussion on the applicability of newly developed ET to civil procedure, it seems to be necessary first of all to clarify the scope of the term 'newly developed ET for communications'. As a matter of fact, there are various kinds of

[5] Harsági, "The Impact of New Information Technology on Civil Procedure," in *Civil Justice Quarterly* 29 (2010): 250–51, pointed out the importance of the consideration of procedural principles.

[6] As to the distinction of ordinal civil procedure, there are some different arguments whether some particular procedure, ex money payment, belongs to the ordinal or summary procedure.

relatively recent developments in ET by which our communication has been greatly modernised, increasing convenience. As the main part of civil procedure is composed of complex mutual actions or communications, formally or informally regulated, between the court and parties and between the parties themselves, one must question whether the introduction of ET into the civil process is necessary. The introduction of newly developed ET for communication acts to reduce some difficulties inherent in traditional civil procedure; in our traditional system of civil procedure there were some difficulties in respect of swiftness or promptness in mutual communications between the court and the litigants; especially if litigating parties live considerable distances from the court, it must be a troublesome and costly duty to attend hearings. Notably, in our business world ET for communication is now a common and prevalent method to, in effect, reduce geographical differences.

So it must be necessary and natural to consider the applicability of some kinds of newly developed communications systems into the civil process to reduce inherent difficulties based on the geographical distance between litigating parties.

Recollecting historical development of civil procedure, there had been an attempt to resolve difficulties in procedural communication by introducing newly-developed communications systems in the society at the time; one of the good examples was introducing the use of the *postal system* into civil procedure. Before the prevalence of a national *postal system*, civil actions could be brought to the appropriate court by the plaintiff or his/her representative in person; for the service of process on the defendant it was necessary that some special judicial authority, such as a bailiff, or *huissiers de justice* in France, be established and available for effectuating civil procedure. The French bailiff had a long history throughout the ancient regime[7] and had competence to serve some relevant procedural documents on the defendant, not through the court,[8] or as an official of the court, but as an independent official engaged by a direct application on the part of litigants. This French system of bailiffs was adapted to the civil process in neighbouring countries but with some different nomenclature[9]; in some systems it was assigned to a court official in charge of service of process. In some other judicial systems the same business has been engaged, other than by private person, by officials such as marshals in the USA.[10] After the *postal system* had prevailed and was commonly used as a daily system of communication and exchanging documents for the purposes of business, it was introduced into civil procedure as a primary or secondary method for avoiding difficulties of service of process or other procedural

[7] See Perrot, *Institutions judiciaries*, 8th ed., No. 458; Schmidt, *Civil Justice in France* (2010): 91.

[8] Art. 651 French Code of Civil Procedure; a notice, *notification*, made by bailiff named as "*signification*", Sub.2.

[9] German Code of Civil Procedure of 1877 adopted French system of bailiff, Gerichtsvollzieher, as an official of service of process, see, Art. 152: "Service of process is performed by a bailiff, *Gerichtsvollzieher*".

[10] Federal Rules of Civil Procedure.

documents,[11] especially for effecting the service of process on a defendant living far from the court. In most systems of civil procedure used today, the postal system is a major method of serving process.

Subsequently the telegram was invented as a rapid system of communication. This system however had some technical difficulties, and was only a simplified form of communication system; it could not transmit detailed information or documents and did not prevail as a common method of exchanging information and documents in daily life. Even so, it was possible to exchange some simple messages in some exceptional cases in civil litigation and there were some discussions relating to its proper use.[12] As a matter of fact, however, due to its technical inconvenience, the telegram did not become a popular communication system in civil procedure.

Until the recent radical changes brought by ET, technical innovations in communications technology did not require a rethink of the fundamentals of civil procedure. Indeed, recent development of ET is described as an "ET Revolution" and had a great success in changing our daily communication; we therefore cannot ignore it with respect to its applicability to civil procedure.

1.2.2 What Are the Newly Developed ETs and Their Functions?

The recent rapid development of the ET for communication system shows various innovations with different functions and utilities. Its development was so radical and diversified that different applications to civil procedure would depend on the aspect of the procedure in question. In order to discuss their applicability to civil procedure, therefore, their particular technical potential or advantages should be specified and focused with regards to each particular stage of civil procedure. We must first of all identify a particular ET instrument, with its capabilities in civil litigation. Regarding its applicability to civil procedure its social function and usefulness as a common communication system in a society must not be ignored.

Some newly-developed ET encouraged our daily mutual communication not only by its highly sophisticated technology, but also by its mass availability due to its low cost.

---

[11] In Germany the unified code of civil procedure of 1877 introduced *postal* system as a supplementary method to service of procedure by bailiff, Art. 176. German law makers considered its introduction to civil procedure on the basis of its promptness, certainty and fairness: see, Hahn, Mugdan and Stegemann, *Die gesammten Materialien zu den Reichsjustizgesetzen*, Bd. 2., Materien zur Zivilprozesordnung, Abt. 1, 1881, 221.; In the United States a step was taken in Ohio in 1917 by the statute, see, Millar, *Civil Procedure of the Trial Court in Historical Perspective* (New York: The Law Center of New York University for the National Conference of Judicial Councils, 1952), 88.

[12] In Germany there have been cases relating to the validity of procedural requirements of formal activities, especially the functions of signatures on procedural documents by telegram, see Vollkommer, *Formstrenge und prozessuale Billigkeit* (1973).

A remarkable example of recent and epoch making technical innovation in ET was the *email system*. Email has made startling progress in the last two decades and is now surprising both in terms of its quality, and in terms of being well equipped in respect of its capacity to exchange information. As it is available on compact equipment while maintaining high quality, and with the reasonable availability of PCs, email can be used widely amongst ordinary people. It is used nowadays in our daily life for exchanging mutual communication and sending detailed documents officially and privately and has been accepted rapidly all over the world. Such a rapid development of ET has been recognised ever since in our civil procedure and it has been admitted nowadays as reliable supporting equipment in court administration and also in civil proceedings; it has already been introduced or proposed for in-house communication and judicial administrative tasks in the court; it has been applied also in some respects in our civil procedure in many jurisdictions. One of the current issues as to the application of email in civil procedure is not whether it can be used in civil procedure at all, but its procedural functions in earlier phases of civil procedure. Indeed, civil procedure is not a mere matter of the mutual communication between the court and litigants or their representatives, but is composed of some procedural principles on which our modern civil procedure can perform its social function for resolving private disputes. If ET will be applied to civil procedure, as an indispensable tool, our considerations must be focused on its relationship with the fundamental principles of civil procedure. In order to consider such relationships with ET and procedural principles, it must be necessary to identify particular forms of ET and their function and the procedural phases of civil procedure in which they will be applied.

Email must be regarded as a main part of newly developed ET for communication. *Email* systems play a significant role; the innovation of email has been estimated sometimes as an "ET revolution". Indeed, it is true that email systems have many advantages over other systems of communication. In addition to technical advantages, due to its reasonable cost performance, it is now prevalent in the daily life of ordinary people in much of the world. Email should be a main focus for the following discussion on the applicability of newly developed ET to the modern civil procedure.

The *facsimile* system was developed as a compact device for exchanging copies of documents before the widespread use of email systems in the world. This system made it possible to exchange the copy of *original* documents by telephone line, without any transforming them by specially equipped electronic technology; in particular it has a clear advantage for exchanging documents written by the complex linguistic methods composed of countless different kinds of characters, like Chinese characters or Japanese *kanji*. In such linguistic cultures most documents have been written generally by hand, due to the inherent technical difficulties in such linguistic systems with mechanised typing,[13] that is until the recent development of software

[13] In Japan the typewriter for Japanese was developed. But it was very complex and difficult to use; only those specially trained can use the machine. In the court practice the judgments were written for many years by Japanese type writer. But due to such difficulty they were written only by trained specialists, and took a long time to write.

able to render these characters. Because of the technical advantages of *facsimile* for exchanging documents, it was used popularly for the first time in Japan. *Facsimile* has been admitted and partly introduced in the newly-established Japanese procedural system for exchanging some particular procedural documents.[14]

Video-conference or link system is an ET by which between people located in far different places can see the picture of another person and talk together for conference as if face to face. By using this equipment people do not need to travel long distances. A conference, otherwise very difficult due to travel time and cost, could be rendered possible by this equipment. It is possible to protect a person from seeing another person, of whom he is afraid, in open court hearings. Virtual courts, or the virtualization of justice could be realized by the introduction of the video conference system, and its possibilities have been discussed.[15] By the introduction of this system, however, one of the main principles of traditional civil procedure, hearings in open court immediately before a judge, could be more or less restricted.

In addition to such mentioned typical ETs, there are some other ETs, for example, the *trio phone system*, a telephone system by which between three or more places can be linked and can converse simultaneously. In this system normal telephone lines can be used, without financial expense. By this simple system for mutual communication between the court and both litigants simultaneously, some fears or doubts of unfairness possibly caused by using a normal telephone would be reduced.

Electronic data processing systems can be also appropriate for disposing with huge numbers of cases by telecommunication lines in short times without becoming burdensome; in our mass consumer market society sometimes large numbers of cases for collecting money claims may be petitioned for disposal by the court. In our business world most business activities and their records may be operated and disposed by a computer through data processing systems; it would be extremely convenient and effective if petitioners could add their petition to the proceedings by telecommunication technology.

1.2.3 Different Structure of Civil Procedure and ET

For our following discussions on the applicability of newly developed ET for communication to the civil procedure, we must clarify some different structures and particular civil proceedings in each country. From the comparative aspects of civil procedure, procedural structures in respective countries are not unified, but rather differently formulated. Surely such differences in procedural structure could be a cause of different attitudes as to the applicability of ET to the civil procedure in each jurisdiction.

---

[14] Amended Japanese civil procedure of 1996 introduced *facsimile* as a regular method of communication between representatives themselves to exchange their preparatory documents for plenary hearings, see Sect. 1.4, below.

[15] See Sect. 1.6.3, below.

One of the differences in formulations of civil procedure relates to the position of the court in the course of proceedings. On the one hand, there are procedural systems in which the courts are given a positive and extensive managing power for pending cases in the whole course of procedure from beginning to end; in such types of procedure the court begins management from the of filing a case with the appropriate court.[16] In this system a filed case will be managed by the court and proved mainly in open hearings in the court room. The main duty of civil procedure will be to maintain procedural fairness mainly in the formal proceedings instructed by the court; in this type of civil procedure the main communication in civil procedure will be focused on the relationships between the court and litigating parties or their representatives. Indeed, plenary hearings before the court will be the main event and fair opportunity in the civil procedure; but the proceedings before that point are controlled mainly by the court and communications between the court and the litigants must themselves be conducted fairly.

At the other end of the scale, there is an extremely different type of procedure in which the court dealing with pending cases has only restrictive powers to manage cases, and only during the restrictive phases of procedure. In this type of civil system, cases will be arranged and prepared mainly by the autonomous cooperation of representatives for the respective parties before proving the formal trial in the open court room.[17] Such a fundamental procedural construction has an impact on other procedures, without trials. In this type of civil procedure, the main part of civil procedure will be the last stage of trial or plenary hearings, and the pre-trial phases for preparations of main hearings could be regarded as informal private matters and it is not necessary to regulate formally their mutual communication.

Of course, such categorisation of civil procedure should be regarded as extremely exaggerated, suitable only for noticing some typical differentiation between procedural systems, as may be relevant to the issue of procedural communication.

1.2.4 Relevant Stages and Fundamentals of Civil Proceedings

We now turn to the course of civil procedure. The regular course of proceedings through which a pending case will be proved and decided is not harmonised between all jurisdictions. But, generally speaking, the normal cause of the most

---

[16] Recently there are some common tendencies to give the court case management powers, especially in common law countries, see Andrews, *The Modern Civil Process* (Tübingen: Mohr Siebeck, 2008), **2**.17, 3.13–3.22.

[17] American civil procedure is based fundamentally on this type due to the jury system; according to the American Federal Constitution Amendment VII, 1791, jury trial is a fundamental procedural right of citizens in common law cases. In such a system, cases must be tried at the final stage of proceedings before jury on the concentrated factual issues. For the trial, the case should be well prepared by discovery and recently by the pretrial management by the court, Louisell, Hazard and Leubsdorf, *Civil Procedure*, 5th ed. (2001), 283.

civil procedure, from the point of view of mutual communication between the court and litigants or their representatives, could be described as following different stages in which some different considerations or fundamental principles should be taken into account.[18]

The first stage: filing the claim with the court and the service of process on the defendant; this is the first contact of a plaintiff with the court and with the opposing party. This action is composed of a voluntary activity to make an application to the court, to require an adjudication of the case by the court. In most jurisdictions the filing of a claim is not regarded as mere information or some transition of the document to the court, but as a step with some procedural effects. The filed claim should be served on the defendant. As civil litigation is constructed as a dispute between two opposing parties, the claim once filed (or in some systems, before being filed), should be served on the opposing party, who should be given an opportunity to respond. Service of process is an indispensable part of civil procedure with some concrete effects: service of process should be performed even in the case of refusal of the defendant to receive it. The result of service should be reported to the court by a receipt and should be recorded for avoiding any possible doubt on the validity of service. There are many opportunities in civil procedure to exchange information, but the service of process has a significant role; other methods to transfer procedural documents might be differently regulated.

The second stage of civil procedure could be a preparation for the following plenary hearings as the main event of the civil procedure. In most systems, pending cases will be prepared previously by some special procedure for identifying the real and current issues for deciding the case. As the preparatory stage, there are, however, some different attitudes to the proceedings in each jurisdiction; whether requiring a separate or integrated procedure, oral or documentary proceedings. In these proceedings cooperation between the court and both litigants and their representatives is essential. To encourage effectiveness and convenience of civil procedure, the impact of the new ET for communication could be considerable; there could be laid down some different principles from those of the plenary hearings.

The third stage relates to access to evidence. Some cardinal factual issues, identified in the preparatory proceedings, should be proved and decided on the basis of evidence. To decide a disputed case, the court has available fundamentally only the evidence presented before the court by both litigants: the principle of party presentation.[19] In these proceedings, the activities to present documentary evidence could be made by using ET; and there are some regulations relating to such evidence presented to the court.

The fourth stage is the plenary hearing which will be heard, generally, at a court open to the public. In most jurisdictions, plenary hearings are formal with ceremonial elements, in a dignified atmosphere. Nevertheless, there remain some

[18] ALI/UNIDROIT, Principles of Transnational Civil Procedure, 2006 points out three main phases of the proceedings, 9.1: the pleading phase, the interim phase and the final phase, see *op. cit.*, p. 28.

[19] In Germany: *Verhandlungsmaxime*.

differences between plenary hearings in different systems: whether the trial is by jury or judge, possibility of documentary proceedings or oral hearings of witnesses and other evidence etc. For proving significant evidence, there are some fundamental principles; principle of *orality, principle of public hearings, principle of immediacy* of the judges.

Finally the court renders its final judgment on the case. The judgment should be delivered, sometimes by way of service of documents on each litigant; there is a possibility to appeal to the appellate courts against the judgment. In this stage there are some possibilities of the communication between the court which rendered the judgment or appellate court and the litigants; serving the final judgment and resorting an appeal. Service of judgment is a formal and usual way to deliver a judgment, but it might be allocated differently to the service of process in respect of the applicability of ET.

In addition to the above mentioned ordinal civil procedure, there are some special proceedings, in which the applicability of ET for convenient communication and for disposing of large-scale multi-party cases mechanically has been developed.

1.3 First Stage – Commencement of a Civil Claim

1.3.1 An Action of a Party to Commence a Civil Claim

Normal civil procedure will be commenced by an act of the claimant requesting a particular judgment by the court; by filing a complaint with the proper court[20]; or by issue of a claim form by the court[21] at the request of a litigant. In most jurisdictions such activity should be performed by filing a documentary, formulated complaint or application according to the regulation of the appropriate court where the case will be commenced, proved and decided by the judgment of the court. In some jurisdictions, for example in France, however, a procedure will be commenced by bringing an originating application by one of two possible ways: one is by writ of summons (*assignation*), served by bailiff (*huissier*), to the defendant for appearance before the judge; the other is by joint petition of the parties, *requęte conjointe.*[22]

[20] In the *United States* FRCP 3: "A civil action is commenced by filing a complaint with the court"; in *Germany*, Art. CPR 253 provides: "An institution of an action effects by the service of procedure"; in *Japan* Art. CCP 133 provides: "An institution of an action should be made by filing a complaint with the court".

[21] In *England* CPR Part 7.2 provides; "(1) Proceedings are started when the court issues a claim form at the request of the claimant".

[22] Art 54 French New Code of Civil Procedure provides: "Subject to cases where proceedings are instituted by way of a petition or by way of a declaration to the clerks office of the court and those where cognizance shall be taken by a voluntary presentation of the parties before a judge, the originating application shall be brought by way of summons or by the filing of a joint petition at the clerks office of the court."

Between both systems there can be observed some slightly different aspects in the proceedings for commencing a civil procedure; one is a system in which, first of all, the plaintiff should file a complaint with the proper court and then the submitted complaint with summons will be served on the defendant; the other is a system in which the complaint will be served on the defendant not by the court, but by the bailiff directly to the opposing party before the filing of the complaint, and then the plaintiff must submit a copy of the writ of summons at the clerk's office, commencement by *assignation* in France. Such differences could affect the suitability of applying ET to these proceedings.

1.3.2 Filing of Claim with the Court

1.3.2.1 General Aspect of a Claim

The act of a plaintiff to bring a claim to the appropriate court for demanding a court judgment is regarded as the first step of proceedings. Since the complaint is served on the defendant, s/he can have a chance to contest the claim. A court can not initiate a civil claim without a claim from a claimant. According to the understanding common to most jurisdictions, there is no judgment without an action.

Filing a complaint with the court to demand a specified judgment is a voluntary action of the claimant and the court will render a judgment corresponding to the request of the plaintiff. This action is not, however, a mere voluntary action to show his demand to the court, rather it has procedural significance given by the procedural regulations; the action, combined with the service of process with the complaint on the defendant, is regarded as a fundamental of the consequent proceedings. Some failure in the commencing action could be a cause of confusion in the consequent proceedings and lead to severe harm to the procedural positions of the parties. Due to such possible significant role of the action to commence following proceedings, the action ought to be made by a document with some formulation able to secure and confirm the validity of an action.

A documentary complaint should be brought to the court for instituting an action. This process could doubtless be made more convenient and efficient through the use of ET. But in order to allow that method, it must be considered that the filing of a complaint by email would satisfy the requirements for commencing a claim.

1.3.2.2 Formality and Some Additional Requirement for the Claim?

A complaint which should be brought to the court for commencing an action should be in the form required by the relevant procedural rules. As to the application of ET to the process for filing a claim with the court, there are some different nuances between national legislation, depending on each jurisdiction's traditional system of civil procedure. The differences reveal some divergent attitudes in respective systems of civil procedure as to the possibility of electronic filings of claim; in some jurisdictions an action of filing a claim is regarded as a formal activity with relatively strict requirements of form; in other jurisdictions the process is more liberal and only the fundamentals are required.

German civil procedure has a relative liberal position as to the form of complaint. There is no prescribed standard form for written submissions.[23] In 2001 the German parliament authorized the use of electronic procedural documents by the newly provided Art. 130a of the German Code of Civil Procedure, the detail being left to subsequent federal and state-level regulation. Although the text does not expressly refer to complaints,[24] another provision, Art. 253 Sub. 5, provides a power for electronic filing of complaints.

With regard to the action of commencement of a civil claim, Art. 253 Sub. 1 of German Code of Civil Procedure provides that "the institution of an action is concluded by the service of process of the complaint". So the institution of a commencement of a civil claim is composed of two activities; it is not composed of only a sole activity to bring a complaint to the court, but also of the subsequent service of process.[25] Service on the opposing party arguably establishes legal relationships between the court and both parties. But since the activity to commence a procedure is composed of two different activities with different natures, the applicability of ET for communication should be considered differently.

On the other hand, English civil procedure requires the issue of a claim form with some formality in order to commence a civil action (CPR 7.2)[26]; service of process is not regarded as a part of commencement of an action. As far as concerns applicability of ET to this stage of proceedings, English Civil Procedure Rules Part 5.5 refers to a practice direction to regulate the provision for documents to be filed or sent to the court by facsimile or other electronic means. And according to such general provisions, PD 5.3 provides for filing by facsimile; as to filing by email, PD 5B refers to a 'Pilot Scheme for communication and filing of documents by email', in force 1. May 2004. In it there are general provisions as following: '1.1 Section I of this practice direction provides for parties to claim in special courts to: (1) Communicate with the court by email; and (2) File specified documents by email. 1.2 Section II of this practice direction provides for parties to claims in specified courts to file specified documents electronically via an online service'.

The Japanese code of civil procedure was amended, and introduced in 2004 a general provision relating to the electronic method for petition. Art 132-10 provides as following: "In the case of a petition or any other statement to be filed or made in

[23] Murray and Stürner, *German Civil Justice* (Durham: Carolina Academic Press, 2004), 192.

[24] The text refers to the preparatory document, petition and explanation by parties, information and explanation by the third person. Murray and Stürner, *German Civil Justice* (Durham: Carolina Academic Press, 2004), 193 argues, 'Although the specific statutory provision authorizing electronic documents does not specifically mention complaints, it appears clear from the statutory language as a whole that complaint can also be in electronic form.'

[25] Rosenberg, Schwab and Gottwald, *Zivilprozessrecht*, 17th ed. (München: C.H. Beck, 2010), § 95. Rn. 1.

[26] For money claims there are some special provision; money claims online, PD 7E.

the civil procedure (hereafter referred as a "petition, etc."), which shall be filed or made, subject to the provisions of this Code or other laws and regulations concerning such petition, etc., by means of document, etc. (meaning a document, a transcript, extract, authenticated copy or duplicate of a document or a duplicate of a bill or note, or any other paper or other tangible object on which information recognizable to human perception such as characters and shapes in stated: the same shall apply hereafter) to the court specified by the Supreme Court (including one filed or made to the presiding judge, authorized judge, commissioned judge or court clerk of such court), notwithstanding the provisions of said laws and regulations, as provided for by the Rules of the Supreme Court, the petition, etc. may be filed or made "by means of an electronic data processing system wherein the computer (including input–output devices; the same shall apply hereinafter) used in the court is concerned, by way of telecommunication lines, to the computer used by the person who files or make the petition, etc. or person who receives a notice of a disposition under the provision of Article 399 (1) provided".

The provisions regulating the electronic filing of documents are general and basic; for its operation it is necessary to establish a computer linked netting system with each court and others. In Japan a system is being prepared and established which will be completed soon, and its use will be regulated by the Rules of the Supreme Court.

1.3.3 Service of Process

1.3.3.1 General

Generally, the service of process or documents is a method to inform the person addressed of procedurally significant matters. But when considering the application of ET, other matters should be considered. One is the procedural attribution of service of process; whether it is performed as the official responsibility of the court or is the private job of the litigant.

In the latter case the task of service of process should considered independently. In *France*, bailiffs (*huissier de justice*) have a statutory monopoly of service of process within France. The bailiff is a separate official, responsible for the service of procedure on account of its long tradition. In this system service of process does not belong to the official duty of the court. In the *United States* the service of process is provided fundamentally as a private matter[27] and Federal Rules of Civil Procedure provides; "Any person who is at least 18 years old and not a party may serve a summons and complaint", FRCP 4 (C)(2). Subsidiary FR4 (C) (3) provides: "Service of process was provided as a private business".

[27] For the historical development of the system of service of process in the United States, see, Millar, 85 et seq.

On the other hand, in the *German* system of civil procedure, the service of process belongs fundamentally to the official duty of the court and private service is an alternative method which can be chosen by the claimant in some special proceedings.[28] Originally the Imperial Code of Civil Procedure of 1877 adopted the French system of party engagement of service of procedure, and in 1890 the system was changed to the system of its official engagement or responsibility of the court.[29] The Austrian Code of Civil Procedure has a same system of official service of process.[30] The Japanese system also belongs to these groups.

Generally speaking, the service of procedure is a special proceeding by which the complaint is served on the opposing party in order to give information regarding the claim and argument of the claimant. In the system of official service of process, whole proceedings are regulated strictly by law and the court official is charged with them. The particular documents, like complaints and summons, should be served normally by using the postal system or some other officially admitted method. The served documents should be addressed and handed directly to the addressing person and this fact should be assured and redressed to the court to avoid subsequent possible disputes regarding its validity. In the case that the addressing person would not receive the served document, the service officer can leave them at the place of the person addressed. Such facts should be reported to the court and will be recorded. Even in the system of private service of procedure, the fact that documents have been sent to the opposing party must be reported to the court. The procedure has a function not only to transmit particular documents to the addressing person personally, but it has a function to be secured by some neutral person, as an official duty or even as a private business.

Such a special requirement of the service of process is based on the fundamental values of civil procedure; service exists to give proper information to the addressed person, as an opposing party for future proceedings; it is regarded as a basic and principal premise of subsequent procedures operating fairly and it contributes to secure fundamental procedural rights of the defendant to deny a claim and give him protective position; such protective position of the defendant, and the rights of hearings in the proper court, must be accepted as a fundamental right in all systems of civil procedure.[31]

---

[28] Rosenberg, Schwab and Gottwald, *Zivilprozessrecht*, 17. Bd. (München: C.H. Beck, 2010), § 72. Rn. 10; 379; German law of service of procedure was amended and modernized by Zustellungsreformgesetz (ZustRG) of 25.06.2001 (BGBl 1206), effective from 1.07.2002.

[29] See Hellwig, *System des deutschen Zivilprozessrechts* Bd.1, 1912, 513.

[30] The Austrian Code of Civil Procedure Art. 87 provides official service of process; the service of process is performed by postal service, Art. 88 Sub. 1.

[31] In the United States, Federal Constitution of the US in V of Amendment of 1791 provides *due process,* see Freedman, *Total Justice*, 1987, 80; in France Art. 14 of NCPR provides "No party may have a determination entered against him without having been heard or called", as to "*le principe du contradictoire",* see Guinchard and Ferrand, *Procedure civile*, 28e ed., 712.; In Germany Art 103 Sub. 1 Grundgesetz provides right of hearing before the court, *Anspruch auf rechtliches Gehör.* The regulation of service of process serves to realize the right, Stein, Jonas and Roth, *ZPO*. 22th Bd., vor § 166. Rn. 2.

1.3.3.2 Possibilities of Electronic Service of Process

Service of process is a special method to deliver some particular documents with procedurally relevant matters to an opposing party. It is a fundamental tool for giving the opposing or interested party some relevant procedural information and has a function in securing his/her procedural fundamental right to be heard. A basic premise for keeping procedural fairness is that all participants in civil litigation should be given significant procedural information; misleading information owing to a defective system of service of procedure could seriously prejudice the subsequent proceedings. Ensuring a successful service of procedure should be validated by a return receipt and verified subsequently to avoid any question of defectiveness in the service of process.

In the service of process, some relevant documents, like complaint, summons etc. must be served in person on the person addressed. Only in exceptional cases, can they be handed to some other person who has close relationship with the person addressed. In the official service system the responsibility to ensure proper service belongs to the court.

Service of process by ET has been introduced to some extent in certain jurisdictions; the German Code of Civil Procedure provides special service of process by way of electronic method (Art. 174. German Code of Civil Procedure), but only to certain special persons with a legally authorized occupation. In this case there is no problem that the person addressed would refuse to receive them. In England, the Civil Procedure Rules provide methods of service, 6.20 (1) (e) providing that "a document may be served by... fax or other means of electronic communication in accordance with Practice Direction 6A". The English system of service of process is not a matter for the official duty of the court. In Austria, a dual system of service of process was commenced; in this system service of process may be made by electronic methods, but if that is not possible the postal system can be used.

As it should be possible to perform service of process coercively upon a defendant, it must be performed even in the case that he declines to receive it. Electronic Service is suitable in such a situation. And it is necessary that the receiving party has equipment to receive email; if he is not connected to the internet, then the method will not work. In many systems service of process by the electronic method is not admitted as an alternative method.

1.3.4 Other Communications Between Parties Themselves or with the Court

In the course of civil litigation there are many situations in which the parties (or their representatives) and the court, or the parties themselves, must exchange information and procedural documents. Service of process has been regarded as a main method for their mutual communication. In most jurisdictions, however, the traditional system of service of process or documents demands seemingly unnecessary costs and time.

By introducing newly developed ET for communication this procedural burden could be reduced and procedural efficiency could be encouraged.

In many civil procedure systems, such communications by using email or by other method have been permitted. For example in England, procedural communication may be sent by email, PD 5B 1. Also the German Code of Civil Procedure provides electronic documents, as mentioned.

The Japanese Code of Civil Procedure of 1996 introduced an informal method for using *facsimile* to exchange preparatory documents between representatives of parties and the court. Before this amendment it was necessary to exchange preparatory documents by service of process; the litigants had to present them to the court which would in turn send them by way of service of process. But the procedure involved needless costs and time. The new Rules of the Supreme Court, mandated by the Art. 4 of Code of Civil Procedure, introduced a new method, so-called "direct sending", *chokusou*, between representatives and they must send copies to the court (Art. 84 Rules of the Supreme Court for Civil Proceedings).

1.4 The Second Stage (1) – Preparation for Plenary Hearings

The second stage of civil litigation is the procedure of preparation of plenary hearings. Only by well prepared examination can effective and speedy justice be achieved. These aims have been an overriding objective of civil litigation in every jurisdiction. To bring about an effective civil procedure system, recent legislation and practice has encouraged a detailed examination of disputed cases. To realize such objectives, some measures relating to the preparation of cases have been introduced.

In the normal civil procedure both parties allege many legal and factual arguments resting on complicated legal and factual issues; the live issues should be arranged and clarified by identifying them in advance of trial, by well organized preparation proceedings. As a matter of fact, in the proceedings for preparation of plenary hearings, both parties must communicate with each other and with the court. Traditionally the preparation of the plenary hearings has been accomplished and performed for the most part by documentary exchanges or arranged in some preparatory meetings, such as conference in the court.

For the preparation of a case, litigants or their representatives must get some additional information by exchanging documents or by negotiating in or outside the court. There are different types of procedure; one is a procedural system which provides a preparation of plenary administrated mainly by the court. In such a system there is a separate procedure outside plenary healings in which the preparation of the main hearings will be administrated and managed under the leadership of the court. Preparation could be made by exchanging documents.

There has been a recent tendency to bring about well prepared proceedings through active management of cases by the court, in order to render civil litigation more effective. The German code of civil procedure amended in 1976 introduced two types of preparation proceedings: (a) early first hearing, *Frühe erster Temin* (Art. 275 of German Code of Civil Procedure) and (b) documentary proceedings for preparation, *Schriftliche Vorverfahren* (Art. 276). In the former cases the presiding judge can fix a period for preparation of the early first hearing to submit documentary denial of the claim by the defendant. The early first hearing is normally an oral hearing,[32] but in normal cases the hearing will be preceded by a settlement procedure, arranged by the court (Art. 278 Code of Civil Procedure). But in the case that such attempts were not successful, preparation for the plenary hearings will be ordered. In the latter case, the presiding judge does not fix an early first meeting, and the case will be prepared by documentary preparation (Art. 276 Code of Civil Procedure).

In France, preparation of normal cases proved in the ordinary court of original jurisdiction, *Tribunal de grande instance,* will be held by way of a meeting between the president of the court division in charge of the case and parties' lawyers, known as a *conférence du president.* In such proceedings the preparation will be managed in the meeting by the president of the court where the case is pending. English Civil Procedure Rules 1998 encouraged the preparation of the case by case management, CPR 3; it is possible to use videoconferencing in a case management conference, PD 32, Annex 3, No. 1.

The Japanese Code of Civil Procedure of 1996 encouraged preparation for the plenary oral hearings. Generally oral hearings should be backed up by documents well prepared (Art. 161 Sub. 1 Japanese Code of Civil Procedure). In respect of exchanging preparatory documents for plenary hearings, it introduced a new system of "direct sending of preparatory documents" between representatives of both parties by using facsimile. In this system, representatives should send or exchange preparatory documents directly and also send copies to the court. Before the reform they were submitted to the court and then served by the court; it was a time-consuming and costly process. The code provides for three types of preparatory proceedings: preparatory oral hearings, preparatory proceedings and preparatory proceedings by exchanging documents. The first type of proceedings is a normal hearing, but only acts as a preparation for the following plenary hearings. The other proceedings are not formal oral hearings, but only preparation outside the open court. The preparatory hearings are held in a room with a round table, with an informal atmosphere where a videoconference device is sometimes prepared. In the latter two cases the court can use a *trio phone* by which it can speak with both representatives simultaneously and can avoid any allegation of bias on the part of the court. As the preparatory proceedings are not plenary oral hearings, there is no reason not to use a videoconference system which is linked to another court nearer the other party. For the preparation of the oral hearings the possibilities of ET could be considered more extensively.

---

[32] Stein, Jonas and Leipold, *ZPO*, 2007, § 275. Rn. 22.

1.5 The Second Stage (2) – Presenting Relevant Evidence Before the Court

To decide a case contested by the defendant, relevant facts must be proved by evidence presented at the hearings. Documents play a significant evidential role in modern society. In our daily business, most transactions are made by documents. But nowadays the nature of transactions have been changed by using ET, email or telefax. The potential evidential role of these new methods of communication must be considered.

In the case of mutual communication by way of ET, whole data are kept not as paper formed documents, but in electronic form on a computer. In respect of submission of documents, such electronic data is not regarded as true documentary evidence under the Japanese Code of Civil Procedure, viewed rather as a quasi-document. Such data cannot be proved by the same method as for documentary evidence. It is necessary to use computer-programming to produce the information. But the outputted information can be proved in the same proceedings as documents. It is, therefore, possible to apply the proceedings relating to documentary evidence to electronic information. Art. 231 of the new Japanese Code of Procedure provides specifically; "the provisions of this Section [concerning documentary evidence] shall be applied mutatis mutandis to drawings, photographs, audiotapes, videotapes and any other objects prepared for the purpose of indicating information, other than documents". As to the obligation to submit documents, a Japanese court decision ordered the submission not only of data recorded in the computer magnetic tape, but also of the programming to print the information on paper.[33]

ET can be used for expert testimony. It is doubtless convenient and effective if an expert can submit his report by way of ET. The newly amended Japanese Code permits the use of a videoconference system; the court can allow, subject to special regulation by the Rules of the Supreme Court, the use of videoconference system in a case that an expert is giving an oral report and s/he lives a long distance away or in other proper case (Art. 215-3 of Japanese Code of Civil procedure).[34]

1.6 The Third Stage – Plenary Hearings

1.6.1 Generals

The third stage of normal proceedings is a plenary hearing before the court. This is the central event of civil litigation in the traditional mode of civil procedure, and parties, or their representatives, must join meetings to argue their case and prove

[33] Decision of *Osaka* Appellate Court on March 3. 1978, Koumin 31. 1. 28.

[34] Amendment in 2003 by Law No. 108, 2003.

evidence relevant to deciding factual issues between the parties. In this type of hearing, participants, not only the parties, but also other public observers, can watch the whole trial in the court. This mode of proceedings is regarded, along with other related fundamental procedural principles, as a guarantee of fair proceedings and good motivation for accelerating effective justice. Oral hearings in public are now the dominant form of civil litigation not only in each national system of civil procedure, but also in the international dimension of civil procedure.[35]

Although oral plenary hearings have relevant merits, their inherent problems should not be ignored; to attend a meeting can be burdensome and difficult for some people. If a significant witness cannot join a hearing, the court must decide the case without calling him/her.

1.6.2 Requirements of Oral Hearings and Their Functions

In most jurisdictions, generally speaking, plenary hearings should be held orally and publicly in open court. Such oral hearings in the open court are regarded as a fundamental principle of civil procedure, but their necessity or demand is satisfied somewhat differently in each jurisdiction:

One of the strictest regulations or formulations can be seen in the Japanese Constitution; it requires oral hearings in court open to the public for all litigated civil cases (Art. 82 Japanese Constitution). This constitutional dimension has an impact on the concept of plenary oral hearings. Any changes or extension to the traditional concept of oral hearings might give rise to serious criticism on the grounds of constitutionality.

Generally, plenary hearings are held in an open court room where the pleading takes place, with the judge and all procedurally related people in attendance. Such a form of proceedings captures many procedural fundamental principles: *oral hearings* as a fundamental form by contrast to documentary examination of the old system of civil procedure, *hearings in open* court and the *principle of immediacy* are regarded as a requirement of a fair form of trial by the court. *Immediacy* in the examination has been regarded as a fundamental principle based not only on logical or formal requirements, but it seems to be based on the sensitivities of human perception which have varied influences on the witnesses and the judge.

Generally speaking, the traditional system of plenary hearings with immediate relationship with the judge for oral examination in the presence of witnesses and parties in a court room can bring about a kind of human contact allowing detailed debate and examination of a case. The judge can get information, not only through what is said, but a wider of perception of the situation (such as through the demeanour or body-language of a witness).

---

[35] Art. 6 of European Convention of Human Rights of 1950 provides the right to fair proceedings which includes hearings at an open court.

1.6.3 Possibilities to Use Video Conference System in Plenary Hearings

For the convenience of the parties and other interested persons, some inherent inconveniences of oral plenary hearings should be removed by introducing ET into civil procedure. For a litigating party or witness who lives or is staying far from the court, it is not convenient to appear at plenary hearings, due to long travel times and cost to business. ET used in the form of a video conference system could improve or reduce such difficulties. Possibilities and limits of video conference system have been considered and are partly in use in some jurisdictions. Its applicability to plenary hearings differs in each jurisdiction.

In Germany Article 128a[36] provides for plenary hearings to use video conference systems; it is composed of two provisions relating to the examination of parties and witnesses etc. Sub. 1 provides; "By agreement with the both parties the court can admit the parties and their representatives and supporter on the request while its pleadings to link hearings with other place. This hearing must be transmitted simultaneously on the image and sound to the other place where the parties, their representatives and supporter are staying and to the court room."

Sub. 2 provides; "By the agreement with the both parties the court can admit that the witness, expert or party stay other place. The hearings will be transmitted on image and voice simultaneously to the other place. If parties, representatives and supporter will be allowed by the Sub. 1 to stay other place, the hearings will be transmitted simultaneously on image and voice to the place."

This provision follows immediately after a provision as to the oral hearings, Art. 128 of German Code of Civil Procedure. So observing from the location of the provision, Art. 128a must be understood as a provision providing an exception to oral hearings in civil procedure.[37] These provisions require the agreement of both parties to videoconferencing. It must be noted that the agreements could be an abandonment of the right of oral hearings in the open court provided by Art. 128. In the use of videoconferencing there is no restriction on the place to be linked to the court. Indeed, the provision seems to extend the traditional concept of plenary hearings, flexibly for the convenience of the litigating parties.

On the other hand, the Japanese code of civil procedure recently added a general provision for video conference, which nevertheless has a tenuous position. The new provisions which authorized video conference systems in plenary hearings were inserted as a part of provisions for witness inspection, following after the provision relating to protective measures for the witness. Article 204 of the Japanese Code of Civil Procedure provides as follows, "The court may examine a witness, subject to the Rules of the Supreme Court, in the following cases by a method that enables

[36] This article was introduced by Zivilprozessreformgesetz of July. 27. 2001.

[37] Stein, Jonas and Leipold, *ZPO*, 22nd ed., 2005, § 128 a Fn. 3; Stadtler, Die Zivilprozess und neue Formen der Informationstechnik, in *ZZP*, 115, 2002, 413, pointed out some doubt on the reduction of the sub. 2 provision in the code; she argues that the provision could be regulated through provisions on evidence: Art. 355 ff.

communication with each other while mutually recognizing the other parties statues by audio and visual transmissions simultaneously; (1) the witness stays in the distant place. (2) Because of the nature of the case, the age or psychological situation of the witness, the relationship of the witness with the party or her/his representative and other matters, the witness may be assumed to be pressured or psychologically suffer by the presentation for examination at the place of the judge and parties are staying". The newly inserted provision should be regarded as a special regulation concerning examination of witnesses for their convenience alone. Behind such careful regulations there seems to be a fundamental requirement of oral hearings in open court guaranteed by the Japanese Constitution.

In England videoconferencing is available. There is detailed guidance in PD 32 Annex 3; the guidance explains that videoconferencing can be used for the taking of evidence and also for other parts of any legal proceedings, for example, interim applications, case management conferences and pre- trial reviews.

1.6.4 Merits and Problems of Videoconferencing Systems in Civil Litigation

A videoconferencing system is comprised of equipment which links different places and makes it possible to exchange pictures and sound simultaneously. It can act as a virtual court.

On one the hand, it is obvious that it can render justice more convenient, allowing participation at meetings without having to travel long distances. Parties do not need to travel for plenary hearings held a long distance away. But there are some differences from the traditional method of oral plenary hearings in open court. A judge can view a witness or party only through the technology; s/he can examine them by seeing their present states through the video screen, at the same time as hearing their statements through a speaker. Seeing an image on a screen is not the same as meeting face to face.

On the other hand, without using such ET, it can be difficult to get evidence from witnesses who live far from the court and cannot travel there. It is not reasonable to deny the use of videoconferencing in civil litigation. But precisely speaking there are, of course, some significant differences between face to face examination in the open court and examination by videoconferencing.

English law admits generally the use of videoconferencing in civil litigation. But English guidance for videoconferencing, PD 32.33.3, makes some technical remarks. It states: "When used for the taking of evidence the objective should be to make the VCF (Videoconferencing) session as close as possible to the usual practice in a trial court where evidence is taken in open court. To gain the maximum benefit, several differences have to be taken into account. Some matters, which are taken for granted when evidence is taken in the conventional way, take on a different dimension when it s taken by VCF: for example, the administration of the oath, ensuing that the witness understands who is at the local site [at which the judge sits] and what their various roles are, the rising of any objections to the evidence and the use of documents."

1.7 The Fourth Stage – Service of Judgment and Appeals

1.7.1 General

The last step of civil procedure is the final judgment by which the court renders its final decision on the case and related subsequent proceedings. The judgment must be given by sentence and the judgment document must be served on the both parties. Service of the judgment has in some jurisdictions significant functions for subsequent proceedings of appeals or enforcement.

1.7.2 Service of Judgment

In most jurisdictions service of judgment must be made by the court.[38] The service of a judgment should be performed upon the person addressed. The appellate period starts from the time when the person addressed receives this service; the period and person should be precisely recorded and reported to the court.

1.7.3 Appeals

Appeal is an action which is a petition for reexamination of the case by a higher ranking court. For appellate proceedings, a petition is required from the party who suffered as a result of the judgment and wants a relief from it. In most jurisdictions the appeal should be brought within a fixed period: the appellant must bring his/her petition before the expiry of the period. As the petition should be brought within short time, a petitioner's intention to appeal must be promptly communicated to the court. In many jurisdictions where ET for communication is employed, especially when electronic filing is introduced in the civil procedure, its usages in appellate petitions are or will be admitted.

1.7.4 Judgment as an Authorized Title for Civil Enforcement

A judgment can authorize civil enforcement in some jurisdictions. In the German system of civil enforcement, a judgment can be a publicly authorized documental instrument for civil enforcement; upon its submission to the enforcement organization by the claimant the enforcement proceedings will be commenced automatically. For filing an enforcement proceeding, the successful claimant should submit

[38] Art. 317 German Code of Civil Procedure; Art. 255 Japanese Code of Civil Procedure.

an officially certificated copy of the judgment. Its delivery to the defendant is regarded as a fundamental requirement for civil enforcement; the service of judgment to both litigants should be adequately performed. In some jurisdictions service of the judgment is regarded as an official responsibility. The possibility of electronic service can be considered, but it will have to meet special requirements.

1.8 Summary Proceedings

1.8.1 Generals

Besides the normal procedure with formally required procedural fundamental principles, in some jurisdictions there are special procedures for summary proceedings. In such proceedings, communication between the court and the applicant has been improved mainly in terms of the ease of making an application. In such proceedings, there is a wide range of possibilities for using ET for communication, and indeed, you can see already some examples of such intensive applications of ET in summary proceedings.

1.8.2 Money Judgments

In some jurisdictions, there are special provisions relating to money claims, due to their formality and the relatively simple process at trial. In those proceedings, the use of ET to some extent must be considered. Indeed, the English Rules of Civil Procedure provide special types of money claims which may be started by using the specially established "money claim online", PD 7E. In these cases, a claimant seeking a money claim within the set categories may request the issue of a claim form by completing and sending an online form by electronically paying the appropriate claim fee, PD 7E 5.1.

1.8.3 Warning Procedure for Collecting Money Claims

Warning procedure for collecting money claims in some jurisdictions is a specially provided summary procedure by which a claimant files a petition to send a warning against his judgment debtor. In the case that the debtor admits his debt or does not bring a motion to the warning, the claimant can apply for enforcement. This is a summary proceeding without normal procedure.

Regarding the warning procedure (*Mahnverfahren*) in Germany, preparatory general provisions for introducing electronic filings were promulgated, and introduced by the Law of December 1. 2008.

In Japan the warning procedure for collecting money claims as summary proceedings has been frequently used in recent years, due to prevailing consumer credit business.[39] Most of them were petitions from consumer companies. It is normal for a consumer credit company to bring thousands of petitions at once to the court. It was therefore necessary and urgently demanded that there be some form of mechanism to deal with this, especially by a data processing system. Dealing with such huge numbers of cases by small number of court clerks in the summary court was a major problem.[40] The credit companies dispose of such cases by computer. By introducing an ET in the proceedings, it must be possible to handle huge numbers of cases automatically and easily and to reduce burdens on court clerks and to reduce the number of mistakes.

The Japanese warning procedure for collecting money claims was amended and special provisions were added by the amendment of Japanese Code of Civil Procedure in 2004 to introduce ET in the proceedings; according to the newly introduced provision (Art. 397 of Japanese Code of Civil Procedure), a petition for the warning procedure for collecting money claim may also be filed by means of an electronic data processing system with a court clerk of a summary court, specified by the Rules of the Supreme Court as a court that shall handle a warning procedure for collecting money claim by means of an electronic data processing system". The courts authorized to deal with the cases have wide jurisdiction; Tokyo summary court was named as the first court and many summary courts in big cities have been subsequently added within a short time.

1.9 Final Remarks

Encouraging effectiveness and convenience of civil procedure must be one of the leading objectives for promoting further access to civil justice, and our civil procedure should be improved in all ways to realize such overriding objectives. In particular, the traditional system of civil procedure has struggled to deal with issues where one of the parties lives or is staying a long distance from where hearings are to be held. In our widely connecting and dynamically integrating society there are many transactions between people living and staying in far different places. Such difficulties caused by geographical distance have been improved with respect to our daily business and private life by the rapid development of ET for communication. On the other hand, in our traditional system of civil procedure, exchange of procedural communications has been accomplished by some time-consuming methods. Exchange of procedural documents by the postal system required a long time for delivery. Litigants and witnesses living and staying at a remote place must travel to

[39] In Japan, in 2007 number of petitions of the warning procedure for collecting money claims was 364,665; number of normal cases to the summary court was 475,624.

[40] In Japan, warning procedures were charged with the court clerk of summary court.

the court to attend hearings. Rapid developments and the prevalence of newly developed ET for communication can assist in reducing many such difficulties. The introduction of new ET into civil procedure is now in progress in many jurisdictions. But it brings some new problems which could or might undermine some procedural fundamentals. Civil procedure is a social instrument by which civil disputes between private people should be resolved. As the disputes or conflicts are, however, human affairs, the traditional system of civil procedure had many humane elements. Especially in its main event of plenary hearings in the open court before the public, there can be observed some human and psychological fundamentals which are enshrined in procedural principles.

Introducing some ET for communication, especially by introducing a video conference system for plenary hearings, contributes to reduce some difficulties in traditional civil procedure, but its unlimited introduction through the use of a virtual court would change in some fundamental aspects the traditional civil procedure based on face-to-face relationships in civil litigation. Surely, it must be necessary to introduce ET into civil procedure to some extent, but it must be questioned whether it would be welcomed by litigants, and it must be considered whether the use of monitor technology, eliminating the human element from litigation, is a necessary development. Its introduction must be, on one hand, a financial problem and a cost-benefit analysis might be necessary,[41] but the factors of personal contact civil procedure would not be ignored. We are now standing at a significant crossroads for considering the future of civil procedure!

---

[41] Japanese central and local governments introduced some electronic administration system, but its use is not popular.

Chapter 2
The Impact of Digital Information on American Evidence-Gathering and Trial – The Straw That Breaks the Camel's Back?

Richard L. Marcus

Like everything else, the pervasive impact of digital communication has profoundly affected the legal profession.[1] Rather than attempt to assess these broad effects, this paper focuses on the specific question of computers' impact on the conduct of litigation.

Drawing in large measure on which one might call "insider" experience,[2] I consider the extent to which digital data may affect the enduring American "exceptionalism" in procedure.[3] To do that, I must begin with a brief reminder of those exceptional features – mainly broad discovery and jury trial – including a chronicle of some of the ways in which the breadth of American discovery was constrained in the last quarter of the twentieth century. Even before the advent of digital media as a central feature of legal work and litigation, American exceptionalism was moderating, just as other countries were gravitating toward reforms that in some ways resembled the American model of litigation.

Against this background of moderate evolution, I will turn to the emergence of E-discovery and the potential of a "virtual trial" as realities and possibilities for twenty-first century American litigation. Whether these possibilities are realized

---

[1] For discussion of the general effect of computer technology on the American legal profession, see R. Marcus, "The Electronic Lawyer," *DePaul Law Review* 58 (2009) 263–310; R. Marcus, "The Impact of Computers on the Legal Profession: Evolution or Revolution?" *Northwestern Law Review* 102 (2008): 1827–68.

[2] Since 1996, I have been Associate Reporter of the Advisory Committee on Civil Rules of the Judicial Conference of the U.S., the body that develops proposals to change the Federal Rules of Civil Procedure, which govern conduct of civil litigation in the U.S. Federal Courts. I write here solely in my personal capacity, and not as a representative of this Committee or anyone else.

[3] For discussion of this exceptionalism, see R. Marcus, "Putting American Exceptionalism into a Globalized Context," *American Journal of Comparative Law* 53 (2005): 709–40.

R.L. Marcus (✉)
Hastings College of the Law, University of California,
San Francisco, CA, USA
e-mail: marcusr@uchastings.edu

M. Kengyel and Z. Nemessányi (eds.), *Electronic Technology and Civil Procedure*, Ius Gentium: Comparative Perspectives on Law and Justice 15,
DOI 10.1007/978-94-007-4072-3_2,

remains uncertain, but it could be that the challenges of E-Discovery will prove to be the force that curtails what had seemed unbounded American discovery, and that the flexibility of a virtual trial will reinvigorate what has become a remarkably rare event in U.S. federal courts – a trial in a civil case.

2.1 American Procedural Exceptionalism – Discovery and Trial

2.1.1 Expanding and Constraining Discovery

American discovery has long been distinctive.[4] For example, in the 1870s American discovery efforts provoked German protests, and various countries have more recently adopted "blocking" statutes designed to prevent American discovery on their soil.[5] Until the 1930s, however, discovery was available in American courts only on a spotty basis. As late as 1911, for example, the U.S. Supreme Court inveighed against a "fishing bill" by which a party sought "to pry into the case of his adversary to learn its strength or weakness."[6]

The adoption of the Federal Rules in 1938 produced a revolution because these rules vastly expanded discovery: "If one adds up all of the types of discovery permitted in individual state courts, one finds some precursors to what later became discovery under the Federal Rules, but… no one [American] state allowed the total panoply of devices. Moreover, the Federal Rules, as they became law in 1938, eliminated features of discovery that in some states had curtailed the scope of discovery and the breadth of its use."[7] And the courts rapidly came to favor the broad discovery. The Supreme Court – which had denounced "fishing expeditions" in 1911 – changed its tune by 1947: "No longer can the time-honored cry of 'fishing expedition' serve to preclude a party from inquiring in the facts underlying his opponent's case. Mutual knowledge of all the relevant facts gathered by both parties is essential to proper litigation."[8] In significant part because discovery was crucial to American reliance on private enforcement of public law,[9] aggressively curtailing

[4] See generally R. Marcus, "Retooling American Discovery for the Twenty-First Century: Toward a New World Order?" *Tulane Journal of International & Comparative Law* 7 (1999): 153–99.

[5] G. Born, *International Litigation in United States Courts*, 3rd ed. (The Hague: Kluwer, 1996), 849, 856–71.

[6] Carpenter v. Winn, 221 U.S. 533, 540 (1911).

[7] S. Subrin, "Fishing Expeditions Allowed: The Historical Background of the 1938 Federal Discovery Rules," *Boston College Law Review* 39 (1998): 691–745, 719.

[8] Hickman v. Taylor, 329 U.S. 495, 507 (1947).

[9] For discussion of American reliance on private enforcement of public law, see S. Fahrang, *The Litigation State* (Princeton/Oxford: Princeton U. Press, 2010) (chronicling reliance on private rather than public enforcement of employment discrimination law in the U.S.).

discovery raised great concerns.[10] Some said that broad discovery itself prompted expansion in the substantive law of products liability, employment discrimination, and consumer protection.[11] Moreover, discovery's centrality could be seen to make limitations on discovery suspect: "Broad discovery is thus not a mere procedural rule. Rather it has become, at least for our era, a procedural institution perhaps of virtually constitutional foundation."[12]

Nonetheless, growing clamor about the cost and burden of American civil litigation led to numerous efforts to constrain excessive discovery.[13] Various numerical limitations were imposed on use of certain forms of discovery. Potentially more central was the introduction in 1983 of the concept of "proportionality" – that discovery requests should be proportional to the needs and stakes of the litigation.[14] In addition, the rules now direct the parties to confer and develop a discovery plan before they embark upon formal discovery.[15] Perhaps most significantly, the rules now rely on judicial control of discovery as a means of curtailing over-discovery.[16]

2.1.2 Declining Rate of Trial

The American trial is also distinctive. It relies on a lay jury selected in a fairly random manner to decide in accordance with instructions on legal principles provided by the judge. Because it relies on jurors, the American trial operates under fairly elaborate rules of evidence; the judge is empowered and expected to prevent the jury from receiving certain suspect information that might mislead it. Thus, hearsay evidence is generally excluded when offered to prove the truth of the matter asserted.[17] Perhaps

---

[10] Consider the views of a leading judge: "Congress has elected to use the private suit, private attorney-general as an enforcing mechanism for the antitrust laws, the securities laws, environmental laws, civil rights and more. In the main, the plaintiff in these suits must discovery his evidence from the defendant. Calibration of discovery is calibration of the level of enforcement of the social policy set by Congress."

P. Higginbotham, "Foreword," *Alabama Law Review* 49 (1997): 1–6, 4–5.

[11] See J. Friedenthal, "A Divided Supreme Court Adopts Discovery Amendments to the Federal Rules of Civil Procedure," *California Law Review* 69 (1981): 806–20, 818 (arguing that discovery itself had fueled growth in substantive law in these areas).

[12] G. Hazard, "From Whom No Secrets Are Kept," *Texas Law Review* 76 (1998): 1665–94, 1694.

[13] For a discussion of this effort, see R. Marcus, "Discovery Containment Redux," *Boston College Law* Review 39 (1998): 727–84.

[14] See Fed. R. Civ. P. 26(b)(2)(C) (setting forth grounds for judicial limitation of discovery that is disproportional).

[15] See Fed. R. Civ. P. 26(d); (f) (providing for moratorium on formal discovery until the discovery conference is held).

[16] See R. Marcus, "Reining in the American Litigator: The New Role of American Judges," *Hastings International & Comparative Law Review* 27 (2003): 3–30.

[17] See, e.g., Fed. R. Evid. 801; 802 (providing that hearsay not generally admissible unless it fits within an exception to the hearsay rule).

the most prominent recent example of judicial control is the judge's "gatekeeper" role in limiting the testimony of party-sponsored expert witnesses.[18] Because the American trial involves lay decisionmakers and relies on evidence provided by the parties, it ordinarily is a single continuous event distinctive from the multiple hearings that characterize similar decisionmaking efforts in much of the rest of the world.

Although the American trial is very popular with filmmakers in Hollywood, it is not so popular in American courtrooms, at least if measured by the frequency of civil trials. For decades, scholars have been tracing and analyzing the declining frequency of trial in the federal courts.[19] There is at least some room to debate whether comparisons with trial rates a half century ago are meaningful, since the nature of civil litigation in the federal courts has changed greatly, but it is surely true that a conventional trial is now a very unusual event in a civil case. Whether this development can be linked to changes in rules is uncertain; some regard increased use of summary judgment beginning in the mid 1980s as playing a role in the declining trial rate.[20] Perhaps more significant is the growing emphasis judges place on encouraging settlement. Except for that effort to promote settlement, however, it cannot be said that there has been a conscious American judicial effort to curtail trials comparable to the conscious effort to constrain overdiscovery.

2.2 The Emergence of E-Discovery

Electronic discovery is the hottest topic in American litigation today. Some suggest that it might be "considered a specialized substantive expertise in the same vein as, for example, patent law," and argue that mishandling such discovery could become a fertile source of legal malpractice claims.[21] There is at least one casebook designed specifically for law school courses on E-Discovery,[22] so that students can be prepared to handle E-Discovery upon graduating. Largely to deal with concerns about the difficulties of E-Discovery, an electronic discovery industry has emerged that reportedly generated over $4 billion in revenues in 2009. Within law firms, the advent of electronic discovery has prompted the creation of staff attorney positions to handle

[18] See, e.g., Daubert v. Merrell Dow Pharmaceuticals, Inc., 509 U.S. 579 (1993) (directing the judge to act as a "gatekeeper" with regard to expert scientific evidence).

[19] For a collection of such studies by a number of authors on the common theme "The Vanishing Trial," see *Journal of Empirical Legal Studies* 1 (2004): 459–984.

[20] For discussion, see M. Redish, "Summary Judgment and the Vanishing Trial: Implications of the Litigation Matrix," *Stanford Law Review* 57 (2005): 1329–59.

[21] J. Kwuon and K. Wan, "High Stakes for Missteps in EDD," *New Jersey Law Journal*, December 31, 2007, E2.

[22] See S. Scheindlin and D. Capra, *Electronic Discovery and Digital Evidence* (St Paul, MN: West Publishing Company, 2009).

the activity. The costs of this form of discovery have mounted quickly; some lawyers assert that the cost of preparing a privilege log identifying requested digital materials withheld on grounds of privilege routinely exceeds $1 million.

2.2.1 Handing E-Discovery Within the Existing Discovery Framework

Almost none of the problems just mentioned were important a decade ago. But that does not mean that E-Discovery did not exist before 2000. In 1970, the Federal Rules were amended to accommodate discovery of computer-based information.[23] That discovery began to occur. In 1985, a federal district judge wrote that "[c]omputers have become so commonplace that most court battles now involve discovery of some type of computer-stored information."[24] The third edition of the Manual for Complex Litigation, published in 1995, said that "[c]omputerized data have become commonplace in litigation."[25]

Nonetheless, the enormous increase in use of the Internet, and particularly of e-mail, made a major difference in the frequency and magnitude of discovery regarding electronically stored information, which has generated an acronym for this material – ESI. These changes began to come to the fore in the U.S. in the late 1990s, but it was unclear how to evaluate them. The main concerns seemed to be like the concerns that had begun appearing frequently about hard-copy discovery in the 1970s – burden and cost.[26]

By the late 1990s, the American discovery rules had already been changed to include provisions that seemed potentially suitable for the challenges of E-Discovery, particularly the concept of "proportionality" and the power of the court to manage discovery and grant protection when needed to guard against undue burden. Thoughtful judges candidly recognized that the existing rules sufficed for these purposes.[27] For an extended period, therefore, it seemed that no rules specifically tailored to E-Discovery

---

[23] See Fed. R. Civ. P. 34 (1970 version) (authorizing discovery of "data compilations").

[24] Bills v. Kennecott Corp. 108F.R.D. 459, 462 (D. Utah 1985).

[25] Federal Judicial Center, *Manual for Complex Litigation*, 3rd ed. (Washington, DC: Federal Judicial Center, 1995), § 21.446.

[26] For discussion of these points, see R. Marcus, "Confronting the Future: Coping with Discovery of Electronic Materials," *Law & Contemporary Problems* 62 (2001): 253–81.

[27] See, e.g., Thompson v. United States Dep't of Hous. & Urban Dev., 219 F.R.D. 93, 98 (D. Md. 2003): "Under Rules 26(b)(2) [regarding proportionality] and 26(c) [regarding protective orders], a court is provided abundant resources to tailor discovery requests to avoid unfair burden or expense and yet assure fair disclosure of important information. The options available are limited only by the court's own imagination and the quality and quantity of factual information provided the judge by the parties to be used by the court in evaluating the Rule 26(b)(2) [proportionality] factors."

would be needed. For example, when the photocopier was introduced in the 1950s and 1960s it had a large impact on discovery, but no special rules were developed to respond to this impact. Perhaps the same would be true for E-Discovery.[28]

2.2.2 *Tailoring Rues for E-Discovery*

Gradually, confidence that the existing rules sufficed waned. E-Discovery *did* seem to present particular issues that deserved a specific rule response. One distinctive feature was the volume of material involved. Much as American discovery has (particularly since the invention of the photocopier) long involved quantities of information that leave the rest of the world agog, the shift to E-Discovery produced much more information. This feature alone seemed to set this form of discovery apart even though computers could to a substantial extent assist in managing and analyzing the mountain of data.

Other recurrent problems also emerged. Too often, failure to focus on the potential difficulties of E-Discovery from the beginning of litigation could result in serious problems that could have been avoided by careful attention at the outset.[29] Too often, ESI was produced in a form or format that the receiving party could not use, and the producing party then refused to provide it in a different form. Too often, demands for retrieving ESI from backup tapes or other disaster-recovery sources generated disputes. Too often, computer systems themselves altered or deleted data – as they are supposed to do in normal operation – leading to claims of spoliation of evidence. Too often, difficulties locating all privileged material resulted in claims that privileged ESI had been produced, resulting in loss of privilege protection for all related information.[30]

After much consideration, therefore, rule changes were adopted to address discovery of ESI. Briefly, the Federal Rule changes included the following:

1. *Requiring the parties to confer*: The rules now require the parties to confer about E-Discovery and preservation of discoverable information at the outset.[31]

---

[28] For discussion of these points, see R. Marcus, "Only Yesterday: Reflections on Rulemaking Responses to E-Discovery," *Fordham Law Review* 73 (2004): 1–21.

[29] See, e.g., GTFM, Inc. v. Wal-Mart Stores, Inc., 2000 WL 335558 (S.D.N.Y., March 30, 2000), in which outside counsel for defendant was informed by the client's contact person that ESI requested by plaintiffs was no longer in existence and so informed the court. A year later, during the deposition of a member of defendant's Information Technology staff, it was revealed that the information had indeed existed at the time the representation to the court was made, but had subsequently been deleted. Sanctions were imposed on defendant.

[30] On privilege protection in the pre E-Discovery era, see R. Marcus, "The Perils of Privilege: Waiver and the Litigator," *Michigan Law Review* 84 (1986): 1605–55.

[31] See Fed. R. Civ. P. 26(f) (directing the parties to confer regarding a discovery plan and to discuss preservation of discoverable information and any issues regarding discovery of electronically stored information).

2. *Providing default rules for form of production of ESI*: The rules now permit the requesting party to specify the form for production of ESI, and provide directions about how form is to be handled, in an effort to avoid disputes arising later on this subject.[32]
3. *Providing for "inaccessible" ESI*: The rules now specifically address the problem of "inaccessible" ESI by absolving the responding party of searching such information so long as it identifies the information and explains why it is not reasonably accessible.[33]
4. *Protecting against sanctions for good faith operation*: The rules now provide that a party may not be sanctioned for loss of ESI resulting from its good-faith routine operation of an electronic information system.[34]
5. *Guarding against privilege waiver*: The rules now provide a process for recalling privileged information mistakenly produced through discovery,[35] and the Federal Rules of Evidence have been amended to insulate "inadvertent disclosure" through discovery from being held to waive privilege protection.[36]

State courts do not operate under the Federal Rules, but most have adopted similar rules for their judicial systems.

2.2.3 *The Straw That Breaks the Camel's Back?*

Despite these changes, concern about the burdens of E-Discovery remain, and may be growing. Indeed, some suggest that these burdens may finally cause even the U.S. to retreat from its long-standing willingness to authorize potentially unlimited discovery. Whether that might happen is not possible to forecast. Indeed, recent work by the Federal Judicial Center raises doubts about whether E-Discovery is as widespread as some contend, and also about whether it generates the costs suggested by some who are raising alarms about the expenses generated by this form of discovery.[37] At least some observations seem in order, however.

[32] See Fed. R. Civ. P. 34(b) (permitting a responding party to object to a specified form for production, and directing that in any event the responding party specify what form it will use before producing ESI, while providing further that it need not produce in a different form if it does so).

[33] See Fed. R. Civ. P. 26(b)(2)(B).

[34] See Fed. R. Civ. P. 37(e).

[35] See Fed. R. Civ. P. 26(b)(5)(B).

[36] See Fed. R. Evid. 502.

[37] See T. Willging and E. Lee, "Their Words: Attorney Views About Costs and Procedures," in *Federal Civil Litigation* 3 (Washington, DC: Federal Judicial Center, March 2010) (reporting that most attorneys interviewed had little experience with E-Discovery).

2.2.3.1 The Importance of ESI Will Not Decline

E-Discovery will not go away. To the contrary, it will become more essential. For one thing, it may provide the only source for information that was formerly available in hard-copy format. Medical records of the future, for example, may be available only in electronic format. This information is crucial in personal-injury litigation, an American staple. Most information generated by organizational litigants (governmental and private) exists only in electronic format; estimates are that only a very small percentage is printed to paper. Unless we are going to abandon proof by "documents," therefore, American lawyers will have to use E-Discovery.

Moreover, the material available through E-Discovery is often highly important. Probably the first case to bring that home was *United States v. Microsoft*, the monopolization case in which the government used Microsoft e-mail messages to devastating effect during the trial. As all litigators know, getting a witness to acknowledge or "recall" things often depends on getting the relevant documents; documents don't forget. But hard copy materials disappear altogether. Indeed, for some types of litigation hard copy materials may be disappearing. Consider medical records as an example; hard copy is probably on its way out for such records. Nowadays one follows the emails, not the money.[38] As the *Wall Street Journal* observed in 2003, "[c]orporate investigations used to mean following a paper trail, but these days many follow an electronic one."[39] The *ABA Journal* similarly reported, regarding an investigation by then-New York Attorney General Eliot Spitzer, that "[a]s with many damaging revelations produced in court during the last decade, these statements did not come from wiretaps… but from e-mails produced in the normal course of discovery."[40] In the words of a Houston attorney, "What I've found is that when you've got the e-mails, people remember lots and lots of things."[41]

Email may even make some discovery unnecessary. For example, an experienced trial lawyer recently suggested that email enables a lawyer to do a "blind" cross-examination at trial without having first taken the deposition of the witness: "The advent and indeed overuse of emails has made blind cross-examination through the use of documents even simpler, and the use of depositions superfluous. Witnesses

[38] This is a reference to a reported comment by "Deep Throat," who leaked details about the Watergate scandal to Washington Post reporters in the 1970s, and told the reporters to "follow the money."

[39] E. Byron, "Computer Forensics Sleuths Help Find Fraud," *Wall Street Journal*, March 18, 2003, B1. See also Dusting for Digital Fingerprints, *The Economist*, March 12, 2005, 32.

[40] J. Krause, "Discovery Channels," *American Bar Association Journal*, July 2002, 49, 49–50; see also A. Berenson, "Once Again, Spitzer Follows E-Mail Trail," *New York Times*, October 18, 2004, C1 (describing investigation of Marsh & McLennan); Dusting for Digital Fingerprints, *The Economist*, March 12, 2005, 32 (describing use of forensic computer techniques to solve crimes); O. Kerr, "Searches and Seizures in a Digital World," *Harvard Law Review* 119 (2005): 531–85 (exploring Fourth Amendment complications caused by increasing importance of digital evidence in criminal cases).

[41] P. Geier, "A Defense Win in Enron Country," *National Law Journal*, January 23, 2006, 6.

do not talk to each other anymore. They merely email each other. Hundreds of emails. Thousands of emails. We are drowning in a flood of emails, and they are all a part of E-Discovery. The overuse of emails has made blind cross a much less risky task. Each witness's testimony is set forth in his or her own words through a series of "instant messages" and endless email exchanges. People are disarmingly candid in emails. Admissions that would never appear in a letter or a formal memorandum permeate the email universe. People spontaneously write emails without hesitation and, more importantly, without reflection or thinking. All you have to do for effective blind cross is to have the witness repeat what he or she wrote and then merely ask if the witness believes it to be a true statement at the time he or she wrote it".[42] So the growth of E-Discovery may make the use of depositions (one of the features of American exceptionalism) unnecessary, or at least much less necessary.

Because of the importance of E-Discovery, there may be increasing clashes between the American appetite for E-Discovery and the greater privacy concerns in other countries. For a long time, many countries have had "blocking" statutes to prevent American discovery on their soil. But with electronic access to information, saying when discovery is on another country's "soil" becomes difficult. At the same time, many countries have adopted stringent digital privacy regulations. Electronic access via discovery therefore increasingly raises issues under those privacy regulations.[43] Increased globalization will magnify these strains.

2.2.3.2 E-Discovery May Change Corporate America and American Law Firms

Initially, Corporate America did not seem to focus on E-Discovery; a decade ago, some corporate litigators urged that the rules be amended to make clear that ESI was discoverable so they could get their clients to pay attention to it. Under the impact of *United States v. Microsoft* and a lot of other cases in which email evidence proved very forceful in court, American corporations have awakened to the reality that email could be viewed as the "corporate equivalent of DNA."[44] One reaction has been to urge employees to be more careful in what the write in email.[45] But the reality is that too many people are willing to write offhand messages in email, and that Instant Messaging and texting seem even more informal, and these activities increasingly are the subject of discovery as well. Another reaction has been to adopt an

[42] M. Neubauer, "Mastering the Blind Cross-Examination," *Litigation* 35 (2009): 23–6, 25.

[43] For discussion, see R. Marcus, "E-Discovery Beyond the Federal Rules," *University of Baltimore Law Review* 37 (2008): 321–47, 339–40 (describing the issues).

[44] N. Varchaver, "The Perils of E-mail," *Fortune*, February 17, 2003, 96.

[45] See E. Ben-Yahuda, "Sending Unwise E-Mails Can Be Hazardous to Your Career," *San Francisco Daily Journal*, October 11, 2004, 4 (describing formal efforts by companies to teach their employees to be guarded in what the write in email).

email program that has a self-destruct feature; after being read it will obliterate itself. One version of this form of email program is called Kablooey Mail.[46]

American corporations also spy on their employees' email. In 2001, it was reported that about three-quarters of U.S. companies monitored employee use of the Internet and spied on employee e-mail.[47] "Snoop" software has been developed to assist companies in doing this surveillance.[48] It may be that failure thus to monitor employee activities could itself expose a company to liability for failure to detect and prevent workplace harassment and similar claims; at least it seems that companies are regularly using electronically stored information to detect employee misbehavior. In part, this effort is prompted by an appreciation that this information could be obtained by plaintiffs through discovery.

American law firms seem also be affected. Indeed, one might say that whatever becomes a big deal for corporate America is likely to become a big deal for American law firms also. E-Discovery surely has become a big deal for law firms. A number of law firms have created E-Discovery departments. It may be that this is necessary as a matter of self-preservation for firms. According to one vendor, "[w]e have already observed… many companies changing counsel because of the lack of expertise of certain law firms regarding electronic discovery."[49]

This self-preservation by law firms may go beyond keeping clients; legal malpractice concerns loom in the background. According to two lawyers writing in the *National Law Journal* in December, 2007, "[i]n the context of electronically stored discovery, the skills and legal knowledge that might be deemed an essential part of 'competency' are rapidly changing with technological advances," and as a result it is "highly probable that malpractice claims will largely center on competency in advising clients as to preservation and production of E-Discovery."[50]

E-Discovery may affect the organization of law firms. The Chicago-based firm McDermott, Will & Emery, for example, reportedly created a new tier of attorneys – called permanent contract associates. Part of the explanation is that regular associates have become very expensive, and "electronic discovery has dramatically increased the amount of basic work that usually goes to these high-priced associates."[51] E-Discovery, then, may be an important stimulus in creating this new law firm employment position. And for all litigators, it may transform document review. Formerly conducted in client quarters, perhaps in remote locations, it may now involve sitting for days or weeks before computer screens while reviewing electronic documents.

---

[46] See A. LaVallee, "This Email Will Self-Destruct," *Wall Street Journal*, August 31, 2006, D1.

[47] See K. Livingston, "Battle Over Big Brother," *San Francisco Recorder*, August 30, 2001. 1.

[48] See J. Schwartz, "Snoop Software Gains Power and Raises Privacy Concerns," *New York Times*, October 10, 2003. A1.

[49] M. Arkfeld, "Growing Pains for the Amended Federal Rules," *The New Horizons of E-Discovery* (2007).

[50] J. Kwuon and K. Wan, "High Stakes for Missteps in EDD," *National Law Journal*, December 14, 2007.

[51] K. Schmitt, "McDermott Plans to Fill Cheap Seats," *San Francisco Recorder*, November 1, 2007, 1.

Law firms may also be more inclined to consider outsourcing because of E-Discovery. A January, 2008, article in the *San Francisco Recorder* reported, for example, that "[h]igh rates and the increasing bulk of e-discovery have pushed the associate general counsel of San Francisco-based Del Monte Foods to seriously consider using sources outside his law firm for the grunt work of litigation."[52] In February, 2008, another article reported that the Washington-based law firm Howrey was opening an office in India that "will handle document management in litigation."[53]

Finally, lawyers find themselves in the curious and unnerving situation of being at the mercy of E-Discovery vendors. Certainly vendors try to frighten lawyers from trying to do E-Discovery themselves.[54] As noted above, outside E-Discovery services have become a billion-dollar industry in the U.S. Whether or not lawyers really need all these expensive services is uncertain.[55] But it is somewhat delicious to contemplate the supposed predicament of the lawyers. "E-discovery has brought about a kind of role reversal in the legal profession: Now it's the lawyers who find themselves surrounded by circling sharks. Once an e-discovery vendor identifies an attorney or law firm as a potential client, there's often no end to the sales pitches, product demos, complimentary mouse pads, and follow-up emails from perky PR reps."[56]

2.3 The Impact of the Digital Revolution on Trial

2.3.1 Problems of Admissibility

As noted above, American trials operate under fairly elaborate and restrictive rules limiting what can be considered in reaching a judgment. Thus, evidence must be admissible under the rules to be considered by a jury, or even a judge if the case is tried to the court.[57]

[52] See Z. Elinson, "GCs Embracing Outsourced Work," *San Francisco Recorder*, January 24, 2008, 1. See also "Inside Out: Working the Split Shift at an Indian Legal Outsourcing Company," *American Lawyer*, March 2008, 20 (reporting an estimate that the number of people working at outsourcing firms in India tripled from 1,800 to 6,000 lawyers between March, 2005, and the end of 2006, and that document review projects are typically billed at $15–25 per hour).

[53] D. Eviatar, "Howrey Opens India Office for Document Management," *San Francisco Recorder*, February 11, 2008, 3.

[54] See J. Noble, "Dangers in E-Discovery," *Legal Times*, June 3, 2002, 15 (identifying "trying to go it alone" as the most common mistake in regard to E-Discovery).

[55] See E. Kircher-Allen, "Electronic Expertise," *California Lawyer*, October 2007, 9 (reporting that "even some E-Discovery consultants caution against the overuse of outside experts. Except in complex cases, 'a paralegal who has been sent to a workshop and trained on a piece of software can probably handle e-discovery' ").

[56] T. McNichol, "The E-Vendors Cometh," *California Lawyer*, February, 2008, 37.

[57] As a practical matter, however, the rules of evidence are applied in a less restrictive manner in court trials.

Although similar evidence rules apply to all evidence, admissibility of ESI raises special problems. Already, the American Bar Association has published a 450-page book on how to present digital evidence in court.[58] According to the judge who wrote the Foreword to this book, the author is "suggesting an entirely new way of looking at how digital information should be considered by the court, and the consequential need to create new rules of evidence based not on tradition but on how digital information actually comes into creation."[59] Whether that revolution occurs is presently uncertain, but at least three areas of concern can be noted.

At the outset, it is worth emphasizing that there is a wide variety of electronic materials that may be important as evidence. E-mail has already been mentioned frequently. But it is important also to note that most business and government records are also maintained on computers, often in dynamic databases. If one wants to prove a financial obligation or prove title to land, electronic records will probably become necessary. Beyond that, various surveillance-type digital devices may produce information that bears on a case. The card reader used to access premises may be used to show that somebody using a party's access card entered or left at a certain time. The automatic fare collection device on a bridge or a freeway may be used similarly to establish where a person was at a given time, as might a GPS reading for a cell phone. And the images captured by a video camera at an ATM or other location may be used as evidence of what happened there. But to be used, all these various items must be received in evidence and therefore qualify under the evidence rules.

Indeed, the availability of such materials may sometimes make trial unnecessary. For example, in 2007 the Supreme Court held that a videotape of a speeding motorist taken by pursuing police officers established as a matter of law that the officers were justified in ramming the speeding vehicle, causing it to crash.[60] Summary judgment in favor of the officers was therefore justified in a suit for injuries by the driver. Although this decision can be criticized,[61] it suggests the potentially central role of such material in resolving cases. The Supreme Court was sufficiently impressed by this evidence that it provided an electronic link to enable citizens to view the video themselves.

"Originals": Once known as the "best evidence" rule, the "original writing" rule requires that when one wants to prove the contents of a writing one must offer the original as evidence.[62] This can apply in a wide variety of circumstances; in most of

---

[58] See G. Paul, *Foundations of Digital Evidence* (Chicago, IL: American Bar Association, 2008).

[59] Id. at v (Foreword by Hon. John M. Facciola, U.S. Magistrate Judge, District of the District of Columbia).

[60] Scott v. Harris, 550 U.S. 372 (2007).

[61] See D. Kahan, D. Hoffman and D. Braman, "Whose Eyes are You Going to Believe? Scott v. Harris and the Perils of Cognitive Illiberalism," *Harvard Law Review* 122 (2009): 837–906 (arguing that jurors might have reacted differently to this videotape).

[62] See, e.g., Fed. R. Evid. 1002 ("To prove the content of a writing, recording, or photograph, the original writing, recording, or photograph is required, except as otherwise provided in these rules or by Act of Congress.").

the instances used as examples in the introductory material to this section above there would be an argument that proof of the contents was being offered. The contents would be used to demonstrate a debt, or prove title to land, or show that somebody accessed a building or crossed a bridge or used an ATM at a given time. And given the dynamic nature of electronic databases, producing the "original" in court could prove difficult.

Fortunately, this requirement should not often be a barrier to admissibility of evidence in U.S. courts, because usually a duplicate can be used in place of the original unless there is a genuine question about the authenticity of the duplicate.[63] Because a "duplicate" can include something created "by mechanical or electronic re-recording,"[64] there should usually be an easy way to satisfy this requirement.

Authenticity: But there is a considerably more important difficulty showing that the electronic evidence is authentic. Although ESI may last a long time, it may also be altered fairly easily, and sometimes it may be difficult to detect the alteration. Indeed, computers may alter or delete records as part of their normal operation, which gave rise to the protection against sanctions for such alteration even when it destroys information. For purposes of admissibility, the proponent of electronic evidence must show that it is a record of what the proponent claims it to be.[65]

Ordinarily, as with hard-copy materials, authentication is not much of a challenge. For example, testimony of a witness with knowledge that the item is what it is claimed to be should suffice.[66] But since the proof offered in electronic form is often not something that a witness can identify just by looking at it, the process can be much more elaborate. Indeed, the recent ABA book on such evidence spends four chapters covering 80 pages on how to achieve authentication with digital evidence.[67]

Hearsay: Hearsay is a "statement... offered in evidence to prove the truth of the matter asserted."[68] Some pieces of digital evidence may not be regarded as "statements." For example, a film of users of an ATM, or a machine-based reading on the speed at which a car was traveling might be regarded of free of hearsay dangers because (like a clock that a witness consulted to determine what time an accident happened) they do not depend on the veracity of a person.

But most digital evidence does depend on the veracity of a person who input the information sought to be proved in court. In any event, the proponent of evidence would usually need to establish the reliability of the electronic device upon which

---

[63] See, e.g., Fed. R. Evid. 1003 (providing that "[a] duplicate is admissible to the same extent as an original unless (1) a genuine question is raised as to the authenticity of the original or (2) in the circumstances it would be unfair to admit the duplicate in lieu of the original.").

[64] Fed. R. Evid. 1001(4).

[65] See Fed. R. Evid. 901(a) (providing that authentication is "a condition precedent to admissibility" and must include "evidence sufficient to support a finding that the matter in question is what its proponent claims").

[66] See Fed. R. Evid. 901(b)(1).

[67] See G. Paul, *Foundations of Digital Evidence* (Chicago, IL: American Bar Association, 2008), 33–113.

[68] Fed. R. Evid. 801(c).

the probative value of the evidence depends. In addition, the proponent will usually have to find an exception to the hearsay rule to surmount a hearsay objection. Very often digital evidence is offered as a "business record."[69] Other possibilities also exist,[70] but this foundation will often require advance planning.

Enduring and growing importance: Despite these admissibility issues, it is certain that ESI will prove increasingly important as evidence at trials. For one thing, much that was formerly in hard-copy format is no longer printed; only ESI can be used. For another, many important sources of information, such as ATM videos or building access monitors, are available only because of the invention of digital methods of recording information. And these various sources of evidence are becoming central to much litigation – both civil and criminal.

2.3.2 Modifications in the Method of Trial

The American trial has long resembled the Hollywood version. But it need not continue to do so.[71] In 1998, Dean Carrington undertook to peer into the possible future for trial in the age of "virtual civil litigation."[72] He concluded that "[t]he traditional trial is becoming obsolete."[73] He assumed that the role of the jury should remain sacrosanct, and that the jury should therefore be assembled in the courthouse to observe the "trial." But he saw no reason for the "trial" itself to be dependent on live testimony in the courtroom. To the contrary, as witnesses were likely to be dispersed over wider and wider areas, it would become more and more important to replace live testimony with recorded testimony, perhaps itself the product of discovery conducted by video conference hookup. That way, all the evidence could be recorded in advance and all evidentiary issues would also be resolved in advance. Meanwhile,

---

[69] See Fed. R. Evid. 803(6): "A memorandum report, record, or data compilation, in any form, of acts, events, conditions, opinions, or diagnoses, made at or near the time by, or from information transmitted by, a person with knowledge, if kept in the course of a regularly conducted business activity, and if it was the regular practice of that business activity to make the memorandum, report, record or data compilation…"

For discussion of the application of this exception to digital evidence, see Paul, supra note 67, at 113–50.

[70] E.g., Fed. R. Evid. 801(d)(2) (for statements of a party opponent, as for an e-mail from a party offered against that party); 803(1) (present sense impression for a statement describing something being observed by the declarant, as in an email message); 803(3) (present sense impression for a statement by a person about his or her present physical or mental condition, including the intention to do an act in the future); 803(4) (statements for purposes of medical diagnosis, such as e-mail messages to a medical provider); 803(8) (public records).

[71] Much of this discussion is drawn from R. Marcus, "E-Discovery and Beyond: Toward Brave New World or 1984," *Federal Rules Decisions* 236 (2006): 598–638.

[72] P. Carrington, "Virtual Civil Litigation: A Visit to John Bunyan's Celestial City," *Columbia Law Review* 98 (1998): 1516–37.

[73] Id. at 1524.

"trial counsel become co-producers of a multi-media presentation,"[74] and "trial advocacy will more closely resemble the work of a Hollywood film producer and less that of a Hollywood actor."[75] The jurors, having been summoned to the courthouse, would there watch the movie. But before that happened, the court could rule on any motions for judgment as a matter of law. And appellate review of all pretrial rulings (including motions in limine) would be accomplished before the "trial" since there would be no development at trial to await in addressing such rulings. There would thus never be a need for a motion for a new trial.

Other methods to exploit digital technology for the trial suggest themselves. Why not have the witnesses give "live" testimony by video hookup during the "trial" rather than relying on a pre-recorded video?[76] Why require the jurors to come to the courthouse to see that presentation; couldn't they view it online from home? And couldn't the jurors deliberate online – in a jury chatroom?

But will anything like this actually occur? At least some American judges have seemed willing to consider the possibility. Digital technology has become important in trials in recent decades because it permits simulations or recreations of events involved in lawsuits to an extent not previously possible.[77] Revising trial techniques as suggested by Dean Carrington, however, would involve much more aggressive use of digital technology. Some judges have suggested that such a step should be taken. In 1991, a prominent federal judge suggested using videotapes for testimony because "jurors are accustomed to acquiring information from the television screen and thus react favorably to video presentations."[78] Nearly a decade before that, another district judge offered the following vision that corresponds to Dean Carrington's: "If all testimony is by videotape deposition, the 'trial' concept would embrace simply the playing of the videotapes (subject to evidentiary objections) sandwiched between opening and closing statements. That could advance the trial date considerably, because the flexibility of scheduling (involving only counsel and the Court) would permit the trial to be placed in any available open date on short notice."[79]

---

[74] Id. at 1526.

[75] Id. at 1524.

[76] See Fed. R. Civ. P. 43(a) (providing that in "compelling circumstances" the court may permit a witness to testify by contemporaneous transmission from a distant location).

[77] Some see the introduction of these techniques as a momentous development. A law professor, for example, says that "[t]he use of electronic visuals is as significant as the introduction of cross-examination in the 1870s and formal discovery in the 1930s. This will be the greatest change in advocacy in the career of anybody alive or about to be conceived." L. Brennan, "Pitching the Gen-X Jury: As Jurors Get Younger, Law Schools Are Thinking More Like MTV," in *National Law Journal*, June 7, 2004, 1 (quoting Prof. Stephen Lubet of Northwestern University); see also H. Gottlieb, "Plaintiffs' Lawyers Have High-tech Advantage in Courtroom," *San Francisco Recorder*, February 28, 2006, 1 (reporting that plaintiff lawyers are more likely to use a "$1,500 a day technical director hired to spike the presentation with computer-generated graphics").

[78] W. Schwarzer, "Reforming Jury Trials," *Federal Rules Decisions* 132 (1991): 575–96, 588 (1991).

[79] Lucien v. McLennard, 95 F.R.D. 525, 526 n.2 (N.D. Ill. 1982).

Another decade earlier – in the early 1970s – the state courts in one county in Ohio embraced just such a method.[80]

Meanwhile, other innovations have been attempted to speed up trials. In the early 1970s, an Oregon federal district judge endorsed having the direct testimony of witnesses submitted in advance,[81] and a federal district judge from the District of Columbia wrote an article in 1983 urging that all direct testimony be submitted in written form.[82] Carrying the idea of trying the case based on written submissions a bit further, it has been urged that a court could determine from a summary judgment motion that an ordinary trial would add nothing of value, and urge the parties to agree to a "trial without witnesses" rather than summary judgment.[83]

As should be apparent from when these ideas emerged, such innovation in trial methods did not depend on current digital technology. And despite the enthusiastic endorsement of judges who developed these new techniques, they have yet to carry the day with most judges. Thus, even though video trials were introduced in the Ohio state courts in the early 1970s, the Ohio Supreme Court ruled in 1992 that a trial court could not ordinarily require unwilling litigants to have such a trial, noting that "videotape trials have not gained widespread use, and are all but confined to Erie County."[84] Similarly, requiring direct testimony in writing does not seem to have swept the land. Somewhat similarly, "[t]he federal rules have not changed the long-established principle that testimony by deposition is less desirable than oral testimony and should ordinarily be used as a substitute only if the witness is not available to testify in person."[85] Although a rule permits transmission of live testimony from a remote location, moreover, that is allowed only in very rare circumstances.[86]

The traditional format of the American trial has much to recommend it that might be lost were Dean Carrington's (or digital technology's) more aggressive possibilities to be pursued. Some adaptation to the present reality of pervasive digital

[80] For an argument in favor of this technique by the judge who pioneered it in Ohio, see J. McCrystal and A. Maschari, "Will Electronic Technology Take the Witness Stand?" *University of Toledo Law Review* 11 (1980): 239–54.

[81] See G. Solomon, "Techniques for Shortening Trials," *Federal Rules Decisions* 65 (1975): 485–95.

[82] C. Richey, "A Modern Management Technique for Trial Courts to Improve the Quality of Justice: Requiring Direct Testimony to Be Submitted in Written Form Prior to Trial," *Georgetown Law Journal* 72 (1983): 73–93; see also Kuntz v. Sea Eagle, 199 F.R.D. 665 (D. Haw. 2001) (denying plaintiff's motion that he be allowed to present direct evidence orally instead of in writing).

[83] W. Schwarzer, A. Hirsch and D. Barrans, "The Analysis and Decision of Summary Judgement Motions," *Federal Rules Decisions* 139 (1992): 441–523, 471, 474 (1992); see also Acuff-Rose Music, Inc. v. Jostens, Inc., 155 F.3d 140 (2d Cir. 1998) (upholding use of this approach if the parties forgo their right to a full trial).

[84] See Fantozzi v. Sandusky Cement Prod. Co., 597 N.E.2d 474, 480 (Ohio S. Ct. 1992).

[85] C. Wright, A. Miller and R. Marcus, *Federal Practice & Procedure*, 3d ed. (St Paul, MN: West Publishing Co., 2010), § 2142 at 626.

[86] See Fed. R. Civ. P. 43(a) (permitting testimony by "contemporaneous transmission for a different location" only in "compelling circumstances").

communication will be necessary. Increasingly, jurors come to court with the short attention spans that typify those addicted to "multitasking"; they will expect trial presentations to adhere to their expectations. Jurors may also be increasingly tempted to use hand-held devices to gather information beyond what's presented in evidence in court,[87] something that courts will have to try to prevent.[88]

But more aggressive changes for U.S. trials may not be adopted for a variety of reasons. First, as many civil law countries have recognized by shifting to a single continuous hearing, a free-standing continuous trial with live testimony can be a relatively low-budget event that is accessible to all litigants and the attending public. Replacing this simple method with a prerecorded substitute would mean that a party would have to depose its witnesses to produce the prerecorded testimony that would be substituted for their live presentations. And once begun, that prerecording effort might go far beyond a video version of a traditional deposition or even traditional trial testimony. For example, in a federal-court trial in Texas in the 1980s many of the witnesses were beyond subpoena range and the parties and court agreed to creation of presentations for the jury that contained a melange of materials from depositions. In the words of lawyers on the case, this produced "a presentation that closely resembled a television documentary or news report," that "was the creation of a production studio, and not merely the playback tape of a tape made in the deposition room."[89] In somewhat the same vein, it has been reported that "[a] new trend among attorneys is to present the jury with split-screen video deposition testimony… Jurors respond to this type of presentation because they are able to see witnesses testifying about each other at the same time, without rewinding or switching between screens as would be required with a traditional video format."[90] Much as the better-financed litigant has an advantage under the traditional trial format, one might hesitate about a substitute that more overtly makes the quality of the presentation dependent on the resources spent in producing it. As a California court said in declining to shift the cost of a successful litigant's high-powered computer presentation at trial to the losing party, "[i]f costs are routinely awarded for high-powered technology, most parties will be unable to litigate."[91]

Besides costs, the high-tech trial may drain important interactive features of the traditional trial. The human interaction of an American trial is more than window-dressing. Relying on technology to provide pre-recorded testimony could

---

[87] See, e.g. J. Solovy and R. Byman, "Confronting the Fact that Jurors do Research," *National Law Journal*, November 30, 2009, 23 (arguing that trial attorneys should assume that, despite judges' instructions not to do so, jurors will engage in illicit E-Discovery on their own).

[88] See, e.g., C. Miller, "New Bill Targets Web-Surfing Jurors," *San Francisco Recorder*, February 22, 2010, 1 (describing legislation introduced in the California Legislature to provide punishment for jurors who use electronic means to gather improper information or discuss the case in a forbidden manner).

[89] C. Buxton and M. Glover, "Managing a Big Case Down to Size," *Litigation Magazine* 15 (1989): 22–5, 22–3.

[90] A. Lagomarsino, "Strategic Use of Video Depositions," *Nevada Lawyer* 11 (2005): 8.

[91] Science App. Int'l Corp. v. Superior Court, 46 Cal.Rptr.2d 332, 338 (Cal. Ct. App. 1995).

significantly detract from these advantages of the traditional trial. As even one of the strongest proponents of the use of technology during trials recognizes, "recorded testimony lacks the immediacy of live testimony."[92] In addition, adroit cinematography could markedly alter the impact of a witness's testimony from what it would be it would be in the courtroom. As D.W. Griffith was the first to discover, skillful use of closeups can convey and emphasize emotions.[93]

Carrying things a step further and permitting juror hookup to the "virtual trial" from home could produce additional difficulties. For one thing, it would seem to deprive the court of any way to observe and police the jury while it was receiving evidence. It might not even provide an assurance that the jurors had actually watched the evidence that was received. In a somewhat analogous situation, a federal appellate court held that it was improper for the district judge to give jurors deposition transcript excerpts and tell them to read the excerpts at home.[94] Labelling this reliance on "evidence to go" or "takeout evidence," the appellate court objected that "[t]he jury's reading of the deposition excerpts is thus totally outside the supervision of the judge."[95]

Relying on video conference technology might satisfy some concerns, but still seems unlikely to recreate that opportunity of the party or witness to feel that she has fully told her story to the decisionmaker. As one opponent to reliance on video conferences for criminal defendants says, "video conferencing technology cannot replicate normal eye contact."[96] Just as this mode limits what the factfinder can see of the witness, it also limits what the witness can see of the factfinder (if anything).

Finally, some possibilities of juror reliance on high-tech methods to deliberate can be unnerving. Although one could consider permitting deliberation by chat room, that would weaken or eliminate the face-to-face interaction of jurors in reaching human decisions that seems an important aspect of the decisionmaking process. In part, effective deliberation depends on collaborative experience jurors develop before they begin to deliberate, an experience that would not seem equivalent if experienced online. The fact that the jurors usually select their own foreperson as they begin deliberations is a recognition that they depend on some mutual familiarity before deliberations begin. And the course of the actual deliberations depends significantly on the sort of attention and participation that actual presence provides, but that virtual presence might not. At some point, for example, it might be clear from the traditional method that a certain juror is not participating in the

[92] F. Lederer, "The Road to the Virtual Courtroom? A Consideration of Today's – and Tomorrow's – High-Technology Courtrooms," *South Carolina Law Review* 50 (1999): 799–844, 819.

[93] See M. Roth, "Comment, Laissez-Faire Videoconferencing: Remote Witness Testimony and Adversarial Truth," *University of California at Los Angeles Law Review* 48 (2000): 185–219, 202.

[94] Stine v. Marathon Oil Co., 976 F.2d 254 (5th Cir. 1992).

[95] Id. at 267.

[96] A. Poulin, "Criminal Justice and Videoconferencing: The Remote Defendant," *Tulane Law Review* 78 (2004): 1089–167, 1111.

deliberation process, but it is difficult to see how a similar determination could be made if the jurors were in different locations and interacting only via the Internet. Undercutting this feature of the traditional trial would further weaken the decision-making process.

2.4 Concluding Spectaculations

The foregoing is mainly speculative in the sense that few of the more aggressive uses of E-Discovery or "virtual trials" have actually been attempted, and even when they have they have been the exception rather than the rule. The American legal system changes slowly. Although that cautions attitude may frustrate some, it can also be reassuring in light of some of the issues raised by adopting new practices emphasizing digital communication.

At the same time, it is almost certainly true that both evidence gathering and evidence presentation will change substantially due to the increased use of digital methods. To some extent, that could encourage common-law jurisdictions to gravitate to an "electronic dossier" rather than live trial events. And it could also result in greater (or at least less burdensome) access to evidentiary materials in traditionally civil law systems.

The potential impact on established American litigation is becoming apparent. A February, 2010, report on the handling of these issues in the New York state court system provided the following reaction to the various characteristics of E-Discovery: "All of these characteristics make it virtually impossible to be able to locate and produce all potentially relevant ESI in an easy and cost-effective manner. Given these realities, the traditional adversarial approach to discovery can be a recipe for disaster in the context of E-Discovery, fostering delay, driving litigation costs to unacceptable levels, and forcing judges to issue multiple written decisions to advance the case to trial. Practitioners and judges confirm that E-Discovery costs in New York often reach levels that are far out of proportion to the actual value of the matters in dispute. And although these costs are most commonly associated with commercial litigation, where E-Discovery can quickly run into the millions of dollars, it is equally problematic in the full range of civil, family, and criminal cases involving electronic materials."[97]

The report concluded, therefore, that change was essential because "the New York court system's standing as a leading forum of both national and international litigation is at stake."[98] Adversarialism is not the only value, after all, and the American commitment to it may have to bend considerably to cope with E-Discovery.

---

[97] Electronic Discovery in the New York State Courts: A Report to the Chief Judge and Chief Administrative Judge 6 (Feb. 2010).

[98] Id. at 8.

For American exceptionalism, therefore, a key question is whether there will come a point soon when even proponents of broad discovery say "Enough is enough." Certainly a significant number of lawyers believe that the era of demanding production of all materials that "relate or refer" to the topic of the lawsuit is passing. Whether that shift in attitude will become widespread, or will become widespread soon, is quite uncertain. Even now, a surprising proportion of American lawyers seem not to rely on or employ E-Discovery, and they probably would not accept any of the more esoteric uses of digital technology at trial. Until a new generation takes their place, therefore, things may continue more like the past.

Chapter 3
Technology and Civil Litigation in the United States in the Twenty-First Century

Paul D. Carrington*

3.1 Traditions of the Democratic Courthouse

As observed in the other papers presented to this conference, the cost of civil litigation is inevitably a cause of chronic dissatisfaction with any legal system. The concern for cost has perhaps been greatest in the United States for several reasons associated with the distinctive role of civil litigation in the nation's political system associated with the chronic weakness of the political structures erected to govern a large territory. That role of civil litigation is centered in what it pleases me to describe as The Democratic Courthouse. My question for the day is: can the traditions of the American Democratic Courthouse withstand the shock of contemporary and forthcoming technologies bearing on the methods and costs of litigation? I hope so, but there is perhaps cause for doubt.

The American Democratic Courthouse can be said to have originated in the late eighteenth century in the former colonies that had won their War for Independence. To be sure, some of its features originated in the English tradition of trial by jury, an institution designed to afford a measure of protection for weak and mistrusted courts against corruption and intimidation. The right to jury trial necessarily carried with it the right to a public proceeding at which evidence is presented orally. Somewhat inconsistently with the civil juries, local American judiciaries presiding over them were sometimes endowed with the extraordinary powers formerly conferred only on the English Chancellor who had long decided cases in the Court of Chancery as the

*Professor of Law, Duke University. This essay is an updated edition of Virtual Civil Litigation: A Visit to John Bunyan's Celestial City, 98 Colum. L. Rev. 501 (1998), republished as Virtuelles Ziviverfahren in den USA: Ein Besuch in John Bunyans Himmlischer Statt in ZZZ International (Spring 1999) and in American Bar Association Judicial Division, The Improvement of the Administration of Justice 477 (7th ed 2001).

P.D. Carrington (✉)
Duke School of Law, Duke University, Durham, NC, USA
e-mail: pdc@law.duke.edu

M. Kengyel and Z. Nemessányi (eds.), *Electronic Technology and Civil Procedure*, Ius Gentium: Comparative Perspectives on Law and Justice 15,
DOI 10.1007/978-94-007-4072-3_3, © Springer Science+Business Media B.V. 2012

right hand of the monarch, with monarchical powers and discretion to compel conduct by unwilling citizens.

A novel additional feature of the early democratic courthouse was the deregulation of the legal profession in the former colonies. Much of the elite of the colonial legal profession had fled the revolution (mainly to Canada), And the social order of the newly developing western frontier spreading across the continent was especially unreceptive to claims of elite status. So it was that Abraham Lincoln, who never attended a school or received any form of professional instruction, could be licensed to practice law in Illinois by a single judge acting on the suggestion of another lawyer. And, notably, he was, by any standard a very, very good lawyer.

By the middle of the nineteenth century, these traditions of civil juries, the judicial contempt power, and untrained lawyers, had merged in the architecture of the courthouses situated at the center of almost every American town of size. Most of the judges sitting in those central courthouses were elected to their offices by vote of the people. That practice of electing judges had evolved in the nineteenth century as the states' citizens recognized that many consequential political decisions were being made by judges who had no particular qualifications to make them except their acquaintance with elected governors or legislators.

The nineteenth century American courthouse was thus a center of local public attention. Indeed, as recently as the mid-twentieth century, it was unsurprising to find citizens gathering on the steps or in the corridors of their courthouses simply because the conduct of trials in democratic courtrooms was sometimes perceived to be the best show in town.

A reflection of the dramatic role of the American judiciary is the awesome structure in which the Supreme Court of the United States sits. It was erected in 1937, but not without criticism by the eminent Justice Louis Brandeis who thought it too pretentious and likely to cause excessive vanity and self-indulgence on the part of the Justices. He was in my view correct in that prediction.

3.2 Private Enforcement of Public Law

Meanwhile, by the early decades of the twentieth century, the special role of the federal judiciary appointed by the President with the consent of the Senate had begun to acquire a greater political role. This enlargement was most directly the result of the advent of a national economy united by railroads to span the continent. Economic development advanced the interests of large business firms that were not always attentive to the conflicting interests or rights of distant citizens, including their remote workers, consumers, competitors, investors, or tenants, all of whom might be exploited in one way or another by distant corporate managers.

The federal Congress took full notice of the development of the national economy with the enactment in 1890 of the federal antitrust law. That law relied heavily on private enforcement in civil actions brought by competitors claiming to be harmed by the illicit practices of their distant rivals. To encourage such private enforcement and deter harmful business practices by marketplace bullies, Congress assured successful plaintiffs of a reward of treble damages. This feature could be said to be derived from the English common law practice of awarding punitive damages to civil plaintiffs proving their defendants to be guilty of egregious wrongs that the royal government might wish to deter.

While Congress was encouraging private enforcement of its new antitrust law, it also commissioned its new Department of Justice to prosecute offenders. And it also detected a need to regulate other business practices of the railroads doing business on a continental scale. It therefore created The Interstate Commerce Commission to control practices of railroads that were harmful to the interests of shippers, passengers, and local communities. But when Congress faced the need to protect railroad workers injured as a result of dangerous business practices, it left that task to the federal judiciary who would be advised by civil juries informed of the events by lawyers serving for fees contingent upon their success in deterring recklessness by employers by compelling them to bear the full cost of workers' injuries. That model of federal tort law was soon also applied to seamen put at risk by their work aboard ships.

As in many other nations, the twentieth century occasioned the creation of many administrative agencies similar to the Interstate Commerce Commission. These were branches of state or federal governments and were empowered to prevent harmful business practices by close oversight enforcing rules promulgated by the public agencies to prevent specified misdeeds. These public agencies were generally designed to be somewhat independent of political branches of government. Some succeeded in performing their assigned roles as regulators. The United States Food and Drug Administration established in 1906 stands out as an enduring success. But it is a marker of American politics that the FDA was not given responsibility for meat safety because meat producers had sufficient political influence on Congress to confer that duty on the Department of Agriculture that is more responsive to their interests. Similar political efforts succeeded over time in capturing many regulatory agencies by assuring the fidelity of their officers to the business interests that their agencies were designed to regulate. As it became evident that many public agencies were being captured by the firms they were established to regulate, both state and federal governments came increasingly to rely on private plaintiffs and their unregulated lawyers asserting civil claims as a crucial means of protecting the safety of citizens, the integrity of the marketplace, the environment and other public interests.

The insurance industry serves as an example of weak enforcement of American public laws needed to protect citizens and their resulting dependency on private enforcement of their rights. Insurance companies that may be engaged in business on a global scale are regulated only by weak agencies of state governments, and are often permitted to engage in business practices that observers might reasonably deem to be dishonorable to their policyholders, unjust to claimants, and contrary to the public interest. So they are sometimes sued by citizens who sometimes recover punitive damages assessed by civil juries to deter their abusive practices.

A recent example is provided by a case in which a Utah state court, on the advice of a jury, assessed millions of dollars in damages against State Farm Insurance on proof that it had engaged in systematic racial and gender discrimination in settling claims of accident victims. It had instructed the lawyers it hired to defend claims against its insureds to settle claims made by white male plaintiffs, but to go to trial in every case brought by a woman or a citizen of color. The state insurance commissioner had not addressed the problem even though the practices were a clear violation of the state's law. Thus, exposure to civil liability is how Americans are protected from the greed of their insurers.

I should also note the example provided by the internationally celebrated case of the old lady who seriously harmed herself by spilling a cup of coffee purchased at a drive-in window of a McDonald's restaurant in New Mexico. Extensive skin grafting was required to treat her injury. McDonald's refused to pay her medical expenses. It was at trial revealed that she was by no means the first victim of a national policy of McDonald's requiring coffee sold at the windows of its franchisees to be dangerously hot. The policy reflected management's belief that many buying coffee at take-out windows want it to be very, very hot. McDonald's was made to pay not only her hospital bill, but a sum sufficient to deter it from maintaining its dangerous policy. That civil judgment served its purpose; McDonald's coffee was cooled by a few degrees to prevent future serious scaldings. But in the end she received just enough to cover the costs of her injury.

These examples show that the United States and the 50 states have come to rely heavily on private civil plaintiffs to enforce their rights against business interests and thus deter business practices that offend the rights of consumers, workers, passengers, tenants, investors, patients, or insureds. And this dependence on private law enforcement is the reason that American courts extended the reach of their jurisdiction in civil cases over distant defendants who may have caused harms to local citizens or firms. Accident victims in Utah thus do not depend entirely on the public officers of Illinois, or of foreign public law enforcement officers, to protect them from abusive practices but can invoke their local jurisdiction over their civil claims.

It is thus often the threat of such civil liabilities imposed at the democratic courthouse, more than the cautions and restraints imposed by public officials that protect citizens from unlawful practices by big firms engaged in business on a continental scale. This is an important difference from legal systems of nations that can entrust the enforcement of public law to their public officials, and that can therefore regard civil procedure simply as a means of resolving private disputes.

3.3 Discovering Electronic Documentation

Modern technologies may enhance, but also may diminish, the effectiveness of private enforcement of public law. As my two examples illustrate, the most important feature of this system of private enforcement of public law may be the empowerment of litigants to gain access to evidence needed to prove their claims. Private lawyers exercise investigatory powers that in many nations would be conferred only on public officers.

While this discovery procedure had an ancestry in the practice of the English Court of Chancery, the right to examine other peoples' files was noticeably expanded by the Federal Rules of Civil Procedure that became federal law in 1938 and were soon replicated in state laws. At that time, it became the custom to allow civil parties to "pry into the case of their adversaries to learn its strength or weaknesses," prying that they had not previously been permitted to do. "Fishing expeditions" became the order of the day. In 1947, the Supreme Court proclaimed that "mutual knowledge of all the relevant facts gathered by both parties is essential to proper litigation." As Geoffrey Hazard put it, discovery became almost a constitutional foundation of our law. Congress and state legislatures came to depend upon private claimants to investigate possible violations of many laws they enact.

Thus, in the Utah case mentioned, a private lawyer representing a private client for a contingent fee, investigating pursuant to a court order entitling him to search the files of the State Farm insurance company, found correspondence by management at the company's headquarters in Illinois addressed to its local managers in Utah directing them to settle quickly claims brought by rich white males, but to resist vigorously those brought by mere women or members of racial minorities who are more easily intimidated by aggressive resistance. These documents outraged the civil jury and Utah's elected judiciary who awarded the plaintiff many millions of dollars in punitive damages, enough to tell State Farm managers not to do it again. And in the New Mexico case previously mentioned, it was a search of McDonald's national files by the plaintiff's lawyer that revealed that scores of customers had been seriously hurt by McDonald's practice.

The extent of public reliance on discovery by private claimants is confirmed by federal and state laws opening public files to private citizens and their lawyers. Government offices often supply documentary evidence to civil litigants.

There have of course long been protests to be heard from business interests decrying interference with the marketplace by civil plaintiffs invoking the powers of the democratic courthouses to investigate possible wrongdoing by their firms. The development of electronic communication has in the last two decades or so elevated the shrill tone of some of the expressions of business discontent. A reason may be that word processing has made it easier in many situations such as those involving McDonald's or State Farm Insurance for the civil plaintiff to search the files of a defendant to uncover evidence need to win their cases. The ability of private plaintiffs to enforce public law has in this respect been improved by technology. So some of the grief expressed by the Chamber of Commerce is protest over the costs of complying with laws protective of citizens and the public interest, laws that firms might be tempted to disregard but for the existence of electronic business records that are more extensive and more accessible to plaintiffs seeking to enforce the law.

On the other hand, there is the countervailing consideration that electronic information storage may in some situations elevate the ability of a plaintiff to impose high costs of document searches on a defendant. This is so because an excess of information elevates the cost of finding the pertinent or crucial evidence. A tactic sometimes employed is the suffocation of a civil adversary with uninformative documents.

McDonald's or State Farm could perhaps impose a cost on the contingent fee lawyer representing their civil adversaries by responding to discovery requests with files of millions of electronic documents including all their e-mail for the years in question. A 1993 amendment of the Federal Rule directed the parties to make a preliminary exchange of information about the data that might be available and that might be sought by a party. But a judge would then often be needed to impose reasonable limits on the disclosure.

A countervailing and complicating consideration is that an electronic search by a plaintiff in such cases may sometimes reveal embedded information that for some specific public reason should not be disclosed to the civil adversary. The right of access to an adversary's file is not absolute. There are evidentiary privileges such as the right of parties to receive in confidence advice from their lawyers. And there are other rights to privacy. Hence the defendant in a civil action may feel the need to study its electronic files before disclosing them to the adversary, and in some instances this may prove to be very costly. Who should bear the cost of sorting through an excess of data is thus a question now frequently arising in civil litigation. On this account, the cost and consequences of electronic document discovery has become a major concern in the last decade.

American business firms are able sometimes to avoid the lash of public law by avoiding civil proceedings in which their files are subject to discovery by adversaries. A way to pursue that hope is by writing mandatory arbitration clauses into standard form contracts imposed on consumers, workers, investors, franchisees, and their like. An arbitrator is empowered to order a party to produce documents, but unlike a litigant in a civil court, a party in arbitration has no right to conduct an investigation of an adversary's files.

It is possible in some situations that the jurisdictional reach of an American court may be extended as a result of the availability of electronic documents confirming the contacts of the distant defendant with the local court. An illustration of the jurisdictional question was provided in 1982 in an action brought in Pennsylvania by an African firm advancing a claim against its Irish insurance company. The defendant insurer challenged the jurisdiction of Pennsylvania over its person. The African plaintiff moved to discover the insurer's business records to show that it had sometimes done business in Pennsylvania and was therefore subject to its jurisdiction. The Irish insurer refused to produce the documents requested on the ground that they were located in London and therefore unavailable in Pennsylvania. This contention was rejected by the court. When the defendant defied the court order, the court simply inferred that the undisclosed files would have revealed earlier transactions in Pennsylvania sufficient to justify the exercise of personal jurisdiction over the defendant and made a finding of fact to that effect. That those files are now in electronic form eases the African plaintiff's claim to the right of access to the Irish insurer's records and gives comfort to the court in imposing jurisdiction on the uncooperative defendant who resists disclosure. A word search of the insurer's files for "Pennsylvania" would in most instances go a long way toward resolving the issue in such a case.

The relevant Federal Rules of Civil Procedure were revised again in 2006 to address concerns about the cost and consequences of discovery in civil cases of electronically stored documents. The controlling rule now envisions a preliminary discussion between the parties to identify the appropriate limits of useful and appropriate document discovery. And the court is responsible in complex cases for working out an agreed plan. Only documents that are "reasonably accessible" are subject to discovery. But some judges are better than others at performing the duty to make a balanced assessment of situations. And million-document disclosures do sometimes occur.

It bears notice that parties are ill-advised to destroy documents selectively in anticipation of a discovery request. A party is not free to downgrade an accessible database into an inaccessible format. One who does so is exposed to the prospect that the court will, as in the case of the Irish insurance company, draw the inference that the adversary's allegation would have been confirmed by the destroyed data.

It must be conceded that business interests represented by the Chamber of Commerce, and others as well, continue to protest the costs of electronic discovery. No one suggests that the 2006 solution is complete. As noted, those costs can in some cases suffocate the factual truths. But the complaints of Business regarding the cost of document discovery in commercial and public law cases has been substantially overblown. A 2009 empirical study by the Federal Judicial Center confirms that, with the exception of a very few outlying cases, the cost of discovery and related pretrial proceedings remains minor in relation to the stakes in the cases, ranging from 1.6% to 3.3% of the amounts in dispute. The question is what can judges to do in the outlying cases?

3.4 Electronic Imagery of Disputed Events

Electronic imagery, as others have noted, is an even more formidable threat to the premises of traditional civil litigation than is the access to electronic documents. Of course photography has been in use by fact finders for a century and a half. And its uses as evidence of past events have been greatly increased, and will surely continue to increase as more and more of our lives are recorded in visible instruments. It seems reasonable to expect that visible recordings of events will tend to improve the accuracy of judicial fact-finding.

But perhaps that expectation may be somewhat exaggerated. We were provided with a striking example of the problem in a decision rendered by our Supreme Court in 2007. The case involved an automobile accident. The event was recorded by the police car that was involved in the event. The lower courts viewed the motion picture and accepted the injured driver's account of the event. While reviewing the case for procedural error, the Justices of the Supreme Court were drawn to the motion picture and decided that the film clearly confirmed that the fault was not that of the police but of the driver-plaintiff. This was an extraordinary departure from the assigned role of the Court, which is to review only rulings of law, not fact.

An essay published in 2009 in the Harvard Law Review records an empirical study in which many citizens were asked to view the same crucial motion picture and decide who was at fault. A substantial majority of the viewers disagreed with the majority of the Supreme Court Justices. Thus, it seems that even video recordings of events are sometimes less conclusive than we may be inclined to suppose. That decision in the Supreme Court tends to confirm the wisdom of assigning the task of viewing videotapes of accidents to groups of citizen-jurors.

The frequency of the availability of such visual evidence is rapidly increasing as technological advances have put in the hands of many citizens the tool needed to make a motion picture of events at hand. Many retail stores are visually recording all manner of conduct for the purpose of deterring larceny. And many public places are subject to similar recordings made by police to detect and discourage street crimes. Most environmental harms can be amply recorded by complainants. Or disproven by films recorded by firms accused of causing them. It seems likely that experience will teach high court judges to defer to courts of first instance in assessing such visual records. But the noted Supreme Court decision stands as a contrary indicator.

Thus, as a result of the technological development easing the task of making video recordings, it does seem likely that there will be less dependence in the future on the observations of eye witnesses who were present at the time and place of events in dispute. This seems to be a benign trend. Modern science tends to confirm that eye witnesses are often unreliable and that judges and jurors who hear the testimony of such witnesses are often poorly prepared to assess their credibility. We may therefore hope that our courts could be more accurate in applying the law to disputed facts when they are presented with more videotapes recording the events in dispute. But the human failings that limit our ability to discern accurately the facts in dispute are, alas, not eliminated by the presence of the videotape.

3.5 Electronic Imagery of Testimony

Of course traditional litigation almost everywhere has been heavily dependent on oral testimony. Witnesses are summoned by our courts to respond in public at the democratic courthouse to questions asked by the counsel who requested the summons. Their testimony is given under oath to assure that the witness is cognizant of the law of perjury. Adversary counsel can then cross-examine a witness to test his or her credibility and information.

In the twentieth century, it became common to summon a witness to give testimony at a deposition. This is a proceeding conducted outside the court, perhaps in a lawyer's office. But a public official is there to administer the oath ritual to assure that the witness is aware that perjury would be a crime, and then the testimony is officially recorded. Traditionally that record would be in the form of a transcript prepared by an official trained to record testimony in short hand. And adversary counsel is invited to attend and cross-examine the witness under oath, and on the record.

Parties to civil litigation may be summoned to submit to deposition at a time and place prior to trial that is reasonably convenient. Their employees may also be examined. And so may independent witnesses having no connection to the parties, but they may be summoned only to reasonably convenient venues. But within the United States, any citizen is subject to a subpoena to be deposed to give evidence in any civil proceeding anywhere in the United States.

The deposition of witnesses is of course an important method of trial preparation enabling counsel to foresee and respond to the case that will be presented by an adversary. If necessary to secure a witness's testimony, he or she might be summoned to appear anywhere in the world that is reasonably convenient to submit to such an examination. Switzerland and some other nations regard the taking of a deposition within its borders as an offense against its national sovereignty. But citizens of the United States who reside abroad have an unqualified duty to give testimony under oath if summoned by any American court to do so. If the deponent is unavailable to appear at the trial, his or her deposition testimony may be presented in the form of the official written transcript that might be read aloud to the court and incorporated into the official record of the proceeding.

Since 1993, civil litigants have been permitted to record a deposition on videotape to be shown at trial if the witness is unavailable. And if the testimony and examination by counsel is recorded on videotape it might be shown at trial if the witness is unavailable. The video may be as effective as if the witness were actually in the courtroom. Indeed, contemporary technology is now able to record with intimacy the face and manner of the witness so that the ability of the observer to assess the credibility of the testimony may be even greater than if the witness were to appear in person in the courtroom.

Contemporary technology also enables the possibility of presenting testimony in court by satellite. In some circumstances, parties may strongly prefer to have the witnesses testify openly at the very time of trial. Imaginably, one might today be able to conduct a trial in Chicago at which witnesses are heard and cross-examined while they are appearing before cameras in Pésc or Tokyo. And the intimate display of their faces and reactions may be more informative than would be obtained by their actual presence in the courtroom.

3.6 The Possibility of the Virtual Trial

Perhaps these developing technologies will lead us to radical change in the conduct of civil litigation. In my 1998 essay on this subject, I suggested radical reform. Given easy, almost costless, preservation of images in digitized form, and their instantaneous transmission over long distances, there will no longer be sufficient reason to require, expect, or even permit much, if any, evidence to be presented in the form of oral testimony by witnesses in a room in which the judge, jury, and counsel are all present. A trial might normally become a movie presentation.

To be sure, there will be something lost in spontaneity and in the interpersonal contact between witnesses and those assessing their veracity, but those benefits will be overwhelmed by the saving in cost, time, and convenience attainable by means of digitized communication of testimony. People will be unwilling to travel distances at a particular time to await the presence and availability of others when images can be made at almost any place or time, preserved, and transmitted as and when needed, painlessly and almost without cost at any time to any place.

Perhaps it will always be desirable to preserve the authority of the court on special occasions to allow or require a party to appear and testify in a traditional courtroom, or perhaps even to compel an independent witness to do so. That might be most appropriate where two observation witnesses are engaged in a swearing match on a crucial issue in a case involving stakes large enough to warrant the increased cost; on such occasions, demeanor evidence that is conveniently available may arguably be worth the cost. But jurors will always have the opportunity to observe demeanor in the recorded testimony when it is screened. It seems at least possible that what is left out when testimony is observed on the cool screen is the part of the demeanor evidence that is positively misleading, for those radiations of the spirit that cause us to be irrationally attracted to a witness or irrationally repelled by him or her may then be less intense. It may actually be harder to lie effectively on a screen than it is in person.

Indeed, it is reported that scientific evidence indicates that we are more likely to be deceived by lying or mistaken witnesses when observing their testimony in person than if we were to review a paper transcript of the testimony. Witness presence, in other words, may often harm, rather than improve, the accuracy of credibility assessments. It seems that science is not quite ready to tell us whether testimony given on a screen is more or less reliable than that supplied in a document. Nor, it seems, is science quite ready to provide us with a reliable lie detector test, although the technology of brain scanning indicates the possibility that such a device might some day be developed.

There would in any case be no problem in providing public access to the virtual trial. Some federal courts in 2010 have begun to post audio recordings of their proceedings so that they are accessible to the public. Those who appreciate the dramatic value of legal contests could have their needs met without necessity of going down to the democratic courthouse and waiting on the steps for a lively trial. They can be e-mailed a video transcript to view on their television sets in the comfort of their living rooms.

On the assumption that testimony in the future will be presented electronically, we can see that trial counsel may become co-producers of a multi-media presentation. Testimony will be recorded in advance of trial and reviewed by adversary counsel, much as documentary exhibits presently are. All evidentiary issues not resolved by agreement of the parties would be resolved by the court at pretrial conferences. Because all the proof would be unalterably recorded before any of it is presented to a trier of fact, every evidentiary issue could be resolved *in limine* (before trial). This would result in a clean visual recording of all the testimony and arguments of counsel to be presented, with no distractions from bickering among lawyers and judges. The pretrial bickering will have occurred and will have been recorded, but not as part of the trial tape to be seen by the trier of fact. The possibility of tactical surprise at trial would be completely eliminated, and quickness of wit as a professional trait of trial counsel would be much devalued, being displaced by traits valued in television announcers and directors of drama.

The pretrial conference conducted by the judge would also provide an occasion to consider whether the material supplied by counsel is as a whole worthy of consideration by a trier of fact. The issue whether a genuine issue of fact is presented would be presented by a motion for judgment as a matter of law. Granting the motion at this point would have the same effect as the "old-fashioned" directed verdict or summary judgment. Given that the entire trial could be previewed on videotape with complete confidence that nothing can later be added or subtracted and a party's factual contentions are seen as legally deficient, there would be no reason to summon jurors, schedule a trial, and conduct an actual submission of the proof, before aborting the proceeding.

Given the opportunity of counsel in such a system to edit the video evidence, long trials would seldom be required. There would be no interruptions for objections, exceptions, or sidebar conferences of any kind. A long movie would be interrupted for bathroom and lunch breaks, but would otherwise run from eight in the morning to five in the afternoon. It should in most cases be possible to return to the time when almost all civil trials were conducted in a single day, or at most two. Because the value of evidence can be accurately appraised in advance of trial, there is no reason to allow lawyers to use valuable court time with rambling presentations. It would seldom be in the interests of lawyers or their clients to overstay the court's welcome, or risk stretching the attention span of jurors.

Given such a change in the mode of presenting evidence and the resulting ability of parties to know for certain the contentions of their adversaries, the use of expert opinion testimony could be reduced. Courts could more readily and more wisely make pretrial rulings on the merit and pertinence of scientific or technical proof. This would be so in part because the court would not in every instance be required to speculate on what the experts might say at trial, for judges would know this precisely at the time of the *in limine* ruling. This reality might in turn be expected to lead to more prudent and effective use of independent, disinterested scientists or technicians to advise the court in determining the utility of opinion testimony.

It also seems likely that, in some cases, the competing experts might be wisely replaced by the single disinterested witness appointed by the court. Such disinterested testimony would be more readily available because the experts would never need to leave their offices or laboratories to testify, and their consultations could be scheduled to fit their reasonable convenience. Use of such disinterested testimony in lieu of a battle of experts could abbreviate and reduce the cost of many trials.

Once the digitized trial has been produced by counsel and all issues between them resolved by the court, the court in a case to be tried by jury could prepare a charge to the jury that would be presented in both written and video form. Opening and closing arguments could also be recorded in video form, diminishing somewhat the influence of the theatrical flamboyance of counsel. The trial would then be a screening of the complete and edited tape for the jury containing not only the evidence but the closing arguments of counsel and the judge's instructions on the law. There would be no particular reason for a judge to be present during that screening. A judge would be needed to preside over the selection of jurors, an event that would likely occur in the courtroom on the eve of trial, and to provide an appropriate ceremony at the start and end of the trial, but it need not be the same judge for all the events. Counsel and the parties would likely wish to be present (at least virtually present), if for no other reason than to observe the jurors as they observe the screening of the proof and return their verdict.

Because few trials would extend over more than 2 days, jury service would be much less onerous, making it possible for more citizens to participate. Such trials would also be less a burden on the public fisc. It is not difficult to imagine that the jurors could remain at their home or workplace and become a virtual jury. As Boris Bittker suggested, "panels of couch potatoes" could quickly announce their verdicts "on everyone's electronic bulletin board." But if trials are to be as brief as I suggest, it would seem to be unnecessary as well as impolitic and imprudent to leave the jurors at their homes. A virtual jury would not provide the same satisfaction to civil litigants that live jurors do, at least to those who seek emotional gratification from the resolution of the dispute or the public exposure of an adversary. Moreover, the deliberation of the jury would be seriously impeded by the absence of social contact. By retaining the citizen jurors to render a verdict, the virtual trial could retain the elements of drama and some involvement of real people. Therefore, the courtroom of the future might do without a witness box or a bench, but it would need a jury box in front of the screen.

The virtual non-jury trial would, of course, closely resemble the virtual jury trial. But the formal screening of the prepared tape might almost as efficiently be seen by a judge other than the one who ruled on the pretrial issues. This change of judges would avoid any prejudice that might have been aroused by the pretrial bickerings and the judge's familiarity with evidence that has been excluded. In lieu of the recorded instructions to the jury and closing arguments, counsel would prepare proposed findings and conclusions.

Whether presented to a jury or judge, the virtual trial would not only be shorter, but it would greatly facilitate efficient scheduling. The precise length of each trial will be known in advance. And because of the limited function of the presiding judge, the trial will not materially interfere with the scheduling of other necessary conferences with judges. Greatly reduced, if not almost eliminated, will be the time lost by lawyers and witnesses waiting around the courthouse for their turn to be heard.

3.7 Virtual Appeals

The digitized trial would lend itself to expeditious review. Roscoe Pound long ago proposed that post-trial motions be consolidated with the first-level appeal, thereby eliminating a redundant stage in the process. Digital technology allows us to go Pound one better. If there is to be a jury trial, it would be most efficient to conduct the appeal before the jury is summoned. Because almost all the court's rulings will have been made before trial, interlocutory review of the pretrial rulings would be feasible and more efficient.

Pound envisioned that the post-trial review would be conducted by a panel of three lower court judges specialized in the conduct of trials. That panel as he envisioned it would review the record of the trial and order a new trial if for any reason that seemed warranted, including the ground that the verdict was against the weight of the evidence. In the exercise of that power, the panel might reduce a verdict it deemed excessive. Pound's panel would also enter judgment as a matter of law when that was warranted by the record. These actions having been taken by three judges rather than one, Pound thought further review of determinations of facts would be unnecessary and unworthy of the expense. Further review would be permitted only at the discretion of the court of last resort, which would be the only high court in his system.

With digitized records of the pretrial conference and trial, Pound's vision becomes even more efficient. The review panel could with relative ease achieve the same familiarity with the case as that possessed by the judge who made the pretrial rulings. There is little reason to confide the large discretionary powers of the trial judge in a single individual, with the added hazards of human failings that such confidence entails. The jury could be deployed only after the trial tape has been cleansed of error by the court of second instance consisting not of one, but of three judges.

The only issues concerning the trial that could be raised after a verdict would be possible jury misconduct, a very rare event, or excessive damages. One advantage in reviewing the pretrial rulings before the jury trial is conducted is that it would eliminate many trials. The mistrial would be eliminated entirely, as would those trials demanded by a party merely for the purpose of delay. A second advantage is that the jury trial becomes a climactic event in which the citizen-jurors are almost always given the final word, thus giving new meaning in to the constitutional guarantee to the right to trial by jury in civil cases.

3.8 The Role of Counsel Preparing for Virtual Proceedings

The virtual trial also makes it possible to place increased responsibility on the parties and counsel to conduct the preparation and production of the trial record. Competent counsel should have less need of oversight by a judge before trial to manage their cases. Attorney case management could be done efficiently in stages. The plaintiff might be allowed a period of discovery in which to assemble her case. The defendant could await a preliminary presentation of the plaintiff's case as the plaintiff proposes to present it before commencing preparation of the defense. Viewing the plaintiff's proposed presentation in the privacy and convenience of his office, defense counsel could notify plaintiff of any objections made to the evidence proffered. If persuaded, plaintiff might delete objectionable material. Other objections might be marked for later resolution by the court. Or, in appropriate circumstances, the court might entertain an early motion for judgment as a matter of law.

Assuming no such motion is granted, it would then be the defendant's turn to conduct discovery. The plaintiff's counsel could view the defendant's preliminary presentation of the defense and record objections, perhaps resulting in the withdrawal of some of the defense material, with remaining objections to be marked for resolution by the court. Plaintiff might also move for judgment as a matter of law, or might renew her discovery efforts to record additional material rebutting that offered by the defense. The defense might then again review the material and respond to the rebuttal. The plaintiff might then be given the last word, with the possibility of another round if authorized by the court upon a showing of reasonable need.

In many cases, this staging of the trial preparation might result in very substantial savings of costs. The defendant would need to defend only against the case actually made by the plaintiff and not against all the claims that might be imagined to appear in an unpredictable real trial. The plaintiff would likewise be freed of any need to anticipate defenses and gather evidence to refute those that are never effectively asserted in the form of credible proof.

The interrogation of witnesses for a virtual trial could be managed with incomparably greater efficiency. They would generally be interrogated under oath by videoconference, at a time and place convenient to the witness, generally at the witness's home or workplace. The interrogator could be thousands of miles away. The only persons possibly needed to be present are the designated officers of the court administering an oath and recording the testimony. And perhaps even they could be elsewhere if counsel agreed. The videotaped record of the testimony would itself be the material used at trial. There would be no rehearsal deposition.

Moreover, because of the reduced inconvenience, the interrogation of a witness could be conducted discontinuously. Thus, a defendant might elect to wait until he has seen the plaintiff's tentative presentation at trial before beginning to erect a defense by conducting cross-examinations that might be used in the defense's presentation. A re-direct examination might then follow later, when it was again the adversary's turn to rebuild her case. This discontinuity would be feasible because neither the witness nor counsel need leave their offices to reopen an interrogation. The earlier testimony of the witness would be readily available for review by the witness as well as by counsel. Discontinuity thus allows all counsel to be better prepared, making the examination of witnesses more effective and more efficient.

Furthermore, inconsequential witness examination can be deleted from the tape presented at trial, so that there is no useless testimony consuming the time of the trier of fact. Counsel could co-produce the trial tape, assembling sequentially that part of the testimony of each witness that either counsel wished to have presented at trial.

The elimination of the rehearsal deposition suggests a reason to enlarge somewhat the duties of counsel to disclose material such as recorded statements of witnesses. A statement to a lawyer that is recorded or made in writing and signed may, under present law, be viewed as protected trial preparation material. But such material is non-replicable, and the interest of protecting work product by withholding shared access to that kind of material is outweighed by the efficiency gained in sharing such statements.

Because digitization makes retrieval so easy, parties could have full access to statements and recorded interviews conducted in earlier adjudication against adversaries presenting cases involving identical or closely similar issues of fact for the purpose of discovering possible evidence, especially including prior inconsistent statements. All examinations of witnesses and examined documents could be filed with the court in digitized form. The problems of storage disappear because all the testimony given in all cases in the United States in a year can be stored in a single computer occupying very little space. A national index of testimony by any citizen in any court could be maintained so that material could be retrieved with modest effort by counsel. The functions of the local clerk of court would thus be substantially reduced, for if all records of all proceedings can be kept in a single digitized file comprehending all actions in the system, there is no need for autonomous local filings and records.

Finally, adversary counsel would be obliged to participate in a continuing discourse with one another regarding the pretrial investigation of facts at issue. This duty would be performed digitally (e.g., by e-mail) and recorded. This electronic conversation (or is it a bulletin board or a chat room or a face book?) would contain or replace formal notices and requests, interrogatories, answers to interrogatories, and case management conferences with the judge. Because the communications between counsel would be part of the record in the case, there would be meager opportunity to engage in off-the-record incivilities. As now, timely responses to any questions or requests for information would be required, and either party or counsel could at any time seek sanctions against others for causing unjustified cost or delay. The record on which such a motion rests would be readily available at the movement of a switch in the judge's chambers, but case management could otherwise be conducted by counsel rather than by the court. This would result in the more efficient use of the time of both judges and lawyers.

A similar form of communication would generally replace the service of a summons on a defendant as the formal means of initiating litigation. Every government or public agency, federal, state, or local, and every corporation engaged in commerce or owning property would be required to register its electronic address for the receipt of service of process. Individuals would remain free to register such an address or not, but if they failed to do so, they would be required to bear the cost of personal service and, if they are familiar with litigation, would be liable for a penalty for delay.

3.9 Virtual Jurisdiction

Much of the law limiting territorial jurisdiction was formed around considerations of physical access to the courthouse. The virtual courthouse is equally accessible everywhere. Thus, the doctrine of *forum non conveniens* becomes obsolete. Parties, counsel, and witnesses would be able to participate in litigation in Fairbanks, Alaska, without leaving their offices in Durham, North Carolina or Budapest. To the extent that convenience of access is a consideration in applying a due process standard to limit "long arm" jurisdiction, that factor would be almost eliminated. The right of the defendant conveniently to attend the screening event of trial can be readily observed, by digitizing and transmitting it to the defendant anywhere.

Considerations of sovereignty would abide. The fact that a courthouse is conveniently accessible to everyone on the planet does not entitle it to rule the world. Moreover, if physical access and convenience are no longer factors, there is less reason to respect the plaintiff's choice of forum. A plaintiff choosing a forum in a jurisdiction other than the one whose law governs the issues in dispute lacks a justification for his or her choice. Accordingly, the principle of due process limiting the territorial jurisdiction of American courts would merge with the constitutional limits on the territorial reach of legislation. Courts should have jurisdiction to enforce their own laws if applicable, and should be presumptively disabled from taking jurisdiction to enforce the law of a foreign sovereign. In other words, choice of law would become the issue of judicial jurisdiction, and the law of conflicts of law will disappear as a separate topic.

There would remain a problem with respect to the plaintiff's right to choose among the courts of the jurisdiction whose law is applicable to the events. There is no reason to allow the plaintiff to forum shop among such courts for sympathy or influence. An appropriate response to this problem might be to select the judge presiding over the virtual trial preparation from among the whole roster of judges available in the system, much as might be done to assemble a virtual review panel. This would effectively forestall shopping for a judge. This suggests that the local court as an administrative unit managing judicial personnel is as obsolete as its clerk's office or as the doctrine of *forum non conveniens*.

3.10 Conclusion

The secondary consequences of the dramatic reduction in cost and delay that will be possible might result from the establishment of the virtual trial are not easily visualized. One must assume that such reductions would lower the threshold of restraint and result in increased filings. More public laws would then be enforced more effectively by private citizens. To those presently preoccupied with the alleged litigation explosion and excessive enforcement of laws regulating business practices, this may be the ultimate horror.

On the other hand, properly deployed, digitization offers promise of revitalizing our appreciation of litigation as an exercise of individual rights and a means of private law enforcement. One might also suppose that the facilitation of discovery will result in closer observation of the law by those with duties whose defaults will be more easily and more fully exposed to public light. This consideration is obviously less important for those nations able to entrust the enforcement of public laws to their public officials. But in such legal systems, comparable improvements in efficiencies and reductions of costs would seem to be in order.

Virtual litigation is not yet on our doorstep. Neither the profession, nor the courts, nor the litigants, and certainly not this author, are ready for it, nor can they be made ready soon. But those who make rules and laws today may make themselves more useful if they have in mind a sense of the direction of change such as I have tried to supply.

Chapter 4
The Future of the Traditional Civil Procedure*

Petrônio Calmon**

4.1 The Cultural Revolutions as a Product of the Industrial Revolutions

I am addressing an audience already accustomed with the use of the new data processing. Even those who do not share technical knowledge on the theme, are certainly users of the systems and have already noticed that today's life would be impossible without data processing. The reader does not have to be convinced on how deep the way of living of humanity has changed on the last few years because of the universalization of the use of the modern information and communications technologies. Therefore, I should not elaborate an exhaustive report on all those techniques and their application on devices that are now a must for the day-to-day life.

However, maybe it is good bringing up some interesting inputs to this work, as they might help us out to be aware of the real dimension of the theme we are about to cover. That is what we intend to do, especially on Chap. 2. However, I do not mean to just inform, but rather provoke an insight about the meanings of the facts that are being noticed and, above all, about the consequences that those technological new things are provoking and will provoke on the social relationships. We believe that the technological revolution has to be seen for its symbolic contents. After all,

*This work was elaborated to be presented as the General Report of the Colloquium Pecs, of the International Association of Procedural Law, which was organized at the city of Pécs, Hungary, in September 2010

**Attorney at Law of the Federal District and Territories of the Public Department, Vice-President of the Brazilian Institute of Procedural Law, Member of the International Association of Procedural Law, Member of the Ibero-American Institute of Procedural Law.

P. Calmon (✉)
Brazilian Procedural Law Institute, Sé Sao Paulo, Brazil
e-mail: petronio@ibep.com.br

M. Kengyel and Z. Nemessányi (eds.), *Electronic Technology and Civil Procedure*, Ius Gentium: Comparative Perspectives on Law and Justice 15,
DOI 10.1007/978-94-007-4072-3_4, © Springer Science+Business Media B.V. 2012

technology is a determinant factor of the historical process and the shaping of social life. Our premise is that of André Lemos, when he states that technique is the main defining element of the environment we live in. Or, in other words, "*cyberculture* is the significant contents of technology."[1]

If there are deep changes in the way of being and living of people everywhere, on the planet, to such an extent of observing a comprehensive cultural revolution, it is evident that a change in the relationships and conflicts that people experience among themselves is also taking place. If the change also happens on the economy (organizations and finances), in the States and governments, in the education and in health and in every other areas of the human life, that means that the factual fundament on which law is sustained has been altered substantially, particularly the procedural law.

For the right assessment of the procedural law, we ought to know the cultural environment of the time when some jurists established the autonomous procedural law, i.e., the period when the studies of Savigny, Muther, Windscheid, Wach and Bülow, among others, were published. We are talking about the mid nineteenth century, when those magnificent jurists founded the modern procedural law based on their own formation of the Roman right and on the observations that were then possible to them on people, their relationships, the economy, politics and ideology as the driving force during the Century of Light.[2] That was the time of another cultural transformation, on which the pillars of the modern civil procedural lie. On this work we mean to demonstrate that all nineteenth century procedural law fundaments, in spite of having lasted for about 100 years with no greater changes, were totally transformed on the last few years, mainly after 1995. Today the social, economic, political and ideological fundaments are different. We see a different thinking and way of life of people. Just like the creation of the modern procedural law happened as a corollary of the then current cultural revolution, the existence of a new cultural revolution makes it essential the creation of a new procedural law. The procedural law is there to serve society. If society is now different, then the procedural law must also be different.

After we watch cautiously how that cultural revolution is taking place, we will understand better some characteristics of the phenomenon which is the object of our work. But then, which is the object of this work? We are certainly not talking about data processing, nor computer use, nor the electronic process as the simple use of the computer to streamline the legal activity. We are talking about the modern information and communications technologies, a phenomenon much more sociological than scientific, which is changing the society roots in its more important aspects. Data processing is here since the mid of the nineteenth century, but only the technologies

---

[1] André Lemos, *Cibercultura: tecnologia e vida social na cultura contemporânea* (Porto Alegre: Sulina, 2002), 176.

[2] That is the Cândido Dinamarco's very lesson, appearing on the first paragraph of his cathedra work *A instrumentalidade do processo*: "The political and social changes that happened in Europe since the previous century had been capable to modify the relationship formula between the State and the individual, with the rupture of the old structures –and that was responsible for the first concerns in defining the process phenomena" (São Paulo: Malheiros), 177.

of the end of this century and their universalization during the first decade of the century are providing a truly socio-cultural revolution. That is the new industrial revolution, comparable to the 1770 and 1850 British industrial revolutions. Comparable, yes, but probably much deeper.

The first industrial revolution was characterized by the upgrading of the steam engine (James Watt, 1769) and the second for the use of electricity for lighting and moving machines (Thomas Edison, 1879 and Werner Siemens, 1886). But we do know that the period of those revolutions was not characterized just for the scientific and technological progress. As a matter of fact, society has experienced great transformations that interacted among themselves. A new philosophy emerged in that period. One that was marked by the predominance of reason over religious dogmas up to then rigorously imposed. The modern state conception also appeared, with an independent parliament (an experience that originated in England a century earlier) and with the first steps for the definition of a judiciary system. The British (1649), the American (1776) and the French (1789) revolutions are the political facts that became registered in history, but it was the scientific progress that provided the ideology operationalism, with the society, economy and political transformations. Philosophy acted as a thinking element of observation and elaboration of new theories and proposals; and politics was the consolidation tool of a brand new society and a brand new State-society relationship.

Along with the eighteenth and nineteenth centuries industrial revolutions, societies grew and changed. The rural exodus became a constant in every corner of the world. Cities started to grow and swell. On the field, the order was now to increase productivity and industrialize production. It was the start of the technological development, with a core concern on upgrading production techniques and with the invention of machines capable of making easier and multiply man's work. Chemical products were created to speed up production and combat plagues and diseases. The building of efficient plants became a reason for social pride. Every people longed for a greater population, greater cities and larger industrialization. The economy indicators started to exist, all of them addressing growth and the benefits of private initiative.

People left the land labor and the field isolation to adventure themselves in the new cities, dedicating a large part of the working day on scale productions. In the beginning the change of life was hailed enthusiastically. Women now had job opportunities and the life in the city was more joyful and dynamic.

The new economy emerging from the industrial revolution (in its two stages) was totally in the hands of the private initiative. The private agents created industry and held the investing capital. It was not a revolution planned or invented by the State or by any other organized entity. It just surged. It emerged from the hands of private citizens and took all by surprise. A quite pleasant surprise for those who held the capital. An inevitable surprise for the new labor class – and an incontrollable surprise for the State. There was, now, a new power capable of disputing with the State in a much better level than the power of barons and feudal landlords of the middle ages. Back from that era, marked for continuous fights, barons finally won. Not necessarily the same barons of the past, but rather those who managed to adapt to the new economy, along with the new barons coming from the recent bourgeoisie. The winners were those who knew how to manage the economic weapons with much more skills as they had in the past dealt with political and military weapons. The liberalism that emerged to privilege the individual and his freedom originated a dominant class, holding the capital and monopolizing the investment.

The dominant class opened the eyes to a world it had never known, with a chance to enjoy life in a better way. They could now pay for better schools, enjoy better health, make the personal hygiene easier, buy more clothes, move faster, communicate through telegraphic messages or through the voice with faraway people, travel to faraway places and promote all kinds of attractive activities. Things that they could not do prior to the availability of the new technologies. Things that they just wouldn't even imagine earlier.

The industrial revolution transferred individuals from the fields to the cities. Not just those belonging to the people, per se, like peasants, working for the feudal barons, but also those envisioning a more abundant life in the cities and, moreover, thanks to the fact of sharing new ways of investment. Latifundia were then reduced in size, and buying and selling medium and smaller rural areas were increased and, by the same token, invasions, occupations and farming conflicts. A new way of service relationship was created and the countryman was transformed into a plant worker or commerce clerk; and the great land lord was changed into an industry or commerce man. The boss-employee relationship on plants and stores was now utterly different from what they had experienced in the countryside. Conflicts of all sorts surged in the growing cities.

It must be recognized once more: that was the factual basis noticed by the nineteenth century procedural law pioneers and that same basis remained practically untouched since the times of Wach and Bülow until almost the half of the twentieth century. Then technological changes started to emerge every time. It seemed that everything was happening as a constant. Inventions covered the automobile, the airplane, television, the typewriter, the audio recorder and, finally, the outrageous bomb. Nevertheless, although at the time inventions were seen as routine and new things would pop out year after year, as a permanent progress, in the final of the twentieth century a brand new and great technological revolution happened, producing in just a few years much more than what had been produced in decades and in centuries, based on the transforming importance of that recent production.

The embryo of the new cultural revolution started to be shaped soon after the explosion of the nuclear bombs. A few years had elapsed after the end of World War II when once again society promoted the fall of their traditional values. The 1960s of the twentieth century left clear that the world would be different from then on. Nobody would ever imagine that such cultural values would once be broken. The State-church separation became definitive because now values, not just political binds, were ruptured. Sexual revolution imposed itself. Women's freedom followed the same path. Women started to fill in space in the economy, in politics and in the working market (although they are discriminated until today). Human rights were invoked and the civil rights fight imposed the equality among men. The universalization of the vote emerged as the ultimate, definitive stone placed in the pillars of democracy.

Youth was taken over by freedom and alienation movements. Marriage was not insoluble anymore, and abortion started to be recognized. Traditional clothes were left aside. A new and stronger rural exodus emerged, plants were modernized… and communications developed.

While all that occurred, young geniuses worked in their parents' home garage and in university basements, sometimes with financial support granted by companies and the armed forces. Freed from culture, then already dated, and counting on machines and tools, those youngsters started to create new technological apparatuses by using techniques that did not exist up to then, changing completely what is known as *state of technique*. In the same way that the pioneer studies on electricity and magnetism worked as driving springs for the second industrial revolution inventions, the theoretical discoveries of mid twentieth century sustained the creation of data processing and communications through data processing. There was scale production of computers, an item that now became popular. In the early days, it was just a new equipment that made things easier for industries and commerce. Then, in a second stage, it became a new personal equipment. From then on, facts succeeded like in the times of the industrial revolutions of earlier centuries. As a multi-pendular movement, each discovery of a new gadget or a new data processing program drove the discovery of another invention. Like Watt improved the Newcomen steam engine (with greater use of the energy it generated), every new invention of the twentieth century final was used as head start for a next one. The fast and continuous improvements generated the phenomenon that we are calling the universalization of new information and communications technologies. That fact catalyzed (or is still catalyzing) a new industrial revolution and, along with it – like in the Century of Light – a new cultural revolution.

Well – if one changes the culture, one has to change the law and, along with it, the procedural law has to be radically changed.

On this work I present you a negative assessment on the twenty-first century procedural science. A change that was not perceived in its plenteousness, the cultural revolution that took place on the turn of this century. Our working hypothesis is that the distance of the proceduralists from the philosophical debates and those debates with other sciences (not just the human sciences, but also the exact and technological sciences) resulted in a harmful isolation, exacerbating the autistic character, always present in the procedural law, since the consolidation of its autonomy.[3] So that the procedural law regains its importance as science, we must know better and live up with the social, political, economic and ideological observation of nowadays, leaving aside the old dogmas and observations so well elaborated by the founders of the modern proceduralism.

It is not hard to see that, since the modern procedural foundation, society has been completely changed, and with it is problems were also changed. So it's not

[3] Cândido Dinamarco, 24 years ago he starts his proposal for a University of São Paulo Professor, saying that "the procedural law suffers from the natural trend for formalism and isolation." Nothing has changed since then, and almost no proceduralist has turned that speech into practice.

possible that the procedural law science keeps formulating answers to problems that no longer exist. And that it keeps on elaborating solution mechanisms for conflicts that have already vanished. The twenty-first century proceduralist must observe in a better way the twenty-first century man and his behavior. Must know and study the current society, the State and its way of organization and action. When all those factors change, logically, so the procedural law should change. The procedural law is a technique, but it should be a technique working for society. And to be like that, it must supply the answers to the real needs, and not to the needs of a society that no longer is there, namely: the 1850 society.

One and a half century separate us from the Wach and Bülow studies. Ever since, society has suffered great transformations, but nothing comparable to those we are facing for the last 15 years. The contemporary transformation is a lot more intense than the ones in the Enlightenment period. The factors that are changing our society, owing to the lack of diffusion of the information and communications technology, are much stronger and intense than those which changed the society in those times.

How can we take advantage of the conclusions of earlier centuries thinkers, when we know that those conclusions are product of the observations of a society that is over?

Our hypothesis is not limited to the observation that the twenty-first century procedural science is heading the wrong way. It is much stronger and radical, since it proposes the complete destruction of the buildings erected on those pillars, now unsustainable. We verified that the traditional procedural law is not fulfilling the social expectations, because it is no longer the legitimate tool for conflict solutions and for legal security. Its foundations have already been destructed by the time and, as a consequence, all the building is not working as it should. Just the façade is there, on its feet, like a peel with no contents. As the poet would say, procedural law is a sealess boat, like a flowerless prairie.

After the destruction of the old and useless, we must build up a brand new system, with new foundations erected from a deep twenty-first century holistic insight, through a differentiated methodology, characterized by the multidisciplinary tone and through serious research supplying us secure information on the existing contentious and about the thinking and way of life of people, today and tomorrow. The new civil procedure must bury the old values, the Roman and canonic tradition and the Enlightenment ideology. Rome, the Church and the Enlightenment should be saved in the memory and at museums, as fine remembrances of the construction of needed values fit for their respective times. The twenty-first century generation ought to take over the responsibility to reformulate the scientific bases that should sustain a new proceduralism as a science of its own time, turned to current cultural values, supported by the new technology, aiming at responding to the current society, whose values and methods are dictated by and determined by that society.

The new observation starts out pondering on the facts that mark the new industrial revolution, now a more intense and accelerated technological revolution, which is building a new network society that communicates and interacts according to new paradigms.

The new procedural law should be built based on the new technologies and must take advantage of them to provide better results. But the most important is that the new procedural law must focus the new data processed society that changed completely its way of acting and interacting, changed its life and work habits, changed the economy, the financial management, the companies, the commerce and the industry. Just the law and especially the procedural law have not been fully hit in an efficient way by the information and communications revolution that was performed on the turn of the twentieth century to the twenty-first century.

On that observation context, no factor characterizing the contemporary society must be left aside, among them the (society) globalization and the universalization (of the information and communications technology use). Those two factors alone are enough to sustain the thesis of a new industrial-cultural revolution and its direct and profound relationship with the procedural law as a science that cares and disciplines the conflict solution means. Those globalized and universalized society conflicts are different and expressed in a differentiated way.

Such is the case, for example, of a buying and selling contract that can be closed without any words, without any personal contact, with no clear clauses and with no signature, among people who do not know each other, that ignore its characteristics and that are far from one another, in different locations of the planet, speaking different languages, subject to different legislation systems. Thousands of those relationships are happening right now in every second, every one of them with no State or government control, aside any traditional legal protection. What law and what judge could act in cases like those?

People are no longer communicating in the traditional way. They do not do businesses in the traditional way and do not even keep love and family relationships solely in the traditional way. The concept of nation is already being reformulated. The ties that unite the persons now no longer find physical barriers, which could add to a territory and to a power to constitute States. That phenomenon takes place with the same intensity both among individuals and legal entities as well as an individual and a legal entity. Organizations are depersonalized, they do not even are multinational entities, because their owners are not identifiable owing to the interlacing among their stockholders. Frequently a company is a partner of another, which has shares in a third one, that acquires stock from the first company. The interbank relationships are constant and spread all over the globe. Money circulates from New York to Tokyo and from there to Abu Dhabi, then to London and from there to São Paulo, all in a matter of minutes or seconds. Schools hold distant students, books are no longer made of paper. And to complete that brief example, there are videocams everywhere, erasing every trace of privacy from the citizens. Not just their electronic paces are recorded and controlled, but also their human paces, their image and their actions can be seen and recorded anywhere in the planet.

To complete that picture we must emphasize the social and economic relationships gigantism. The communications networks formed and are forming communities so large that they do not find any parallel in the traditional communities. Orkut, Facebook, YouTube, Twitter and other relationship networks make millions of people to relate simultaneously, covering all corners of the planet. When a famous data processing program or a new videogame is launched, millions of copies are usually sold in the first selling second, at the same time, to customers from dozens of countries. Sales are made electronically, and so does the payment. And the product is immaterial: it's a computer program.

Let's just think how the conflicts of interest generating legal processes are right now and will be for the next few years. Would it be possible to imagine that the conflicts of today and the future are any similar with the conflicts characterizing the society lived by Wach, Savigny, Bülow, Chiovenda, Carnelutti, Calamandrei and Plósz? The mass conflict factor, alone, that has originated the so-called serial actions (or repetitive legal actions) is sufficient to demonstrate how today's litigations differs from those of the nineteenth century or the first half of the twentieth century. The conflicts of family rights and urban rents emerged only in the final of that period. Only in the final of the twentieth century started to surge mass conflicts, with thousands (or millions) people moving suits again the same company. Insurance conflicts, product damage reparation, adhesion contract alterations at large communications service providers. Those are just a few examples.

Century twenty-first relationships are lived through computers and the communications among them. Conflicts also result from the use of those technologies. The means of solution have to consider all those factors and be ready to make use of the new information and communications technologies. Then, on the second chapter, we will present a short report on facts that marked electronics in the twentieth century to show that at the final of that time, the third industrial revolution in the history of humanity took place. Being aware of the facts and their timing it will not be hard to conclude that, in spite of the constancy of the previous evolution, the twentieth century starts its path under the impact of a huge transformation whose limits are still unknown. On the third chapter we intend to demonstrate how that new industrial revolution provided a new cultural revolution, i.e.: how the way of life of the twenty-first century society is like. That is the foundation for our proposal for a new procedural law presented in the fourth chapter, which is the expression of our vision about the electronic process.

4.2 The Information and Communications Technologies Universalization

It is important to think how the technological revolution that today presents itself as so close to the social life took place. Up to the 1970 decade, the scientific development occurred through short paces although practically continuous, starting with the second industrial revolution, when the first electric motors were invented. The electronic emerged from the inventions of Thomas Edison back in 1879, which, in addition to the incandescent lamp, improved Graham Bell's telephone and developed a device that he called Edison Effect, and that became known as thermionic valve. Edison invented it by chance when trying to fix the deficiencies of his lamp.

You see that both the lamp and the valve are filaments inserted in a glass ampoule with the production of vacuum.

Edison's valve was improved right in the beginning of the twentieth century with the creation of diode by the Englishman John Flemming. The diode served as the basis for the North-American Lee Forest who in 1906 developed the triode, a key part for the coming of the amplifier and radio broadcast.

Valves provided the birth of a so important electronic industry with the emergence of household, industrial and military items. Thanks to them it was then possible to think on the first electronic brains, machines capable of performing calculations at great speeds. Valves reigned for over 40 years.

The second World War alerted the governments of the dominating potencies of the twentieth century, that concluded that the technological development would be a key factor for future war equipment and started to invest in revolutionary ideas. The rivalry among ex-allied nations that became enemies and competitors contributed substantially, increasing the dispute for technological development, which increased its own importance as a war tool.

From 1939 to 1942, at the University of Iowa basement, Professor John Atanasoff, helped by his student Clifford Berry, built the first electric computer of the human kind, calling it ABC – Atanasoff-Berry Computer. Although it was not programmable, since it was just supposed to solve linear equations systems, the equipment already used binary arithmetic and electronic commuting and, for that reason, back in 1973, a US court bestowed him a patent for the first computer, voiding ENIAC's patent.

Many computers followed the ABC then. Among them the Zuze Z3 (Germany-1941), the Colossus Mark 1 (United Kingdom-1944), the Harvard Mark 1 IBM ASCC (USA-1944), the Colossus Mark 2 (United Kingdom-1944), ENIAC (USA-1946), the Manchester Scale Experimental Machine-Small (Baby) (United Kingdom-1948), the Modified ENIAC (USA-1948), the EDSAC (United Kingdom-1949), the Manchester Mark 1 (United Kingdom-1949) and the CSIRAC (Australia-1949).

ENIAC was the first general purpose computer. It was totally financed by the US Army and cost about 500 thousand dollars, which, in 2010, is worth nearly seven million dollars. It was composed of 17,468 vacuum valves, 7,200 crystal valves (diodes) and about five million welding points. It weighed 27 t, was 2.50 m high, 90 cm wide and 26 m long, equivalent to 680 m^2, consuming 150 kW of power.

Soon after the post-war period, Bell Telephone labs developed a new device that years later would replace the thermionic valve for good. In 1947, John Bardeen, Walter Houser Brattain and William Bradford Shockleym came up with the *transistor*, and won the Physics Nobel Prize for that invention in 1956. The transistor is also known as semi-conductor because it fits in an intermediary spot between the semi-conductor devices and the insulating devices. *Chips* are a set of transistors, and with them the micro-electronic science was born, serving as a basis for all computers until today.

At the beginning, the North-American industries did not care much about the transistor. They would rather keep the thermionic valve in their sets. At that time, people responded favorably to large dimension sets. It was only when Japan launched the fashion of small devices that valves were definitively buried. Starting back in 1954, the transistor was manufactured in silicon, a perfect semi-conductor. In 1958, Jack Kilby, of Texas Instruments, developed the first integrated circuit, formed by a transistor, three resistor and one capacitor. In 1962, the manufacturing process of the integrated circuits was improved to such an extent (plain process) that in just 3 years, between 1959 and 1962, the semi-conductor prices plummeted 85%, and during the ten following years production increased 20 times. Those are facts cited by Manuell Castells, in his work *A sociedade em rede*.[4] The author compares that phenomenon with the times of the British industrial revolution, when it took 70 years (1780–1850) to decrease 85% the cotton tissue price. He adds that "as manufacturing technology advanced and it was possible to improve the design of chips with the help of computers, employing faster and more advanced micro-electronic devices, the average price of the integrated circuit fell from US $50 in 1962 down to US $1 in 1971".

That is an abstract of the new industrial revolution pre-history, but it was in the decade of 1970 that new micro-electronics discoveries provided the making of new machines. With that, the development of languages would become popular in just a few years.

In 1971, engineer Ted Hoff, of Intel, invented the micro-processor, the single chip computer. That fact served as a driving spin for the development that followed. The micro-processor use paved the way to the computing that we are living up to today. From that period until today (2010) the data processing industry applies the very same micro-electronics. And there are already new things that once more changed the "state of technique", although they have not definitively been applied so far. The new era that we still live up started in 1971 and extended up to 1995.

On the following year, *routers* and electronic commuters were developed. In 1969, Bell Laboratories had already made the first electronic commuter in scale, but it was in the decade of 1970 that the *digital commuter* was created, increasing the computer's velocity, power and flexibility, saving space, power and work, as compared to the analogical devices. On that same period the first advances of optoelectronics and the data transmission through packages became possible thanks to the utilization of fiber optics and laser.

At the same time that the history of the modern electronic computing started, with the micro-processor, the pre-history of the *Internet* started. Its origin is in experiences of a research agency of the Defense Department of the United States, ARPA. The objective of that pioneering initiative was the creation of a disperse communications system that would not depend on a central control, which would make the system immune against military attacks. The information would then be disseminated and not concentrated in a single spot. From ARPA, the technology went to the California University. During the Cold War, an operation between US military institutions and universities was rather common. One financed the other

[4] Manuel Castells, *A sociedade em rede* (São Paulo: Paz e Terra, 2002).

and both promoted an interchange of experiences. At the California University, in 1969, emerged the first computing network, called ARPANET.

In 1973, Internet's basic structures were created by ARPA researchers, Vinton Cerf and Robert Kahn. They and other interested parties developed a technology that permitted a computer to talk with others. In the data processing language, a protocol for data transmission.

Soon after, in 1975, MITS Ed Roberts built ALTAIR, which he called a "computing machine". It can be considered the spark that revolutionized the explosion of the new industrial revolution. Its inventors sold the new machine through the mail, delivering the parts with an assembly manual. Thinking on selling tenths of units, hey booked a few thousand in the first month of sales.

The ALTAIR was the basis for two young strangers, Steve Wozniak and Steve Jobs manufacture their first Apple computers, the brand that popularized the personal use microcomputer. They developed two models, the so well known Apple I and Apple II. Launched in 1976, with three partners and a capital of US $91 thousand, Apple's sales exceeded 24 billion dollars in 2007. In 1981, IBM launched its first microcomputer and coined the name Personal Computer (PC) that became the general name of microprocessors. In 1984 Apple launched the Macintosh.

At that same time, emerged the company that would turn its founder the richest man in the world, Bill Gates. He and Paul Allen founded Microsoft, that started to dominate the computer software market.

In 1978, the data transmission protocol, created by Cerf e Kahn, was divided into two parts: one protocol that processed the communications from a server to another, the *Transmission Control Protocol*, or TCP, and a protocol to act internally in the networks, the *Internet Protocol*, or IP. The set of those two protocols, called TCP/IP is utilized up to the present as a world standard protocol, although others have been created, mostly in Europe. The TCP/IP however prevailed.

Also in 1978, Ward Christensen and Randy Suess invented a device capable of transferring data from one computer to another, the *modem*. ARPANET, that was restricted to universities and military bodies, then stopped being the only network and any person could communicate through independent networks.

In 1983, the old UNIX operational system, created by Bell in 1969, started to use the TCP/IP protocol, which made it popular, since the UNIX was financed by public budgets and sold at cost. The UNIX system was adapted to use the modem and its communication protocol originated something utilized universally, the *electronic mail,* thanks to an application created by Ray Tomlinson. In the same year, with the *bulletin board systems* (BBS), developed by Tom Jennings, it was possible to create the FIDONET network, connecting thousands of computers. BBS is a system that allows connections among computers through a telephone line. Until 1990, however, the Internet had not been popularized.

The popularization that today took over the Internet was only possible thanks to the development of some special techniques. The world wide network *www* (world wide web) was developed in 1990 by the *Centre Européen pour Recherche Nucleaire* (CERN), with headquarters in Geneva, under the leadership of Tim Berners Lee and Robert Cailliau. It started to use a format, also created by CERN, called *hypertext markup language* (HTML). They also configured a hypertext transference protocol, the *hypertext transfer protocol* (HTTP) and the *uniform resource locator* (URL), the standard address format used until today in the Internet (for example, the address of Instituto Brasileiro de Direito Processual, is: www.direitoprocessual.org.br). All computer programs used on the www system were distributed free, providing new operational easiness.

In 1992, MOSAIC was created by Marc Andreesen and Eric Bina, of NCSA. MOSAIC, considered the first Internet *browser*, meaning a simple way to operate the communications. The success of that Internet navigator may have happened owing to the fact that it was elaborated not just for the UNIX, but also for the brand new WINDOWS system. Investors soon invited the browser's creators to open a new company that disseminated the well-known NETSCAPE. Therefore, just like the 1970 decade drove the data processing technology, 1995 is the great turning point. I mention as the symbol of that transformation the launch of the Microsoft Internet Explorer browser, occurred on August 23 of that year. In Brazil, in 2010, 53% of Internet users used the many versions of the Internet Explorer. Starting in 1995 the universalization of the modern information and communications took place. It was from that year on that the whole society headed towards the Internet. There were much more paradigmatic changes than what happened with the famous British industrial revolution The key point is this: universalization. Technology provided the universalization and this converted the new technology into a sociologic phenomenon. Social, financing, economic, cultural, psychological aspects and persons from the whole society were transformed. We will cover this matter on the next chapter.

We still have to analyze two aspects of that technological revolution: the technologies of the present and those of the future.

For the purpose of this analysis, the present means the 15 years we are living through the universalization of the new information and communications technologies. On this period, that started with the launch of the Microsoft Internet Explorer, technology has been used always envisioning universalization. For new things to be utilized it was necessary to limit the costs of the devices and the computer programs. Since those costs are not small, the industry would only be economically viable if succeeded in selling in very high scale. It was the concern for the business survival that took companies on looking for a way out through the universalization. From 1995 up to 2010, the computer manufacturing multiplied in an amazing scale and new equipment were invented, aiming at expanding the use of those technologies. Along with data processing we saw the development of the cell phone, laser sets for reading audio and video discs (replacing the old magnetic tapes), plasma television sets, LCD and LED and, ultimately, the electronic book reader, among so many other gadgets.

Not withstanding the emergence of new devices for personal, commercial and industrial use, the new technologies is what amazes us when assessing the last 15 years. They are all focused to the improvement of techniques, size reduction, velocity increase, larger storing capacity, higher transmission speed, better image definition and better manipulation of all those features. When we sum up all that to low cost, we promote the universalization we are pointing out, and that universalization is the reason for the social transformation.

As to the future, it must not be portrayed as an imagination exercise or sci-fic. As future, we should include just the application of techniques that are there already and are still being improved, especially the application in data processing and in the society "technologicalization". On that sense, we point out the development of nanotechnology, the molecular electronics and the holography, the latter particularly important for our study on the procedural law.

4.3 The New Cultural Revolution

In this stage, we must elaborate a close observation in the twenty-first century society, being aware that the universalization of the new information and communications technologies changed substantially (and still is changing) all human life facets, especially their living together. It's about a paradigmatic change of ample and universal character. Although it is evident that the world has not yet become horizontal, and the economic and social inequalities are still huge, we notice a real universalization if compared to the dissemination of preterit technologies. Furthermore, it is already possible to see that the new industrial revolution is a robust factor for reducing inequality. All that because of low cost which, as we have emphasized, is a key component of the virtuous cycle: lower cost implies larger production, and larger production implies lower cost.

The data processing society study is not part of the object of the legal science, not even of its sociology, its anthropology or its philosophy. Those studies are part of those sciences in their new specializations: the technology sociology, anthropology and philosophy. That is why we make use of expert readings, for sustaining our work hypothesis. We do not state that the new technologies are determining a new society, but we do want to propose that the society has appropriated the new technologies to perform its self-transformation and that law, especially procedural law, will rebuild itself, if we intend to respond to that new society.

An important observer of the new industrial revolution, and the consequent cultural revolution, is the already mentioned Catalan Professor Manuel Castells,[5] for whom the new paradigm has five characteristics: (1) The new technologies are meant to act upon the information; (2) The new technologies effect penetrate in the social tissue; (3) The new technologies provide a *network logic* in any system or set of relationships; (4) The new technologies are based on flexibility, because they have an enormous capacity of reconfiguration; and (5) The new technologies have a

[5] Manuel Castells taught at the University of Paris for 12 years, and at the University of Berkeley, California, for 20 years. He now lives in Barcelona, were he teaches at the Open University of Cataluña. From 2000 to 2006, according to *Social Sciences Citatio Index*, he was the fourth scientist more quoted in the whole world.

convergence of specific technologies for a highly integrated system. For this Catalan, "the information technology paradigm does not evolve for its closing like a system, but rather aims at an opening like a multiple access network. It is strong and imposing in its materiality but adaptable and open in its historical development. Comprehensiveness, complexity and disposition like a network are its main attributes."[6]

The new technologies of information and communications change the economy completely and, without any doubt, shape the most essential pillar of the recent globalization. In Castells we find the conclusion that this new economy is characterized by a network informality, globalization and operation.[7] Practically every and any business and labor activity is supported on these new technologies. Therefore, like the governmental activities, the relationship government-citizen and the individual's social life. It's about a new social culture, cyberculture.

According to Francisco Rüdiger, "Cyberculture is the historical movement, the dialectic connection, between the human subject and his technological expressions through which we change the world and so our own interior and material way of being in a determined direction (cybernetics)."[8] For this author, mechanization is the principle of reordering values and rebuilding culture. Modernity emerges with the rupture of the principle that men were organized in heteronomous communities and associations, founded on the belief in common values, promoting an individualism on whose basis the concept of society affirms, where the individual sees himself in a process of social and interior defragmentation.

In the Internet there emerges an innovating communication space, free from political barriers, universal, inclusive and transparent. Like Castells, we believe that we can reach an organized social order so as to satisfy the demand for free interactive expression and for an autonomous creation. "The Internet penetrates all social life dominions and changes them. It is therefore a new configuration, the network society, which is in gestation throughout the planet, although under the most varied shapes between a point and another, with much different effects on the life of populations owing to their history, culture and institutions. Like the previous structural mutations, that complete reversal brings with it just as many new possibilities as new problems. The result from that is undetermined: it will depend upon a contradictory dynamic of eternal fight in the efforts always renovated to dominate and explore, and the defense for the right to live and search a sense to life itself."[9]

It is important to stress that there are many writings criticizing our position, which confers revolutionary traces to the new information and communications technologies, but we cannot overlook that, on this subject, any text becomes dated a few months later after being written, since the universalization that characterizes the present cultural change grows geometrically, if related to time. For us, the twenty-first century technology if a *form of life*. It's a truly cultural revolution, and from that

[6] Manuel Castells, *A sociedade em rede* (São Paulo: Paz e Terra, 2002), 109.

[7] Manuel Castells, *A sociedade em rede* (São Paulo: Paz e Terra, 2002), 119.

[8] Francisco Rüdiger, *Introdução às teorias da cibercultura* (Porto Alegre: Sulina, 2007), 71.

[9] Free translation of Castell's work *A galáxia da internet* elaborated by Francisco Rüdger.

conclusion on, all and any study about society must be reformulated, and all practices and all theories to solve social problems are already dated. As a consequence, everything has to be reformulated for the new society.

Therefore, we are living the Internet culture which, according to Castells, is characterized by a four-tier structure: the technomeritocratic culture, the informational society culture, the virtual culture and the corporate culture. Those tiers shape the Internet culture. Or, in a better way: according to the author, they form the *Internet Galaxy*.[10]

Our proposal addresses the procedural law. We are not prepared to cover great distances in social research. But, in addition to learn with specialized scientists, we must involve ourselves with wisdom to perceive that "our" science is destined to cure the social problems and, therefore, we should try to know the society of our times joining the experts that observe it with other lenses to reach the finality to erect proper solutions that are objectively effective.

As we mentioned above, Dinamarco, since his utterly important cathedra work, already alerted for the procedural law's autism. Until today, however, that science keeps isolated, and that is the first reason for its inefficiency. This is not the time to build temples to formalism, valuing the means as if it were the ends. It's time to search for holistic solutions, starting from a multidisciplinary study.

4.4 The Rebuilding of the Procedural Law

Essentially, the electronic process is formed on the use of the new information and communications technologies for the practice and communication of procedural acts. Utilizing the modern techniques and with the duly legal authorization, all procedural acts (civil, penal and administrative) may be practiced through electronic means. And there is more: the acts practiced through electronic means and those that are still being practiced via traditional means, may be communicated through electronic means. That is the core of the electronic process. As a matter of fact, all practical consequences of the procedural changes result from those two permissions: to practice and communicate acts. It is evident that behind those two labels we find a huge variety of activities, including those practiced today, in the process, as well as those that may come ahead, practiced through the electronic procedural.

The acts of the process are practiced by the parts, by the judge and by justice clerks. The acts of those parts consist in a statement of the will, usually covering requests and argumentation. For example, acts of the parts and the initial demand petition, the defendant's answers, replications and other petitions, allegations, production of the evidences and the appeals. All those acts can be practiced through electronic means. Instead of being written, they could be elaborated and stored in a computer, in its original binary language.

[10] Manuel Castells, *A galáxia da internet* (Rio de Janeiro: Zahar, 2003), 34.

As a rule, the judge's acts are the procedural decisions, sentences and process moving orders. The judge also produces the evidence and coordinates the production of the evidences by the parts. In the electronic procedural, the judge will practice those acts in a computer and store them in it accordingly.

The procedural acts are also practiced by justice clerks, experts and witnesses. By the same token, the legal permission for the electronic procedural includes the possibility of all acts being practiced through that means.

The electronic communication of acts is a two-way street. On one side, the parts are authorized to send the parts' petitions to the judge through a computer and the electronic communication. On the other, the judge will summon the parts through that same system. The relevant experts will report electronically, and the witnesses will be able to testify through the same way.

We can't help but emphasizing the use of electronic means for performing the audiences. On that aspect, the electronic procedural is very innovative, shaking the traditional structures. The audience has always been oral. It's the moment where the parts may expose their points of view and dialog among themselves and with the judge. Depending on the procedural law model, the audience is more informal or has more participation by the judge. Anyway, there is a substantial change when the electronic means are utilized. The electronic audience brings in two new things. The first is the hearing's recording in audio and video, without any written record, even if there are appeals. The image and sound are recorded in data processing files and may be seen and heard in any computer and at any time. The second new thing: that hearing may take place without the participating people being at a same place. It can take place with each participant being in his own location, like his workplace or his residence. By the same token, image and sound will be recorded, for any checking at any time.

That would be the visible face of the electronic procedural. However, restrict the electronic procedural to the practices mentioned above would be the same as not considering all the premises presented up to this point, that expose the existence of a great cultural revolution resulting from the technological revolution of the final of the twentieth century. In fact, the electronic procedural is not different it its look, but rather in its soul, for adopting principles and rules quite different from those of the traditional procedural. Up to this point, we saw how the computer and the world wide net could serve as a technological apparatus to ease the process and streamline its execution, aiming at fulfilling the constitutional commandment of the reasonable duration of the process. That, however, is not the electronic procedural's limit. The electronic procedural is plenty more, as it allows that all other procedural principles and all fundamental guarantees are handled in a better way.

Even so, it's important to start in that point and conclude that that step, alone, would already mean a significant advance. After all, the legal activity has always been the most delaying, practiced in a greater disharmony with the advances of the twentieth century society. When managing Justice exclusively through electronic means, we could greatly improve the judgment. Let's check just one example: the settlement of the process's object. Through a truly electronic procedural, the initial petition and the defendant's answer will not be elaborated in a written document, for the proper filing. The manifestation of the parts will be elaborated on forms, where each of the elements will have their own field for a database. Thanks to that, we will be able to know the conflict's accuracy, making the determination of an effective procedure easier and providing in a better way the review and the answers to the allegations of the parts.

The elaboration of the demand through forms will provide justice administrators and the community in general the exact knowledge of all conflicts and work as orientation for a perfect analysis of the social phenomenon of the conflicting. That system will also serve to be aware of the forthcoming demands at the same time, which will foment uniform decisions and collective decisions, thus complying with one of the main characteristics of the present mass society. The joint assessment of the demands will respond to the equality principle in a better way.

The soul of the process, however, will be transformed with the adoption of the electronic procedural. For this study we took one of the nationwide reports presented at the *Colloquium Pecs*. It is a magnificent work by José Eduardo de Resende Chaves Júnior, named *O processo em rede* (The Process in a Network). After a complex analysis of the typology *network society,* with its connections and means, the report brings in an interesting proposal of specific principles for the electronic procedural, confronting them with the traditional principles. For the Brazilian jurist, the electronic procedural principles are *immateriality; connection; intermediality; interaction*, *hyper-reality, instantaneousness* and *deterritorialization.*

Principle of immateriality is the dematerialization of the records, rejecting the idea that this word would indicate a sense of misticity. In the material means, the process tends to restrain and block certain forms and behaviours. On the contrary, the principle of immateriality tends to be proactive. As *principle of connection,* or network process, it reveals the inter-relationship among systems, machines and persons. That principle has two perspectives, the reticular and the inquisitive. The *reticular connection* is printed in the form of the network and, among other characteristics, exacerbates the process's publicity. The process is in connection with the world, and does not have any more the paper's physical barriers. That is why this principle throws down the rule that *quod non est in actis no est in mundo*. The connection increases the responsibility of the parts, in the process, as a counterpart to the very widening of its participation. Democracy increases rights, duties and responsibilities. The *inquisitive connection* potentiates the judge's initiative in the process, re-writing the concepts of public fact and notorious fact, facilitating the discovery of truth, so-called real. The *principle of the intermediality* results from the conjunction, interaction and reciprocating contamination among several media. The process stops being a paper and starts to utilize several media for its communication. The *principle of hyper-reality* is about, too, the search of the truth added with the procedural streamlining. According to Chaves Júnior, it's about the radicalization of the process's orality. He explains that "reality confined in the process's text is a static reality, resulting from the utilized means and conditioned by it, the paper. In the electronic means we can record not the effective reality but a digitalized, coded and virtualized reality, i.e.: *hyper-realized*." The *principle of interaction* is the new apparel of the constitutional guarantee of the adversary system. In the author's imagination, the adversary system in the electronic procedural will be more intense, more extensible, more verisimilar, more authentic and more instantaneous. The *principle of instantaneousness* is explainable by its own name, being nothing more than the extension of one of the main characteristics of the new informatized society. Finally, the *principle of deterritorialization,* according to which one dematerializes the idea of forum and judicial district. It means much more than the mere physical transposition of territories and legal districts and even jurisdictions, meaning the fluency of the rights effectiveness, that cannot be more stopped simply by the physical space material limitations. The judge's *longa manus*, when dematerialized and connected, becomes more extensive.

Proceeding the march by José Eduardo Chaves Júnior, we can envision a much ampler horizon for the procedural law's change. His proper observations should be object of assessment and amplification.

The adversary system principle, today applied in its merely formal aspect, may be, in practice, a permanent and productive dialog, making the access to justice easier. It will also facilitate the response to the citizen's right of influencing in a clear and direct way the elaboration of legal solutions. At this point, it comes to our mind the instant message systems and the email. In the electronic procedural, during all of the time, there must be space for an effective participation and for a dialog. It will be easier to build consensual solutions and it will be feasible to carry on the judicial process in a more objective and productive way.

The production of the evidence is given another dimension in the electronic procedural, mainly if we consider that the process's data processing takes place in an informatization context of the whole society. The electronic procedural is not a pioneering activity and must not be isolated. Together (or even prior to) the use of the modern technologies of information and communications in the legal process, we have to bear in mind that those new technologies are being used (and have to be used) in all society's activities. For example, all contracts have to be elaborated electronically. Not just public contracts, closed through public notaries, but also private contracts. By the same token, both the public and general administration, and also police bodies and all the remaining segments of the public and private activities, should make use of the modern information and communications technologies. For example, if the police investigations are being developed through electronic means, all the elements of the investigation will already be electronically stored when the legal process is initiated. It will not be necessary to attach any document resulting from this activity to the process. The judge's computer will access the required information automatically and instantaneously. We might also add an access to the records of birth, marriage, death, legal entity constitution, as well as data bases of financial, educational and health institutions.

The legal evidence concept production changes radically, which makes us to review all canons on the judge's participation in the production of the evidence and in the quest for the truth to support his decisions and orders. We currently live with the traditional concepts of public fact and notorious fact. With society transplanted from a persons' society over to a network society, we will be living in a brand new reality where information are accessible to the judge instantaneously. We can state that if for the judge there was a formal world for the records, now the real world is at a click's far. In this informatized society, videocams are spread all over, all of them accessible in the Internet. All information about all persons, individuals and legal entities, are recorded in some database, technically always available. The proceduralist is being taken to revise the traditional concepts and it will be hard, if not impossible, to keep a judge formally distant from all the reality that is now at the reach of his finger tips.

The judge will also know the information on the witnesses: who they are, where they live and work. He will be able to analyze in a better way the participation of those witnesses in the facts that are being investigated. And a separate checking must be elaborated about the witnesses, that is: about the communication of the judge with the witnesses. I feel I should open a parenthesis for a deeper analysis, here, It's about a deficiency in the traditional communications system that might be settled through the proper utilization of electronic means. I'm talking about the communication of the judge with the witnesses. The judge has to know how to communicate correctly to obtain a free and frank testimony from the witness, getting the right narration of the facts. In other words, the judge has to know the witness and how to communicate with the witness. Today the witness is formally summoned to be heard by a judge. He/she is obliged to participate by the law. When with the judge, the witness finds himself/herself in a public building, usually built and decorated like a palace court. The witness sits under a vested man, in front of flags and country symbols. The questions to be made are addressed to the feelings of that witness, who is in a situation of exacerbated sumptuousness and formality. The judge, I repeat, has to know the witness and how to communicate with him/her. Is he prepared for that? How to make someone talk openly in a situation like that? That's an important dilemma, nowadays, and it takes us to consider that not all judges have the ability to develop that conversation. Let alone counsellors. Today the technique is to press the testimony to tell the truth, which is something done through taking an oath and persuasion, products of manipulating abilities. In a network society, the witness is already someone already accustomed to deal with electronic gadgets and has already learned how to communicate online better than personally. For the new generation, it is easier to talk through online chatting than at a personal meeting. So, we must add to the legal practice the conversation through instant messages plus, naturally, through audio and video, so that the witness feels himself/herself in a more favorable environment. We are referring to the new orality, an orality fit for the modern life.

Another topic to be emphasized is the timing, in the legal process. One of the pillars of the traditional process is the performing of acts in their proper time, with rigid preclusions for the parts and relative preclusions for the judge. The fundaments of that traditional system are not much dealt with, but it is right that its consolidation has rooted in such an extent that discuss it might be considered an anathema. However, the building of a new procedural requires respect for the freedom of an ample and profound discussion. There are models, like the Brazilian, where the preclusion is stronger, and it is forbidden even the extension of the process's object after its formation has been completed. On the contrary, the Italian model is in trouble with an endless possibility of extension changes of the appeals at any time. As a rule, however, the laws establish rigid times for the defendant's answers, for the production of evidence and for the presentation of witnesses, all in the name of the process's security. No doubt the form must be exalted when its objective is to provide equality and the execution of the process in a reasonable time. However, the internal process should admit a more flexible dialog, with a better opportunity for knowing the truth of the facts and a better understanding of the arguments. For example, in a legal audience there is the right moment for each counselor to talk. Even when allowing replications and rejoinders, the opportunities are determined in an inflexible way. The same applies for testimonials, that are taken in a single occasion and a witness cannot, by his own initiative, return to testify and contradict what was said by the other. That inflexibility practice, I repeat, is justified by the need to keep the order, but it also has some fundaments of more sedimentary contents. Those fundaments, for example, lead to the prohibition, in some cases, that a witness listens to the testimony of another. As we see the process, all that is nothing but fiction and there are no studies proving their fundaments. You cannot prove, for example, that the testimony of one witness may be distorted on purpose in case he/she gets to know what the contents of the other witness said to the other. Fiction resides in the simple fact that, before the audiences, witnesses may know what is already prepared to be said by the others, since it's not rare that the legal testimony be a repetition of what was said by the witness in another occasion. Anyway, agreeing or not with that thesis, it is important bearing in mind that in the traditional process resides a practical impossibility to effect any change in that system. Nevertheless, with the electronic process it will be possible to open the structures of that dialog without jeopardizing the process's duration and, most of all, without harming the equality and the procedural safety. The electronic procedural can establish a maximum time, but on the other hand, it has to allow that, within that time frame, an open and free dialog is operated, providing freedom to speak and not just being subject to a cold and formalized testimony. In the electronic process, the principle of the adversary system receives a more real essence that turns the process into a productive dialog that ends qualitatively, not in its fixed temporal aspect.

Finally, I stress a factor of the traditional process that is considered the most deficient, both by internal critiques and the community: the effectiveness of the legal decisions. That because, after the sentence, most of the times other acts must be practiced so that the well-being of the life that was in litigation is really bestowed to the entitled party of the material right. For that effectiveness to happen, either we resume the legal activity or the acts to be practiced are administrative. On both situations, the means available to the judge, in the traditional procedures, are very restricted, blocking him to impose his decision or weakening its contents with the practical difficulties and/or the time required to execute the right, up to now only certified. The modern culture permits that the possibilities of the effectiveness of the right are much greater, and we here comment again about the data processing of the society in general. The use of the modern technologies of the information and communications allows a complete control over certain assets and services, which can be executed by the judge at the same time he decides about the destinations of the author's request. The judge may transfer the property of an asset, carry on the attachment of money and assets, matriculate a person in a school, contract a citizen for a public position and take all sorts of steps directly which, in the traditional process, would be limited only to issue written orders, that many days later would or would not be accomplished by its addressee. Legal commands like paying, delivering or doing are replaced by a direct execution of the right through the interference, also direct, in the social tissue. The judge can even cease a plant's activities because if its production is commanded by computers, these will be accessed by the judge to change their programming. For example, if the request is for a reduction of the monthly value of the amount of a telephone bill, the judge can execute that command directly on the company's computers. If the command is for paying alimony, the judge can access the payroll, make the discount and an automatic deposit of that amount on the creditor's account. For all that, it is enough that the access to the data processed systems is technically facilitated. In the penal process it could happen, for example, the direct access to a telephone central station to execute an authorized hearing without any employee nor the telephone company's director knowing about it.

So, there are many facilities of the electronic procedural, and they may be at the reach of the judge and the parties involved in the legal process. The time of the studies and of the debates are then opened on those themes, since the new technological things allow us to figure out solutions that were impossible in the past.

Which are the habits of the new culture that may already be incorporated to the procedural law? All of them.

That is why our conclusion must emphasize the important weapon that the electronic procedural is for the development of three utterly significant activities right now: consensus, research and mass processes. The electronic procedural places the litigant in a position of collaboration and changes the formal adversary system into a cooperative and permanent dialog. The electronic procedural makes us know in a better way the phenomenon of the conflicting, since it allows a qualitative analysis in real time. The electronic procedural allows us to know better the mass conflicting, facilitating the collective treatment of the individual demands or their transformation in collective demand, when applicable.

Chapter 5
Electronification of Civil Litigation and Civil Justice – The Future of the Traditional Civil Procedure Facing the Electronification*

Nikolaj Fischer

5.1 Introduction: Electronic Justice as Actual Topic of Civil Procedure Law Worldwide

The topic of this article contains the so-called "E-Justice"[1] as one important aspect of procedural law and of procedural law comparison[2] worldwide. That means that this focus is on the electronic civil justice and electronic civil litigation. Basically one can say that this topic is a very new one – especially under scientific regard. The colloquium of the IAPL in Pécs (Hungary) focused the scientific discussion

*The author teaches as Privatdozent at the Institute for Comparative Law at the Johann Wolfgang Goethe University of Frankfurt/M. See for further details esp. (on line): http://www.jura.uni-frankfurt.de/l_Personal/priv_doz/fischer/index.html (04.04.2011).

[1] Compare here for a choice of articles to this issue: Regarding Germany esp. W. Kilian, *Juristische Entscheidung und elektronische Datenverarbeitung* (Frankfurt/M: Athenäum-Verlag, 1974); A. Rossnagel, *CR* 1994, 498 ff.; M. Herberger, in *Prozeßrecht an der Jahrtausendwende*, ed. P. Gilles, (Baden-Baden: Nomos, 1999), 33 ff.; Stadler in *ZZP* 111 (2002), 413 ff.; P. Gilles, in *ZZP* 118 (2005), 399 ff.; see also N. Fischer, *Justiz-Kommunikation* (Berlin: VWF, 2004); N. Fischer, *Kritische Justiz* (2005), 152 ff.; N. Fischer, in *DRiZ* 2005, 90 ff.; see for Austria here only G. Kodek, in *ZZP* 111 (2002), 445 ff.; see also the speech of the author (topic: comparison Lithuania-Germany) on the International Conference "Business Law in Transition in a Comparative Context"of the Swiss Institute of Comparative Law (ISDC), 03 July 2008 in Lausanne, published in *Business Law in Transition, A. Comparative Perspective on Eastern Europe, Reports of the ISDC Colloquium* ed. E. Lein, J. Skala and L. Heckendorn Urschuler, (Zurich et al.: Schulthess, 2010), 229.

[2] See e.g. P. Gilles, *Prozeßrechtsvergleichung* (Köln: Carl Heymanns, 1996) (also for the different tasks of Procedural Law Comparison); P. Gilles, in *Direito processual comparado – XIII. World Conference of Procedural Law,* ed. A. Pellegrini Grinover and P. Calmon, 826–41 (Rio de Janeiro: Forense, 2007).

N. Fischer (✉)
Institute for Comparative Law, Johann Wolfgang Goethe
University of Frankfurt/M, Frankfurt, Germany
e-mail: Nikolaj.Fischer@jur.uni-frankfurt.de

M. Kengyel and Z. Nemessányi (eds.), *Electronic Technology and Civil Procedure*, Ius Gentium: Comparative Perspectives on Law and Justice 15,
DOI 10.1007/978-94-007-4072-3_5,

on the “Electronic Justice” nowadays and in future development.[3] According to the call for presentations in preparation of this colloquium this paper is forced to give here only an overview and some remarks about comparative aspects about the topic – from a German point of view. A good reason for thinking and writing about this topic in Germany are the recent legislative actions in German civil justice system.

5.2 Development of E-Civil Litigation and E-Justice

Let us first have a look on the development of e-civil litigation and e-justice in Germany as an example for specific actual – and probably future – problems and questions of e-justice (see Sect. 5.3).

5.2.1 E-Justice Law in Germany

One way to approach to the development of e-civil litigation and e-justice law in Germany is to have a look on the reform history of e-justice-law development. Before the “Justice Communication Act” of April 2005, we have had in Germany the following reforms: “Formvorschriftenanpassungsgesetz” (2001), “Zustellungsreformgesetz” (2002) and “ZPO-Reformgesetz” (2002).

5.2.1.1 E-Justice Law Reforms 2001 and 2002

First of all there was the so-called “Formvorschriftenanpassungsgesetz” from 2001 (effective since 01.08.2001).[4] This act has created the legal possibility for to send briefs between the parties of a law suit and to the court in the so-called electronic

[3] Compare for the actual world report for the civil law countries E. Jeuland, in *Direito processual comparado – XIII. World Conference of Procedural Law*, ed. A. Pellegrini Grinover and P. Calmon, 152–85 (Rio de Janeiro: Forense, 2007); for the actual world report for the common law countries J. Walker and G. Watson, in *Direito processual comparado – XIII. World Conference of Procedural Law*, ed. A. Pellegrini Grinover and P. Calmon, 119–51 (Rio de Janeiro: Forense, 2007).

[4] “Gesetz zur Anpassung der Formvorschriften des Privatrechts und anderer Vorschriften an den modernen Geschäftsverkehr” from 13.07.2001, BGBl. I. p. 1542. This act is the legislative execution of Art. 9 of the EC-Directive RL 2000/31/EG (ABl. EG Nr. L 178, p. 1). See also the “Gesetz über rechtliche Rahmenbedingungen für den elektronischen Geschäftsverkehr”(Elektronisches Geschäftsverkehrsgesetz, EGG) from 14.02.2001 (BGBl. I, S. 3721, 3724), effective since 21.12.2001, which has leaded to the modification of § 1031 Sec. 1 and Sec. 5 ZPO.

form – see § 130a ZPO (Zivilprozeßordnung, German Civil Procedure Code). This means that you can send a pleading to the court by e-mail. This requires not only a qualified electronic signature according to § 2 No. 3 SigG (Signaturgesetz, German Signature Act)[5] but also the general admission that the court can participate in the German electronic justice system (the so-called "Elektronischer Rechtsverkehr", ERV). But in fact only a few courts (esp. the High Federal Courts) take part in the German electronic justice system – so we are far away from each court participating in e-justice in Germany nowadays. Along the way, we can see that the well-known difference between law in books and law in practice also exists in the new areas of civil litigation – not only, but also in Germany.

The following reform was the so-called "Zustellungsreformgesetz" from 2002 (ZustRG).[6] This reform enabled the electronic delivery of pleadings. The electronic service of process is allowed by law either it is an electronic document and the receiver is an attorney (or has a similar profession) or the parties have agreed expressly to the electronic delivery (see § 174 Sec. 3 ZPO). Electronic justice system in this regard means the accreditation and admission of electronic transfer between courts on the one hand and between courts and parties on the other hand by justice administration (see § 130a Sec. 2 ZPO).[7] In a wide sense you can also describe it as an electronic information and communication system between the participants of civil litigation.

In a third step there was the generation of the video-conference in civil litigation with the so called Civil Procedure Reform Act ("ZPO-Reformgesetz", ZPO-RG) from 2002.[8] Since this law reform it is possible and allowed to use video-conference for the oral negotiation and the hearing of evidence if the parties agree – see § 128a ZPO.[9] These electronic options are not limited to civil jurisdiction because they are also applicable with respect to similar or familiar rules for labour-, social-, finance- and administrative jurisdiction, see § 46b Arbeitsgerichtsgesetz (ArbGG), § 108a Sozialgerichtsgesetz (SGG), § 77a Finanzgerichtsordnung (FGO) and § 86a Verwaltungsgerichtsordnung (VwGO).

---

[5] "Gesetz über Rahmenbedingungen für elektronische Signaturen" from 16.05.2001, BGBl. I p. 876. See esp. A. Rossnagel, in *NJW* 2001, 1817 ff.

[6] "Gesetz zur Reform des Verfahrens bei Zustellungen" from 25.06.2001, BGBl. I p. 1206. Compare e.g. Rosenberg Schwab, Gottwald, *Zivilprozessrecht*, 16th ed. (München: Verlag C. H. Beck, 2004), 469.

[7] See e.g. the "Verordnung über den elektronischen Rechtsverkehr beim BGH"(Elektronische Rechtsverkehrsverordnung – ERVVOBGH) from 26.11.2001 (BGBl. I p. 3225); also Fritsche in *NJ* 2002, 169 ff., 177; Barth in *ZAP* 8/2004, 395 f.

[8] See for the "ZPO-Reformgesetz" from 27.07.2001 (BGBl. I p. 1887, changed pp. 3138, 3179), effective since 01.01.2002, see here N. Fischer, *Zivilverfahrens- und Verfassungsrecht* (Berlin: VWF, 2002), esp. 19 ff.

[9] See esp. Borchert in *CR* 2002, 854 ff.; Schultzky, *NJW* 2003, pp. 313 ff.; see for the video-conference in German Criminal Procedure Code (StPO) only § 247a StPO.

5.2.1.2 "Justizkommunikationsgesetz" (2005)

Finally the "Justice Communication Act", effective since April 2005,[10] filled the gap between the existing regulations of e-civil procedure law, which governed so far only the input and the output side of the electronic justice system (see § 130a ZPO on the one side, and § 174 Sec. 3 ZPO on the other side). The most important regulations of this "Justice Communication Act" are the following: The judicial electronic document (§ 130b ZPO), an electronic filing system (§§ 298 ff. ZPO) and electronic evidence regulations (§§ 371 f. ZPO). With respect to § 130b ZPO, the basic regulation is the rule about the judicial electronic document (see § 130b ZPO). This norm has the purpose to allow the courts to use electronic documents – which means here again documents with a qualified electronic signature according German Signature Act (see § 2 No. 3 SigG) – so that the traditional written form for court documents can be changed into the electronic form.[11] It is necessary not only in this context to know that the qualified electronic signature differs from the simple electronic signature like the scanned original signature in an e-mail – see here only § 2 No. 1–3 SigG. Regarding §§ 298 ff. ZPO, the "Justice Communication Act" enables the administration of justice to allow and establish an electronic filing system (see especially § 298a ZPO for the electronic file in civil litigation). The importance of this new regulations (§§ 298 ff. ZPO) lies in the justice intern transfer of media – this means the transformation from written form to electronic form and reverse.

And finally with respect to §§ 371 f. ZPO, the "Justice Communication Act" introduces evidence regulations concerning electronic documents: According to § 371 Sec. 1 S. 2 ZPO the traditional rule about evidence by appearance is also applicable for e-documents – if the e-document for itself is the subject of evidence. In case that the private e-document was sended with a qualified electronic signature, the rules about the probative force of private documents (§§ 415 ff. ZPO) are also applicable for e-documents (§ 371a Sec. 1 ZPO).

5.3 Specific Problems and Questions of E-Justice

A critical view on this development of e-civil justice would lead to the following problems and open issues which need to be discussed[12]: First of all, there is a contradiction between pragmatism and conception of e-justice (see Sect. 5.3.1). Secondly, there is a difference between the formalism and the content of e-procedural law (Sect. 5.3.2). Thirdly, you will see the detailed regulations of e-procedural law on the one side and the lack of basic regulations on the other side (Sect. 5.3.3). Finally we have the fixation on IT-instruments in contrast of the lack of theoretical

[10] "Gesetz über die Verwendung elektronischer Kommunikationsformen in der Justiz" from 22.03.2005, BGBl. I p. 837; compare e.g. N. Fischer, in *DRiZ* 2005, 90 ff.

[11] See also Krüger and Bütter in *MDR* 2003, 181 ff., 183.

[12] Compare for further details N. Fischer, *Justiz-Kommunikation* (Berlin: VWF, 2004), esp. 17 ff.

discussion about e-justice (see Sect. 5.3.4). Hence, these aspects of actual and future e-civil litigation and e-civil justice are transnational, because they are probably not only relevant in Germany.

5.3.1 Pragmatism Versus Conception of E-Justice

First, one has to think about the contradiction between pragmatism and conception of e-justice. It is obvious, that the discussion about the need and regulation of e-civil justice took place almost among practitioners – and only a small number of scholars are involved at all.[13] In contrary to the fact that e-justice is worldwide a topic of many procedural law conferences in the last years it is obvious that especially questions of justice organisation and the connection between justice administration and e-justice in general are seldom topics of scientific interest. However, this is to point out because of the important connection between the reform of procedural e-law and questions of justice organisation and administration. It is obvious that we have in Germany a very detailed and huge justice organisation and administration – see e.g. the basic constitutional regulations in Art. 92, 95 GG (Grundgesetz as the legal basis of the German Constitution). But only a few scientific spectators see even the need for a reform of justice organisation in light of the new IT instruments.[14] This is very interesting because this need for reforms of justice organisation had been advised in several empiric studies in the past. As you can see, this claim sounds very new but is in fact an old one: One of the empiric based studies of the "structure analysis of judicature" ("Strukturanalyse der Rechtspflege", SAR)[15] has emphasized the need for a connection between the modernisation of court organisation and electronic service in order to have a more effective justice system.[16] Here you can see one of the problems of the previous and actual development of e-justice in Germany. It is a fact that the existing regulations of e-civil procedural law were created without reforms of justice organisation. This leads to the copy of the existing, traditional civil procedure system in the new e-service in justice and in this way to the electronification of the past in the present and future justice system.[17] In regard of this we should reactivate the analysis of justice organisation in a first step before we plan a basic reform of e-justice in a second step. Not till then we should regulate the e-civil litigation in detail.

---

[13] Compare for further information P. Gilles in *ZZP* 118 (2005), 399 ff.

[14] See for further details N. Fischer, *Justiz-Kommunikation* (Berlin: VWF, 2004), esp. 24 ff.

[15] See for the results here Leutheusser-Schnarrenberger in *NJW* 1995, 2441 ff.; see also Strempel and Renning in *ZRP* 1994, 144 ff.

[16] Compare only H. Fiedler and F. Haft, *Informationstechnische Unterstützung von Richtern, Staatsanwälten und Rechtspflegern* (Köln: Bundesanzeiger 1992).

[17] See esp. Viefhues and Volensky in *DRiZ* 1996, 13 ff., 14; Strempel and Götzel in *DRiZ* 1990, 121 ff., 125 f.

5.3.2 *Formalism Versus Content of E-Procedural Law*

The next step has to be to think about the difference between the formalism and the content of e-procedure law. In the actual development of e-civil justice we can discover a dominant position of procedural formalism in relation to the content of e-civil procedure law. In order to set an example, I just want to mention the different formalised conditions of sending documents to the court by e-mail (see e.g. §§ 130a, 130b ZPO). The central condition is the qualified electronic signature (according to the German Signature Act, see § 2 No. 3 SigG). Just to clarify: It is not my intention to negate the need for a safe and reliable data transmission between parties and the court outside and inside administration of justice.[18] But I want to point out that it is also necessary to realise that the concentration on the electronic form leads to the lack of discussion about the content of future e-procedure regulations.[19] Besides that, we should also challenge the argument that we have extreme dangers of falsification by the electronification of justice. For example it is argued, that a claim could be falsificated electronically.[20] To my mind this argument is not a very realistic one but significant for the clash of IT and civil procedure law as another form of "clash of cultures".[21] We could also question if the procedural law is the right medium to react to the risks of data transmission and not the criminal law. We have therefore to reconsider what relevance the original signature (by hand) will have in the future. This may lead us to a reasonable discussion about the need of a reform of pleadings.[22] For the avoidance of doubt: I am still not sure if the idea of structured party pleading like in the English Procedural Law is a good solution also for Germany, but this is not a strong argument against discussing this question. Under aspects of procedural law comparison the reformed English Civil Procedure Rules can give a useful example also for possible future German law reform plans: In Part 5.3. of the Civil Procedure Rules the IT-reproduction of the signature is accepted whenever a signature is needed.[23] It could be useful if this general rule is transferred also in German Civil Procedure Code in order to avoid unjust differences between traditional and new communication means – especially when documents are necessary in procedural law to keep termins of material or procedural

---

[18] Compare here esp. Viefhues in *CR* 2001, 556 ff., 560; von Lewinski in *BRAK-Mitt.* 2004, 12 ff.

[19] See e.g. Viefhues in *CR* 2001, 556 ff., 560; diff. E. Schneider, in *NJW* 1998, 1844 ff.

[20] Compare E. Schneider, in *NJW* 1998, 1844 ff.; see also Hahn, in the discussion report of Iqbal, *ZZP* 111 (2002), 491 ff., 494; Schwachheim, *NJW* 1999, 621 ff., 622; see for the German judicature e.g. BGH, decision from 18.12.1975, VII fR A23/75 *NJW* 1976, 966 f.

[21] So Hoeren in *NJW* 2000, 188 ff.; see also Hoeren in *NJW* 1998, 2849 ff.

[22] See for diff. positions e.g. D. Ulmer, in *DRiZ* 1991, 280 ff., 283 f.; M. Zöller, in *DRiZ* 1991, 326 ff., 331 f.; Ernesti in *DRiZ* 1987, 129 ff., 137 f.; Herr in *DRiZ* 1986, 374 ff.; see also Werner in *NJW* 1997, 293 ff., 294, with additional references.

[23] See Part 5.3. of the Civil Procedure Rules: "Where any of these Rules or any practice direction requires a document to be signed, that requirement shall be satisfied, if the signature is printed by computer or other mechanical means." See also Stadler in *ZZP* 111 (2002), 413 ff., 418, 435.

law.[24] Consequently the existing duties to use qualified electronic signatures are to think over, especially with respect to the speed and efficiency of procedural information and communication.

5.3.3 Detailed Regulations of E-Procedural Law Versus Lack of Basic Regulations of E-Justice

The third aspect concerns the contrast between the detailed regulations of e-procedural law on the one hand and the lack of basic ideas on the other hand. I tried to point out that the new detailed regulations of e-procedural law are not based on a former deep analysis of the basics of civil procedure law. It is possible to provide you here only with one example, but it is an important one: Even the question about the compatibility of electronification of civil procedure with the main principles of procedural law have not been discussed or raised yet. This means that on the one hand the electronification of procedure is to scrutinise first on the basis of this general principles[25]: It is to ask which one of the new regulations will fit to the basic principles. This means on the other hand that one have to reconsider if the old and traditional principles fit to the future of procedural law. In other words: It is also to ask which of the basic principles will fit to the electronic future of procedure.

Before we should review these basic principles (as a future task for further examinations) we should debate on the requirements of the future for our (sometimes oldfashioned) civil procedure law. And not till than we can work on a modification or change of these basic principles. For an example, I would herewith like to raise the question of the actual and future importance of presence in trial. Especially the traditional needs for the physical attendance of the parties at the court during the hearing seems to be (in some cases) oldfashioned and overruled – if we think of the possibilities of IT. This question has already importance for the comprehension of speech and writing of litigation and therefore is of importance also for the traditional principle of oral speech in civil procedure (see §§ 128, 137 ZPO) in a future e-civil justice. The question about the modification or change of the basic principles in face of the new IT-means is an important one – not only under regard of German law but also under aspects of procedural law worldwide. On the XIII. World Conference of the International Association of Procedural Law (IAPL), which took place in September 2007 in Salvador de Bahia (Brasil) this topic was discussed as well, especially under the aspect of "access to e-justice" – see also Art. 6 European Convention on Human Rights (EMRK) and Art. 47 Sec. 2 of the EC-Constitutional Rights Charta – and it was evident, that this question was of worldwide common

[24] Compare (with further references) Zöller, 21th ed. Greger § 130 ZPO, Rn. 11; E. Schneider, in *NJW* 1998, 1844 ff; see also M. Schmidt, in *BB* 1999, 1125 ff., 1127.

[25] See here only N. Fischer, *Justiz-Kommunikation* (Berlin: VWF, 2004) 59 ff.

interest.[26] If you remember the growing importance of electronic justice in all our countries we should discuss the future importance of an "access to e-justice" due to the also growing duties to use – sometimes expensive – IT-instruments and electronic signatures. These duties are not only relevant for professional lawyers – we think here especially of Austria – but also for natural parties and consumers.[27]

These stated questions lead to the following ideas: Before planning and executing the next (national) law reforms in the field of e-justice we should debate on the general question if and how the e-justice is consistant with the traditional basic principles of procedural law, like the principle of oral presentation (§ 128 ZPO), the principle of directness (resp. immediateness, see §§ 128 Sec. 1, 355 Sec. 1 S. 1, 309 ZPO), the principle of publicity of trial (see §§ 169 ff. Gerichtsverfassungsgesetz, GVG). Last but not least we should not forget the principle of a fair, human and independent trial according to Art. 6 Sec. 1 of the European Convention on Human Rights (EMRK). Due to this Article a public trial is necessary. For a public trial only the oral and therefore direct presentation is senseful – esp. under regard of finding the truth and of efficiency of procedure – and to grant a fair trial with the right of a legal hearing.[28]

As you can see, the question of compatibility is not always a simple one to answer if you remember the traditional principle of publicity of procedure – which will become may be a burlesque in the age of video conference or electronic filing.[29]

5.3.4 Fixation on IT-Instruments Versus Lack of Theoretical Discussion About E-Justice

These questions are also important for the last of the four aspects which is to mention here, the fixation on IT-instruments in contrast of the lack of theoretical discussion about e-justice. You will find the evidence for this contrast in some of the already existing regulations of e-justice in Germany – which are not consistant with

[26] Compare esp. the relevant World Report from E. Jeuland, in *Direito processual comparado–XIII. World Conference of Procedural Law*, ed. A. Pellegrini Grinover and P. Calmon, 152–85 (Rio de Janeiro, 2007); see also N. Fischer, *Justiz-Kommunikation* (Berlin: VWF, 2004), 59 ff.; see for the both German national Reports: P. Gilles, in *Neue Tendenzen im Prozessrecht*, ed. P. Gilles and T. Pfeiffer, 153–77 (Baden-Baden: Nomos, 2008); N. Fischer, in *Neue Tendenzen im Prozessrecht*, ed. P. Gilles and T. Pfeiffer, 85–152 (Baden-Baden: Nomos, 2008).

[27] See N. Fischer, *Justiz-Kommunikation* (Berlin: VWF, 2004) esp. 36 ff.

[28] Compare for the right of a legal hearing in German Constitutional Law (see Art. 103 Sec. 1 GG) esp. BVerfGE 55, pp. 1 ff., 6; see also N. Fischer, *Zivilverfahrens- und Verfassungsrecht* (Berlin: VWF, 2002), 15.

[29] See for another opinion Stadler in *ZZP* 111 (2002), 413 ff., 418, 437; see also Schultzky in *NJW* 2003, 313 ff., 315.

the traditional German procedural law terminology and theory.[30] Some spectators see here also the lack of the art of legislation in this new regulations created almost without a previous scientific discussion.[31] These legislative problems are also relevant for the future reform plans due to the need to have a consistant legal basis for further law reforms. According to another critical remark the obsession with IT-instruments – like the qualified electronic signature as standard condition to use electronic communication in justice nowadays – could cause problems for the future service of IT in civil litigation and civil justice administration. For further reform plans and foregoing discussions we have to reconsider it.[32]

5.4 Conclusions

To put it all in a nutshell: First, it is necessary to gain a basic concept of electronic justice and its embedding in a reform of organisation of justice administration. Second, it is now time to reconsider the authority of thoughts about risks and dangers of the data security and to conquer the hyper-formalism in e-justice nowadays. In this context one has to raise the question if it is really a task of the procedural law to battle against the misuse of data transmission by hackers. To my mind it should be the duty of the better fitted criminal law.[33] Third, it is necessary to get over the contrast between detailed procedural regulations and the lack of interesting in the basics of procedure. For this purpose the basic principles of publicity and oral presence are very important esp. under regard of the condition of an electronic trial, which is compatible with Art. 6 Sec. 1 of the European Convention on Human Rights (EMRK). Therefore the problem, having a bondage using electronic data transmission for procedural purposes, could arise. Even on the above mentioned IAPL-World Conference in Salvador de Bahia I advanced the view that the "access to justice" should not be restricted by the mandatory use of electronic transmission (see for this example especially Portugal or Austria)[34] or even the binding use of

---

[30] Compare for example the wording of § 128 Sec. 1 ZPO (the regulation of the video-conference), see for a critical point of view towards this regulation P. Gilles, in *Neue Tendenzen im Prozessrecht,* ed. P. Gilles and T. Pfeiffer, 153–77 (Baden-Baden: Nomos, 2008), 167, 169; concerning § 130a ZPO see Dästner in *NJW* 2001, 3469 ff.; vs. Fritsche in *NJ* 2002, 169 ff.

[31] See only P. Gilles, in *Neue Tendenzen im Prozessrecht*, ed. P. Gilles and T. Pfeiffer, 153–77 (Baden-Baden: Nomos, 2008), 163 ff.

[32] Compare N. Fischer, *Justiz-Kommunikation* (Berlin: VWF, 2004), 40 ff.

[33] See here §§ 267 ff. StGB (German Criminal Code) and esp. § 202a for the so-called Hacking, and § 263a STGB for Computer-Fraud and §§ 269, 303a, 303b for Computer-Sabotage. Compare to this argument in detail N. Fischer, *Justiz-Kommunikation* (Berlin: VWF, 2004), 41 ff.

[34] Compare esp. the legal situation in Portugal: Since the year 2003 it is mandatory to send pleadings in appeal cases only via data medium (see Art. 150 Sec. 1, 152 Sec. 6 CPC Portugal); see also Stadler in *ZZP* 111 (2002), 413 ff., 425.

qualified electronic signature.[35] From my point of view the "access to e-justice" has also to be guaranteed as a basic procedural right. Hence, we should think about a social electronic justice.[36] Under this aspect the electronification of justice is also a subject of constellation of tensions between a speedy and a social and fair trial. This question is also of interest for the civil procedure law worldwide. Fourthly, we should think about the consequences for the last problematic aspect – the relationship between the fixation on IT-instruments and the lack of theoretical discussion: Before talking about optionally further reform plans and ideas we should talk and negotiate first about the reform procedure in general, before talking about the electronic procedural law in detail. To my mind it is very important that the basic rules and principles of civil procedure – like Art. 6 Sec. 1 of the European Convention on Human Rights (EMRK) – should also be the conceptual basis for the framework of electronic civil litigation and administration of justice. Far from it we have to monitor a new tendency of formalism of civil litigation within the introduction of the new e-justice-regulations. Anyway – and not only for the situation in Germany – it is to claim that the parties in civil litigation should not be excluded from the technical development of modern means of communication.[37]

5.5 Prospects

In order to finish this consideration I want to get back to the early beginning: To deal with the topic "Electronification of Civil Litigation and Civil Justice" as a matter of electronification of civil litigation and administration of justice (as a worldwide movement) one has, before starting to deal with form and security matters, to ask what kind of civil justice is desired and applicable. These matters of form and security could also be designated as "technical law" in its strongest forming – referring here to the famous words of *Friedrich Stein*.[38] As shown here, the problems of e-justice are not really based in detail questions, but mainly in the basic consideration[39]: If the actual development and concentration on formal matters of e-justice is going on, then the new reform acts in Germany – especially the Justice Communication Act

[35] See N. Fischer, *Justiz-Kommunikation* (Berlin: VWF, 2004), 65. According to this ideas a new working group of the European e-justice and e-civil procedure law was founded on this conference with several members from different European countries (about further activities will be reported).

[36] Compare for this claim already N. Fischer, *Kritische Justiz* (2005), 152 ff.

[37] This claim arised already in the year 1899 when the old German High Court of the German Empire (Reichsgericht) postulated this for term keeping pleadings (see RGZ 44, pp. 369 f.: filing of an appeal by telegram).

[38] See F. Stein, in his preamble of the first edition (1921) of his famous "*Grundriss des Zivilprozeßrechts und des Konkursrechts*".

[39] Compare for further details N. Fischer, *Justiz-Kommunikation* (Berlin: VWF, 2004), esp. 64–71.

("Justizkommunikationsgesetz") – will be also only reforms of the form, but not reforms of content. In this context it is not essential in which form the information is exchanged between the parties – especially paper or electronic form – but it is necessary that there is an effective exchange of information at all. It is desirable that an electronic file should help for an efficient collection of information and in order to have a better access to information for all participants of civil procedure and in this way to a better procedural information and communication. Therefore the hearing has the purpose to enable an intensive and concentrated communication between the parties of a trial. The leading idea here should be to create a consensus – which means a compromise between the parties – or at least acceptance, which means acceptance for a judicial decision. Both require an adequate allocation of time.[40] According to this, the electronification of civil procedure should be accompanied by a fortification of the oral hearing and therefore should lead to a refreshment of the old basic principle of oral hearing which is too often misunderstood only as a principle of adduction.[41] In this way the content and not the form of procedure is decisive and at the same time the civil litigation is recognised as a sophisticated system of information and communication.[42] As a matter of fact, these questions are not only national ones, but also tasks of European and worldwide interest – and especially a challenge for procedural law comparison.

---

[40] See N. Fischer, *Kritische Justiz* (2005), 152 ff., 161.

[41] A famous example you will find in O. Jauernig, *Zivilprozeßrecht*, 26th ed. (München: C.H. Beck, 2000), 84.

[42] See also P. Gilles, in the discussion report of Iqbal, in *ZZP* 111 (2002), 491 ff., 491.; B. Hahn, *Anwaltliche Rechtsausführungen im Zivilprozeß* (Köln: Dt. Anwaltverlag, 1998), 430 ff.

Chapter 6
Proceedings on the Web

José Eduardo de Resende Chaves Júnior*

The just are only effective,
can only maintain the existence of a community,
establishing a collective intelligence.

Pierre Lévy

6.1 Introduction

The new doctrine[1] on the computerization of court proceedings is still too highly conditioned to the aspects of the nomenclature[2] of these novelties of the legal world, which is natural in the process of affirmation and construction of all new disciplines; however, this discussion must not, on any account, leave out what appears to us to be the true *quid* of the issue: the reticular character that the new technologies of communication and information bring to the court proceedings, that is, their virtual insertion on the network, especially on the world computer network, the Internet.

*José Eduardo de Resende Chaves Júnior, President of the Latin American Judges Network – REDLAJ; Principal Judge of the 21st Court Circuit of Belo Horizonte, State of Minas Gerais, Doctor in Fundamental Rights by the Universidad Carlos III de Madrid; Coordinator of GEDEL – Justice and Electronic Law Study Group of the Judicial School of TRT-MG. Member of the Brazilian Institute of Electronic Law – IBDE.

[1] Carvalho Leal, for example, quotes the following terms: electronic proceedings, electronic procedure, e-proceedings, digital proceedings, virtual proceedings, cybernetic proceedings, telematic judicial proceeding and tele-computerized proceeding. Cfr. LEAL *in* Jus Navigandi – Available at: http://jus2.uol.com.br/doutrina/texto.asp?id=9296 [01.04.2011].

[2] Professor José Carlos de Araújo Almeida Filho supports the terminology *Material Law and Electronic Procedures* in his manual. Cfr. ALMEIDA FILHO, (2007), 54–6.

J.E. de Resende Chaves Júnior (✉)
Principal Judge of the 21st Federal Labour Court Circuit of Belo Horizonte, Brazil
e-mail: rede.pepe@gmail.com

M. Kengyel and Z. Nemessányi (eds.), *Electronic Technology and Civil Procedure*, Ius Gentium: Comparative Perspectives on Law and Justice 15,
DOI 10.1007/978-94-007-4072-3_6, © Springer Science+Business Media B.V. 2012

The dematerialization of court records, that is, their passage from the analogical world of atoms to the digital world of *bits* – and even the *qbits* with the imminent arrival of the quantic computer – is undoubtedly of itself comparable to the arrival of the printing press to culture. But the changes resulting from this dematerialization are expanded exponentially when it is understood that the electronic process is, and can be, above all, a *networking process*.

The specificity of the network phenomena brought about, in the field of the exact sciences, a new science called *network sciences*. The typical formalization of mathematics and physics brought the migration of these concepts of the new sciences to sociology and economics.

The challenge that appears now is precisely to catalyze this new knowledge to the legal world, and, especially, to court proceedings. In this paper we will seek to give at least the first hints so that these new tools may be applied to the electronic mesh of procedural rights and duties.

6.2 Connection and Legal Circles

6.2.1 Connection

An initial observation that can be made is that, from the conceptual point of view of pure decidability, the e-procedure profoundly alters the relation between court records and the world. On the paper proceedings, the records are the very material incarnation of the division, distance, that is, the separation between what is decided and what is in the society-world, synthesized in the so-called principle of writing: *quod non est in actis non est in mundo.*

In electronic proceedings, the principle is – or intends to be – precisely the opposite, that is, the principle is that of approximation, connection between the records (virtual) and the network-world, in that the frontier records-world is dematerialized, since both are in so-called *data space*.

The access of electronic court records to the real-virtual world through hypertext (link), although it does not allow access to the material world, brings to the records another world of information, evidence and radically changes the very reasoning of procedures.

This new records-world connection is in fact the connection of the proceedings subjects, judge, plaintiff and defendant with virtual society. The proceedings, therefore, are no longer a merely angular,[3] segmented and isolated flow and become a reticular (networked) and collective flow.

But why is so much said about networks? According to the scholars of the School of Networks,[4] especially Augusto de Franco[5]: *"networks are systems of nodes*

---

[3] Angular in the sense of angular legal relation.

[4] www.escoladeredes.org

[5] Augusto de Franco, in Carta Rede Social n. 171, available at: http://augustodefranco.locaweb.com.br/cartas_comments.php?id=260_0_2_0_C [01.04.2011].

and connections. In the case of social networks, these nodes are people and the connections are the relations between these people. The relations here are marked by the possibility of one person issuing or receiving messages from another. When this happens, we in fact say that a connection has been established".

The first steps of the so-called network theory were taken in the work of mathematician Ëuler, who formulated "graph theory". A graph is a representation of a set of nodes connected by edges.[6] Erdös and Rényi were the first to relate graphs to social networks. There is a lot of work about complex networks, later applied to social networks, including virtual networks. Here we may quote the models of Barabási, Watts and Strogatz and Erdös and Rényi.[7]

It is important here to underline the cumulative and expansive character of network, as shown by Barry Wellman[8] and Barabási.[9] In a network, everything tends to grow on gigantic proportions and scales, and even out of apparent control.

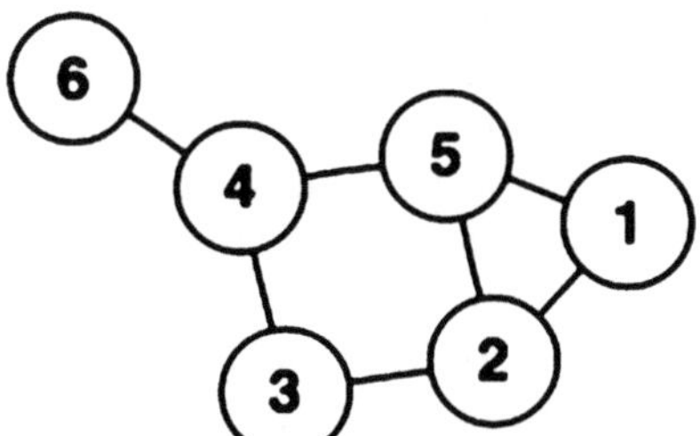

[6] A graph with six vertices and seven edges. "*A graph G is an ordained triple (V(G), E(G), Ψg) which consists of a set V(G) of vertices, a set E(G) of edges without intersection with V(G), and a function of incidence Ψg that associates to each edge of G a non-ordained pair of vertices (not necessarily distinct) in G.*" BONDY, MURTY, 1976, p. 01.

[7] The analysis of social networks begins with two branches: (i) of whole networks and of (ii) personal networks. The first branch focuses on the relation of the group with the network; the other, on the individual with the network. The concept of multiplexity is involved in complex networks. This means the degree of multiplicity of the flow of social bonds found in a given social network. The novelty in the study of networks is in realizing the network structure not as determined and determining, but as changeable in time and space. Another concept of network theory is cluster, which is a group of social groups connected coherently (nodes). Cfr. RECUERO, last access on June 14, 2009

[8] Barry Wellman speaks of the rule 'the more, more', which applies in the interaction between networks on the internet: the more the social-physical network is used, the more the internet is used; the more the internet is used, the more the physical network is reinforced. Barry Wellman and Mena Gulia, "Netsurfers Don't Ride Alone: Virtual Communities as Communities," in *Networks in the Global Village*, ed. Barry Welmman, 331–66 (Boulder, CO: Westview Press, 1999). *apud* Castels (2002), 444.

There are many papers by Wellman and his group available at his virtual page at the University of Toronto, Access on 05/09/2008.

[9] The model 'non-scale networks' was formulated by Barabási. This model is based on the rich get richer phenomenon, as in Wellman. This means that the more connections there are to a node, the more opportunities there are of having others. Therefore, the networks are not equal, since there is a preferential link to the most widely used. Cfr. Barabási (2002), 79–82. The name 'no stop-overs' comes from the mathematical representation of the network, which follows a curve called power-law, also known as the 'Pareto law' or the '80/20 rule, which refers to a proportion that frequently occurs in network phenomena. Cfr. Barabási (2002), 66–71.

Augusto de Franco also notes that "*there are many kinds of network, amongst which the better known and quoted are biological networks (the neural network, for example, connecting the neurons of the brain of animals, or the web of life that assures the sustainability of ecosystems, connecting micro-organisms, plants and animals and other natural elements) and the social network (although there are also machine networks – such as the world wide web which we call the Internet – which are social networks in that they connect people.) There is a similarity between these organizational patterns so that, studying them, it is possible to understand the multiverse of hidden connections which make up what we call social*".[10]

These network phenomena can be clearly seen in the very evolution of the economy of new technologies. Bill Gates confirmed his retirement in mid 2008, announcing the second digital era, the era of connectivity – others prefer the term 'connectibility'.[11] His retirement underlines this new era, since all the fame and money o Microsoft were generated from the idea of the PC, the personal computer, which gave individuality a central position.

The value of the brand *Google* overtook *Microsoft*[12] in the market, which explains this race after the so-called virtual social networks. We have, in effect, entered the era of the common, of the Spinosean crowd, the Deleuzean rhizome, in detriment to the traditional ideas of collectives, such as people, public, class, proletariat, mass.

In the second era, there is the decline of the personal, the individual, the private: the *common*[13] crosses over the public and the private. The third sector with the greater potential of networking than the public and the private sectors has the energy to swallow up the State and the company in the future.[14]

We believe that we may, in fact, think of a new digital and political era, the network era. If IBM was unable to see in the 1970s that the decisive novelty of

[10] Cfr. *in* www.augustodefranco.com.br – Carta Rede Social n. 171 [01.04.2011].

[11] In Portuguese, the "Vocabulário Ortográfico da Língua Portuguesa – VOLP" accepts both forms.

[12] Microsoft announces the launching of its competitor to Google: BING: www.bing.com.br

[13] The idea of common has roots in Negri and Hardt and is linked to Aristotle's concept of 'common place', as described by Paolo Virno: "*Today when we speak of "common place", we usually understand it as stereotyped speech, almost devoid of any meaning, banalities, dead – "his eyes are headlights" – , repeated discourse. However, this was not the original meaning of commonplace. For Aristotle, the topoi koinoi are logical and linguistic forms of general value, as if speaking of the bone structure of each of our discourses, that which allows and ordains each particular utterance. These 'places' are common because no one – neither the refined orator, nor the drunk who blubbers meaningless words, nor the salesman or the politician – can escape them.*" Cfr. Virno Paolo *Gramática de la multitud* (Madrid: Traficante de Sueños, 2003), 34–5.

[14] It is not by chance that Bill Gates retired to dedicate himself to his foundation, after donating 30 billion dollars of his personal fortune to it. Warren Buffett, owner of the Wall Mart chain, also one of the richest men in the world, did the same and donated 44 billion dollars to his foundation. This common fund begins with a budget larger than that of most countries in the world – about 10% of Brazilian GDP.

computer economics was the immaterial software, today Microsoft, on the other hand, is losing ground to Google because it was slow to understand the wealth provided by the so-called network *externalities.*[15]

Besides the network *externalities*, in the immaterial economics of computing, there arises the idea of the nullification of two classic laws of economics: the law of scarcity[16] and the law of decreasing income.[17] In the new immaterial economics, what is valid is the law of abundance and growing income.

On the other hand, production based on network externalities institutes a new form of economic production which is decentralized, collaborative and which can escape market rules – *commons-based peer production.*[18]

Giuseppe Cocco points out that in reticular production, the terms *netwares* and *wetware*[19] *"are mobilized to supplement hardware and software and to apprehend the new forms of work and/or production interaction in the spheres of virtual cooperation networks*".[20] Moulier-Boutang adds that the hegemonic goods of so-called cognitive capitalism are made up of four simultaneous factors: (i) *hardware*;

---

[15] *French economist Yann Moulier Boutang, observes that: "When an economic operation between two agents A and B has an effect on a third agent C without a monetary transaction or convention of Exchange between A and C, or B and C, it is said that it creates an externality. If an externality created operates in detriment of C, that is, if it reduces his current wellbeing, or deprives him of enjoying a good, a potential service, it is said that it is a negative externality or an external diseconomy. If, due to the transaction between A and B, agent C sees an increase of his wellbeing, his wealth, opportunities for action, knowledge or improvement of his environment, it is said there is the creation of a positive externality. Economist Alfred Marshall introduced the idea of technologically possible externalities, for a company C which, through its geographic implementation, benefits the setting (transport, accessibility, market proximity) beyond its fiscal or mercantile contribution. For Marshall, the growth of the company that does not depend on the accumulation of capital and labor, but of technique, is explained by technological externalities*". Cfr. Moulier-Boutang (2004), 147.

[16] Traditionally, economics is seen as the science which seeks to balance human needs, which, by nature, are unlimited, to the resources which are always limited and scarce. One of the purposes of economic activity is the fight against scarcity. Cfr. Cotta (1978), 168.

[17] The so-called law of 'decreasing income' (or non-proportionals) was formulated by classical economists. According to this law, the relation between the quantity of the product and of one (or several) production factors has a tendency to decrease when production increases. In other words, this law says that duplicating inputs in a physical process does not duplicate production. Marx criticizes this law, stating that it applies more to agriculture than to industry, because technical growth itself stops the decrease. Cfr. Cotta (1978), 361–62.

[18] Cfr. Benkler Yochai, *The Wealth of Networks: How Social Production Transforms Markets and Freedom* (New Haven/London: Yale University Press, 2006), 60.

[19] *Wetware* and *netware* are correlate terms. The first refers to the individual capacity to operate hardware and software systems, a capacity that is developed from the point of view of the user or consumer, interactively, in production. The emphasis here is work and innovation, from the point of view of consumption. Netware is the collective outlook of this same interaction with consumption from the network. Cfr. Cocco (2003), 9–10.

Cfr. also Moulier-Boutang (2004), 54–5.

[20] Cfr. *Ibid.*, p. 9.

(ii) *software*; (iii) *wetware* and (iv) *netware*. He states that *netware* has a hegemonic, that is, determining role among them, however, the four factors cannot be reduced to one of them. He also notes that capital cannot have complete control of any of the four factors.[21]

Transporting these outlooks to the special case of electronic court procedures, what seems decisive is to understand it as a phenomenon of interaction between the judge, the parties and society, that is, as a phenomenon of a social, economic and political network and not a mere IT structure for court cases.

6.2.2 Legal Circles

There is a great deal of resistance (which we might call ergonomic) to the use of the computer in legal circles. It is often stated that it is easier to handle court records on paper.

Some difficulty at the current technological development cannot be denied, however, we all know how difficult it is to handle, for example, records which run into 40 volumes; in electronic proceedings, internal search mechanisms will greatly facilitate searching into the records.

There are also technologies such as multi-touch and e-ink available on the market, and even computational vision and voice recognition, which will soon provide more ergonomic solutions to electronic proceedings. There is also the so-called e-paper, of *oleds* or organic *leds*, which make it possible to produce very thin monitors, the thickness of paper, and which are also flexible. There is even an anecdote that clearly shows the overcoming of concerns with the comfort of electronic handling, because, for those who prefer the old paper proceedings, it is possible to foresee the e-paper proceedings…

This wordplay is interesting also to undo another mistaken idea, that is, that electronic proceedings are the simple virtual transposition of court records, without any intervention in the principles and traditional science of proceedings.

The great media scholar of the twentieth century, the Canadian Marshal McLuhan, summarized in the well-known phrase that "the medium is the message"[22] – that is, the idea that the means of communication and transmission of the message is not neutral, since it conditions even the content – the importance of the means of communication and information for rationality itself.[23] The means are seen as an extension of human beings.[24] This idea is very useful for the current discussion of electronic proceedings on the dematerialized and virtual court records.

---

[21] Cfr. Moulier-Boutang (2004), 55.

[22] Cfr. Marshall McLuhan, *Os meios de comunicação como extensão do homem* – trad. português Décio Pignatari (São Paulo: Cultrix, 1979), 54.

[23] Cfr. McLuhan, op. cit. (1969), 69.

[24] Cfr. Marshall McLuhan, *Os meios de comunicação como extensão do homem* – trad. português Décio Pignatari (São Paulo: Cultrix, 1979), 21.

There is a profound interaction between the actual and the virtual, every actual is surrounded by a cloud of virtuals, as Gilles Deleuze had noted.[25] The great Internet philosopher, Pierre Lévy, a disciple of Deleuze, rightly notes that it is important to understand that virtualization:

> is not a de-realization (the transformation of reality into a set of possibilities), but a mutation of identity, a dislocation of the ontological center of gravity of the object considered: instead of defining it mainly for its actuality (a solution), the entity finds its essential consistency in a problematic field.

And he continues:

> To virtualize an entity consists in discovering a general issue to which it is related, to mutate the entity towards this interrogation and redefine the starting actuality as a response to a particular issue.[26]

Cândido Dinamarco, on the other hand, stated that proceedings are 'medium',[27] instrument of validation not only of material rights, but also social and political values, that is, he underlines the importance of proceedings also to assure meta-legal scopes. For Dinamarco, the instrumentality of the proceedings is double, negative (the instrumentality of forms) and positive (instrumentality for the implementation of rights).[28]

Drawing together McLuhan and Dinamarco: if, on the one hand, these legal circles cannot become an end in themselves to the delight of proceedings scholars, on the other hand, the medium is not exempt, nor neutral, since it influences and contaminates the very course of proceedings, the way the parties participate and even the content of the judge's decision, as they are conditioned in this way by the hypertextual and reticular dynamics of the new decidability procedure, that is, the Levynian redefinition of "starting actuality as a response to a particular issue".

Therefore, the electronic medium, besides greatly conditioning the jurisprudence content, will maximize the very instrumentality[29] of the proceedings, that will have

---

[25] Cfr. Gilles Deleuze, "*O Atual e o Virtual*", in *Deleuze Filosofia Virtual,* ed. E. Alliez, 49 (Rio de Janeiro: Editora 34, 1996).

[26] Pierre Lévi, *O que é o virtual?,* Trad. Paulo Neves (São Paulo: Editora. 34, 1996), 15–20.

[27] "Every instrument, as such, is the medium; and every medium is only such and is legitimated according to the purposes it was intended for (p. 206) (…) In other words, the instrumentalist view of proceedings is, by definition, teleological and the teleological method invariably leads to a view of proceedings as the instrument predisposed to carry out the chosen objectives". Cfr. Cândido Rangel Dinamarco, *A instrumentalidade do processo* (São Paulo: Editora Revista dos Tribumais, 1990), 207.

[28] "This has in common with the instrumentality of forms its negative direction, that is, the function of warning of the functional limitations (of the forms there, here of the same procedural system). The negative side of the instrumentality of proceedings is a current methodological achievement, an awareness that it is not a purpose of itself (…) The positive direction of instrumental reasoning leads to the idea of effectiveness of proceedings, understood in the social and political legal context." Cfr. Cândido Rangel Dinamarco, *A instrumentalidade do processo* (São Paulo: Editora Revista dos Tribumais, 1990), 379.

[29] Cfr. Sebastião Tavares Pereire, *O processo eletrônico e o princípio da dupla instrumentalidade* [Electronic Proceedings and the principle of Double instrumentality]. Jus Navigandi, Teresina, year 12, n. 1937, October 20, 2008. Available at: http://jus2.uol.com.br/doutrina/texto.asp?id=11824 [01.04.2011].

far fewer material links and limitations, allowing the increase of its *deformalization* and widening the possibilities of proof. Finally, the electronic medium reinforces that the proceedings is a medium and an instrument, therefore enabling the privileging of the social and political scope of the demand in the proceedings.

6.3 The Judge and Connectivity

Redefining the actuality of the process of judicial decision-making will be, primarily, conditioned by the alteration of the individualized way these proceedings occur in making up the conviction of the judge, since the judicial decision, even when proffered in a collegiate way in the courts, flows from a viscerally monocratic conviction process.

Setting aside the complex mathematical formalizations of the new network theory, Catalan sociologist Manuel Castells transformed a simple insight into a monumental trilogy about the information era. His great perception was that '*the power of flows is more important than the flow of power*'.

This statement is far more revolutionary than appears, because it means, in practice, to highlight all the power of interactive connections of collectively articulated intelligence.

However, how does this affect contemporary judges? The answer is simple: it allows a view of the magistrate not as Power, but as power and counter-power, that is, before representing Power, constituted and static Power, contemporary judges can catalyze the flows of the dynamics of the bonds of collectivity.

On the one hand, a sentence is an act of intelligence. On the other, the feeling is imbedded in the depths of its own Latin etymology (*sententìa,ae*, 'feeling'). But feeling and intellect are the two spheres most affected by the power of flows and inflows of the new technologies of information and communication.

The solitary genius is no longer valid today. No one can compete with the speed and creative wealth of the flows of knowledge which irradiate over the Internet. No one, in isolation, holds even the knowledge available of a single area of learning.

The solipsist judge who disconnects the proceedings from the world, who does not interact with the parties and socio-cultural context, finds it hard to operate with social appropriateness.

The sentence is no longer an isolated feeling, resulting from a private legal rationality, an individual justice. The contemporary feeling of justice is eminently collective, collaborative and showing solidarity. This feeling, crystallized in the very etymology of the sentence, rather than individual, is *pro indiviso*, common and shared in its entirety.

The contemporary sentence tends to be 'common place', not in the sense of stereotyped expressions, but in the Aristotelian sense – *tópos koinós* – that is, a discourse contrasting to 'special places', specialized discourses, private knowledge.

This 'common place' is not the text signed on paper by the individuality of the judge, but the hypertext, the communal mesh which does not cease to communicate, construct and reconstruct through the virtual proceedings.

Therefore, we must be alert so that the stupendous advance brought by the law of electronic proceedings, which asserts open source code, free software, the Internet and virtual proceedings as a rule, is not channeled in the opposite direction, that is, to the *bonding verticalization* of proceedings.

Experience shows that closed and opaque computerized proceedings of judiciary systems in the hands of a few specialists condition and imprison the freedom and independence of the judge in guiding the judgment which, in effect, results in a brutal loss of procedural and instrumental phenomenality of legal access and, consequently, the loss of effectiveness of the material rights of citizens.

6.4 Specific Principles of Electronic Proceedings

This moment of implementing virtual justice in Brazil and in the world may be a privileged occasion, in which doctrine and jurisprudence can channel the flows of emancipation provided by the new technologies of information and communication, or it could mean a conservative option, the choice for the simple '*computerization of inefficiency*'[30] of today's proceedings.

We have, in fact, two paths before us. The first would be to think of the *scanned proceedings*, and not refer to the digitalization of the paper proceedings, but, in fact, the mental process of copying the vices of writing, the inefficiency of paper proceedings and export them to the electronic proceedings. This would mean to think the electronic proceedings with a paper head.

The second path, which seems more promising, is to effectively exploit the potential of the new information and communication technologies, of connection, the law of abundance of *bits* and *qbits* and the law of growing income, of the so-called positive externalities of the internet, and channel these outlooks to new proceedings, a new proceedings rationality that can make people's rights more effective and court decisions fairer and more adequate.

It would be naïve to imagine that a sheet of paper has the same political and social potential as the computer monitor. The press brought down a 1,000 year hegemony of the culture of writing, the iron fist of the Church, giving way Gutemberg's universe. The new technologies, in the same way, are re-ordaining the shape of power.

Therefore, to mechanically import the classical principles of the paper proceedings to the electronic proceedings seems absolutely inadequate and, for this reason, we disagree with those who interpret Law 11.419/2006 as mere procedure.[31] It is

---

[30] This term was coined by Judge Antônio Gomes de Vasconcelos, during the debates in the Thematic Workshops of the *I Congresso Mineiro – Justiça Digital e Direito do Trabalho (1st Minas Gerais Conference on Digital Justice and Labor Law)*, held by the Regional labor Tribunal of Minas Gerais and its Judicial School in the town of Caxambu, State of Minas Gerais, Brazil in August, 2008.

[31] "[…] the extrinsic medium in which proceedings are initiated, carried out and concluded: it is the extrinsic manifestation of this, its perceptible phenomenological reality." Cintra Antônio Carlos de Araújo, et al., *Teoria Geral do Processo* – 17.ed (São Paulo: Malheiros, 2000), 277.

essential to develop a specific legal technology in order to optimize the potential that these new communication and information technologies can provide for the resolution of legal conflicts.

The scope of this challenge is not the work of a single person, nor is it within the scope of this paper. What is intended here is to suggest the first steps of this path.

Seven new principles are suggested here, which are clearly connected to the traditional principles of proceedings, but which, given the new nuances brought by the new medium, take a quantum leap or undergo a topologic twist which differentiates them from the traditional view. In the current phase of our research, we may present the following list[32]: the principle of *immateriality; of connection; of intermediality; of interaction; hyper-reality, instantaneity;* and, finally, the principle of *de-territorialization.*

Let us examine each of them more carefully.

6.4.1 The Principle of Immateriality

The first characteristic of the electronic proceedings is the very dematerialization of the court records.

Therefore, in Portuguese, *autos* (court records) and *atos* (acts), which have a common etymology,[33] come even closer to the idea of pure movement, impulse, activity. Both have a more intense duality in that they do not crystallize, they are certified in an immaterial, digital form. However, the court records continue to be the pure immaterial certification of the acts of proceedings. Therefore, one can no longer see certification as the mere materialization of the procedural acts.

To a certain extent, the verb to *document* and the noun *document* recover the etymological meaning in the electronic medium, which is not linked to the idea of the materialization of the procedural acts. Document derives from the Latin *documentum,*[34] which means teaching, lesson. The meaning is, therefore, more abstract than material. Teaching is an activity, and not a material object (*res*).

[32] Chief Judge Fernando Neto Botelho, one of the greatest Brazilian authorities in legal computing, also follows the line of developing new and specific principles in electronic proceedings. Cfr. Botelho (2009), available at http://www.amatra18.org.br/site/Index.do [01.04.2011].

[33] Latin etymology: *actus,us* 'movement, impulse, right of passage, action, acting of a play', noun derived from adj. *actus,a,um*, past participle of *agère* 'place in movement'. Cfr. Antônio Houaiss, *Dicionário Eletrônico Houaiss da Língua Portuguesa Mutiusuário 1.0.20* Editora Objetiva, Junho de 2003.

[34] Latin etymology: *documentum,i* 'teaching, lesson, announcement, warning, model, example, clue, sign, indication, proof, sample, evidence that brings faith, document', from Latin verb *docére* 'teach'; see *doc(t)-* Cfr. Houaiss, Antônio, *Dicionário Eletrônico Houaiss da Língua Portuguesa Mutiusuário 1.0.20* Editora Objetiva, Junho de 2003.

Following the same line, the ideas of 'proceedings', 'procedure' and 'court records' are also drawn together, since they are not distinct in the electronic proceedings, in the pure materialization of procedural acts, that is, these three concepts come close to the idea of flow, impulse and movement.

To dematerialize here does not obviously mean the passage to a mystical, spiritual or similar world, but simply the passage from the world of atoms, of matter, to the world of *bits*, that is, to the logical or formal world, to the world of language – the language of machines.

Of course this passage is not neutral, because the analogical world of matter is not the same as the logical, formal world of language.

The world of *bits* is the world of language, of binary language. And language has a very unique feature which is the communication-information duality, that is, language is the content of information and, at the same time, it is communication, transmission, connection.[35]

Therefore, the principle of immateriality of electronic proceedings reinforces the idea that the new proceedings is, above all, a linguistic process which links the subjects of the proceedings – the judge, the plaintiff and the defendant – essentially through the language of men and of machines. In other words, the stabilization of judicial demand is done through language of pure logical form, and no longer by the material form (on paper). Here writing and language take on very different meanings.

Also, the dual character of the immateriality of the new proceedings emphasizes, on the other hand, that the proceedings include both the load of content (information) of the material rights at issue, as well as the very discussion, the debate, the communication, the transmission and the traffic of acts and data.[36]

This new linguistic and immaterial concept of proceedings better balances its democratic-formal aspect – proceedings as the formal presupposition of democracy – as well as its material aspect, the social effectiveness of the rights assured by the democratic constitutional order.

From this point of view, immateriality emphasizes the instrumentalist current of proceedings by dematerializing formalisms in favor of a social adaptation of material rights.

The principle of immateriality is not opposed to actualized reality. The virtual is not opposed to the actual, as was shown. The virtual privileges more the *might* than the *act*, and, in this way, invites to a stance which is more transforming of reality – (actual). Electronic proceedings are not proceedings that crystallize an actuality, the *status quo ante*, and in this way tend to seek the unending actualization, the power of update.

---

[35] Décio Pignatari, *Informação. Linguagem. Comunicação* (São Paulo: Ateliê Editora, 2003), 13.

[36] This duality is well highlighted in the very dogmatics of electronic proceedings, as can be seen, *in verbis,* clause I, § 2o, Article I of Law 11.419/2006: "*For the subject of this law, it is considered: I – any kind of electronic medium for* ***storage*** *or* ***transfer*** *of digital documents and files;*". (my highlight).

Electronic proceedings are, therefore, able to perform as being the expression of *might*, than the *Power* of State, reiterating, therefore, the concept of law as the limit to power. Here the distinction by Spinoza in presuppositions 34 and 35 of his Ethics, between might *(potestas)* and power *(potentia).*[37] For Negri, the separation between *potestas* and *potentia* was the center of the fundamental logical struggle of Spinoza's *Ethics. Potestas* was understood as the capacity to build things and *potentia* as the force that actualizes, that is, the force that makes real.[38]

The proceedings in the material sphere tend to repress and contain forms and conducts. The principle of immateriality, on the other hand, tends to be proactive.

It is well known that principles are not as rigid as norms. They are more flexible, indicative, precepts of optimization and indicate tendencies and new paths; they do not impose necessary behavior or provide only a single correct answer. From this point of view, the principle of immateriality is a permanent invitation to doctrine and to jurisprudence and, mainly, to the daily practice of proceedings, to find the most pragmatic and fair medium, to seek a fairer solution to the demand. They are not casuistic proceedings, but of concrete case proceedings.

On the other hand, if the principle of immateriality indicates flexible proceedings, on the other hand, the workflows of electronic proceedings systems will tend to contain and balance possible excesses and discretions of the judiciary. Conditioned to workflow and challenged by immateriality, the actors of the proceedings will mold over time a more constructivist and more democratic concept of proceedings.

6.4.2 Principle of Connection

Electronic proceedings are, above all, a network process open to connection from the (i) technological point of view and from the (ii) social point of view, that is, a process of connection between systems, machines and people.

The idea of a networking connection makes all the difference. The connected proceedings are very different from the *unplugged* proceedings from various points of view. We may systematically organize them from two main points of view: the first is *reticular* connection and the second, *inquisitive* connection. However, it must be stated, and this could not be otherwise, that both views are connected to each other, both of which benefiting from what Pierre Lévy called 'collective intelligence'.[39]

[37] "*Propositio XXXIV: Dei potentia est ipsa ipsius essentia*". "*Propositio XXXV: Quicquid concipimus in Dei potestate esse, id necessario est*". Cfr. Spinoza (1913(a)) 66.

[38] Cfr. Negri Antonio, *Spinoza Subversivo – Variaciones (in)actuales* – trad. esp. Raúl Sánchez Cedillo (Madrid: Ediciones Akal, 2000), 43.

[39] "*It is an intelligence distributed everywhere, incessantly valid, coordinated in real time, which results in an effective mobilization of competencies*'. *We ad to our definition this indispensible supplement: the base and objective of collective intelligence are the mutual recognition and enriching of people, and not the worship of fetished or hypothesized communities*" Cfr. Lévy Pierre, *A inteligência coletiva – por uma antropologia do ciberespaço* (São Paulo: Edições Loyola, 2003), 28–9.

6.4.2.1 Reticular Connection

Reticular, of course, is an adjective that describes everything with the shape of a network. This adjective describes and emphasizes not only a connection, a linear connection, but a qualified connection in a network.

A linear connection is only the approximation between two adjacencies. A reticular connection presupposes a change of scale, of level, of logic. A linear connection brings a predictable and stable flow; a network connection has a complex unstable flow. There is no rigid linearity in the sequence of online electronic procedural flow. There are not even *numbered pages* in virtual court records, but *flowing events*.

Electronic proceedings are not different only because of dematerialization, but, above all, by the possibility of this dematerialization enabling incessant transmission, in real time, of the content of the proceedings acts and practices. With electronic proceedings it does not make sense to ask for time out for individual examination of the proceedings, since the proceedings are connected to the parties and to society 24 h a day 365 days a year.

Publicity in paper proceedings was a mere possibility, the physical and material distance made publicity into mere presumption; with virtual proceedings, however, it is much more than presumption, it is a reality, that is, publicity is virtual, but not in the sense of possibility, but of a virtual and effective reality,[40] since, as has been shown, the virtual is not opposed to the real.

The so-called principle of writing – *quod non est in actis non est in mundo* – ended the oral phase dating from Roman proceedings[41] and even the Germanic medieval proceedings,[42] entering into the Canonic Code. The principle of writing sought to give legal security and stability to procedural acts, but, at the same time, separated the court records from the world.

This records-world disconnect tended to shape all the argument strategy and the performance of the parties and the judge in the proceedings. Not even the later recovery of orality, five centuries later, was able to change the profoundly structuring nature of the principle of writing, since the orality of the paper medium did not break the idea that what was not in the records was outside the proceedings.

With the arrival of new communication and information technologies and the extended possibilities of connectivity provided by them, there is finally a breakdown of the rigid separation between the world of proceedings and of social

---

[40] This is the only reason why Law 11.419/2006 (art. 11, § 6°), considered allowing access to *outside network* of the private documents only to the parties, those with power of attorney and the public prosecutor's office.

[41] During the period of procedure of the *actions of law,* Roman proceedings were totally oral. Only when proceedings started to follow a formula did they become partially written. Cfr. José Rogério Cruz E Tucci and Luiz Carlos Azevedo, *Lições de história do processo civil romano* (São Paulo: Editora Revista dos Tribunais, 2001), 78.

[42] Germanic barbarian proceedings, in the high Middle Ages, were essential oral, although in the Iberian Peninsula there still retained aspects of the former mixed Roman proceedings. Cfr. Jefferson Carús Guedes, *O princípio da oralidade* (São Paulo: Editora Revista dos Tribunais, 2003), 21–3.

relations, since the electronic medium transcends the material limitations of the paper medium. Hypertext – the link – the so-called language of technological jargon – allows the drawing together between records and truth (actual and virtual) contained on the network, without bringing about a chaotic degree of legal destabilization in the mediatic structure of the proceedings.

Also, the principle of reticular connection makes judicial proceedings into a less segmented and sequential phenomenon. It makes the acts less deductive and syllogistic. It swaps the departmentalization of acts by the instantaneous, by logical time, actual time. Deadlines are no longer a separate concept, but take on a more dynamic, concrete and real meaning which reaches all hours of the day, but also reduces and shapes the concrete pragmatics of the acts.

From *logical preclusion* we strike towards *preclusive induction*, that is, *induction* is emphasized to the detriment of *deduction* in procedural logic. Reticular preclusion is not conditioned to a rigid process of formal contradiction between acts. The incompatibility of acts is not only logically deduced, since it can also be induced in a much stronger way in the concrete and specific case. Clearing up formal nullities no longer depends only on the inertia of one of the parties at the first opportunity to manifest themselves in the records. The principle of network connection imposes on the parties the need for permanent, real time surveillance.

Connection increases the responsibility of the parties in the proceedings, as the counterpart to the very expansion of their participation. Democracy increases rights, duties and responsibilities. As has already been stated with regard to the sentence, the principle of reticular connection leads proceedings to the 'common place' – *tópos koinós* – to the virtual Agora, where specialized discourses and procedural technicalities tend to give way – to some extent, we may think in terms of the *technology of deformalization* of proceedings.

6.4.2.2 Principle of Inquisitive Connection

On the other hand, the principle of connection, naturally makes proceedings more inquisitive. For proof, the classical principles of writing – *quod non est in actis non est in mundo* – has always been decisive. This separation between what is in the records and what is in the world is also a mechanism of rationalization and organization of the production of proof. In the paper proceedings, this principle is also intuitive, since there is no way to demand that the judge should know anything outside the reality materialized and established in the records.

In virtual proceedings, this separation is literally dematerialized. The frontiers between court records and the world are no longer so clear, because both belong to the virtual world. The virtuality of the connection – the hypertext – profoundly alters the limits of the search for proof, because, as is known, the links allow indefinite navigation through the virtual worlds of information, one link always leads to another and so on… The so-called s*emantic Web*[43] carries this radiation of information to unimaginable levels.

The theory of proof used the open concept of 'public and notorious fact' to deal with public facts in proceedings. In the world of the Internet, the scale of facts of public knowledge increase to gigantic proportions, since what is decisive is not knowledge of the fact, but the possibility of access to it, the *connection*. It is certain that doctrine, jurisprudence and legislation will over time establish the limits for virtual navigation, in order to avoid bringing chaos into the flow of proceedings. But this regulation shows that in fact reticular proceedings place the actors of the proceedings in another world, in another kind of proofs logic.

What has to be kept in mind, however, is that this possibility opens interesting possibilities in the search of the so strongly desired actual truth – *rectius*: virtual truth – and also greatly transforms the scope of the proceedings calculation of the parties with the burden of proof. This possibility also makes the proceedings into a more ethical instrument, since the increase of the possibility of the search for the actual-virtual truth will be proportional to the reduction of the allegation and negation of virtually verifiable facts.

For electronic proceedings, better than speaking about a 'public and notorious' fact, will be, therefore, to operate with the idea of a 'common and connectable' fact. Here 'common'[44] is understood also as a noun, extra-state, non-government fact, with open access by the worldwide web. This will be the possibility of connection by the judge – inquisitive connection – the decisive criterion of insertion of information in the proofs stage of networking proceedings.

[43] Also known as the World Wide Web Consortium (W3C). The Semantic Web is a data web. There is a great amount of data used everyday that is not part of the web. The view of the Semantic Web is to widen the principles of the Web from documents to data. It allows humans and machines to work in true interaction. After all, the idea is to transform the web from a sea of documents into a sea of data. There is an excellent FAQ at http://www.w3.org/2001/sw/SW-FAQ#What1 [01.04.2011].

[44] The concept of 'common' has been used today in a post-structuralist political trend. The concept is formulated mainly by Negri and Hardt and Paolo Virno. The idea of 'common' as a noun is connected to the Aristotelian concept of 'common place'. "Today, when we speak of 'common place', we usually understand stereotyped speech, almost deprived of meaning, banalities, dead metaphors – "your eyes are two beacons" – artificial conversation. However, this was not the original meaning of the expression 'common place'. For Aristotle, the *topoi koinoi* are the logical and linguistic forms of general valor, as if it were the bone structure of each of our discourses, that which allows and ordains any specific utterance. These places are common because no one – neither the refined speaker or the drunkard who utters meaningless words, nor the trader or the politician – can set them aside.". Cfr. Virno Paolo, *Gramática de la multitud* (Madrid: Traficante de Sueños, 2003), 34–5.

6.4.3 Principle of Intermediality

"Intermediality" is a concept under construction formulated by theorists of information, communication and literature, which means the process of reciprocal conjugation, interaction and contamination between various media.[45]

This idea is relevant to mark the passage from a strictly fixed proceedings, materially recorded on paper, to a dematerialized, fluid proceedings, recorded only linguistically as binary language.

At first sight, electronic proceedings would mean only the passage from one communication medium – paper – to another, *electronic medium.* However, from the immateriality of electronic proceedings the result is that the electronic medium does not stabilize in a univocal medium, a single form of communication and information, since the scientific miracle of computing makes it possible that virtual records transcend written language, aggregating sound, image and even sound-images in movement.

Reading a novel is very different from seeing the film from the novel, which, in turn, is also different from its corresponding stage play, which is different from a soap opera. Although the theme is the same, the medium changes and even conditions the way the transmitted message is perceived and understood. Therefore, the medium transforms the very content of the message. It is not neutral. As McLuhan said, *the medium is the message.*

The possibility of interaction between these different media within virtual proceedings make it, without doubt, much more complex than the traditional proceedings that are recorded almost entirely in written form. In the Brazilian tradition, it is possible to incorporate into the paper records the recordings of electronic, sound and image, but this is precarious, departmentalized, segmented, because, in order to be integrated to the proceedings, these media always challenge a transposition to writing. Image without movement, photography can interact in the paper records without transposition to writing, but this only operates in the paper proceedings in an extraordinary way, not as a rule. It is an exception and it is very limited.

This greater freedom of writing implies, on the other hand, the potentialization of proceedings as a medium, an instrument which assures material rights, since,

[45] The term "intermediality" is a concept under construction and may appear as a synonym of terms such as 'intermedia', 'intermedias', drawing close, within the scope of literary studies, of ideas such as 'intertextuality', 'inter-semiotic transposition', 'interart studies'. "*In this discursive sphere, intermedial refers to the text which intentionally feeds itself from the conjugation of principles which guide different esthetic propositions and media definitions in the plane of a work, producing a multiple context within the specific textual unit.*" Cfr. Gustavo Silva Saldanha, A leitura informacional na teia da intermedialidade: um estudo sobre a informação no texto pós-moderno. *Perspect. ciênc. inf.* [online]. 2008, vol.13, n.1, 55–66. ISSN 1413-9936. doi: 10.1590/S1413-99362008000100005., 2008. In Brazil, in the Literary and Language Graduate Studies Program of the Federal University of Minas Gerais, "intermediality" is the term chosen by the intermedia research group coordinated by Professor Thais Flores Nogueira Diniz. Cfr. http://www.letras.ufmg.br/poslit/13_projetos_pgs/projetos002.html [01.04.2011].

besides increasing the chances of finding real truth, its intermediality, that is, the greater interaction between the various media, ends up deformalizing the proceedings, making them more pragmatic and less subject to the rigid rules of a single medium. This deformalization enables a stronger channeling of the media towards the social benefits of the proceedings.

Finally, intermediality shows the interdisciplinary aspect of electronic proceedings. It reaches over disciplines, since it applies to civil, penal and labor proceedings. It is not, therefore, a simple procedure; on the contrary, it is much more a transverse process.

6.4.4 The Principle of Hyper-Reality

Another important aspect of electronic proceedings, both from the point of view of the search for actual truth, and the aspect of procedural agility, refers to the radicalization of orality in the proceedings. The principle of orality was recovered at the beginning of the nineteenth century, with the French *Code de Procédure Civile* of 1806, followed by the proceedings code of Klein, in Austria (1985), besides the enthusiastic defense of orality in the proceedings carried out by Chiovenda at the beginning of last century and, finally, with its most recent reaffirmation by Cappelletti in the 1960s.

However, traditional orality was always greatly mitigated since, in the end, it always challenged some degree of writing. In electronic proceedings, orality can be completely preserved – even radicalized – because the hearings can be certified in the court records in their pure sound verbalization, through electronic voice files.

More than simple orality, we may think also of full hyper-realization[46] of procedural acts, a hyper-reality which recreates and simulates not only sound data, but also image data.

It is worth remembering that orality has always been valued, not only because of its capacity to seek the actual truth – as opposed to the old apothegm that '*paper accepts everything*' – but also in view of the potential agility that the oral concentration of the acts provides. If the principle of oral concentration in the paper proceedings already provides agility, imagine its potential from the instantaneous intermediality of electronic proceedings.

[46] The concept of hyper-reality was formulated by French sociologist Jean Baudrillard, from Borges' fable about the cartographers of the Empire Who draw such a detailed map that it covers exactly the whole territory that was mapped. "*Today abstraction is not a map, of the duplicate, of the mirror or of the concept. The simulation is the simulation of territory, of a referential, of a substance. It is the generation of models of an actual without the original or reality: hyper-real (…) The real is produced from miniaturized cells, matrices and memories, from models, from command – and can be reproduced an undefined number of times after this. It no longer has to be rational, because it no longer compares to any instance, ideal or negative. It is only operational. In fact, it is not the actual, since it is not involved in any imaginary. It is hyper-real, the product of synthesis irradiating combining models in a hyperspace without atmosphere. It is only operational. In fact, it is not the real, since it is no longer involved in any imaginary.*" Cfr. Baudrillard (1991), 8.

From the Internet we find the very relevant remarks of the French Internet philosopher, Pierre Lévy:

> The arrival at writing accelerated a process of artificialization and exteriorization of memory which, without doubt, started with hominization. Its massive use transformed the face of Mnemosyne. We end up thinking of a memory as a record.[47]

Roman proceedings were essentially oral, but this tradition changed, as can be seen, from the thirteenth century, with the 1,216 Decree by Pope Innocence III,[48] that established the canonic code of the principle of writing – *quod non est in actis non est in mundo*. In fact, the principle of writing in proceedings, which then showed the longing for safety and stability in the proceedings, over time resulted much more in the distancing from reality, the crystallization of the imminent dynamics of the world, rather than anything else. We moved from the "*lettres passent témoins*" to the then dominant "*temoins passent lettres*".[49]

The reality imprisoned in the writing of the proceedings is a static reality, resulting from the medium used and conditioned by it, that is, paper. In the electronic medium, what can be recorded is not actual reality, but digitalized, codified and virtualized reality, that is, 'hyper-realized'.

The 'hyper-real' is not a representation of the actual, but its *presentation*, translated into binary language, bits; it would be better to think in terms of *transpresentation* of the actual, simulating the actual, since the proceedings themselves are a performance, a show. The records are already a representation of this performance, that is, the representation of representation, the *precession of the simulacrum.*[50]

From one point of view, it is necessary, therefore, to align the records to the order of social, concrete reality, because, in this way, hyper-reality brought into play by the new technologies of proceedings can insert us into a chain of oniric and virtual realism. In the same way that the culture of paper imbues in us a frame of mind of security and formalism of writing, with the loss of gross phenomenality, virtual proceedings can also move us away from reality because virtual reality tends toward simulacrum.

The resurrection of the principle of orality in the nineteenth century had the objective of recovering the actual truth in the proceedings, which had been distanced by the regime of writing. Also, the idea was to seek the lost celerity. Orality meant, therefore, the search for actual truth, associated to procedural agility. The limitation of the medium meant that it occurred through rigid schemes of representation.

---

[47] Cfr. Lévy Pierre, *A inteligência coletiva – por uma antropologia do ciberespaço* (São Paulo: Edições Loyola, 2003), 2009: http://caosmose.net/pierrelevy/nossomos.html [01.04.2011].

[48] Cfr. Jefferson Carús Guedes, *O princípio da oralidade* (São Paulo: Editora Revista dos Tribunais, 2003), 23.

[49] Moacyr Amaral Santos, *Prova Judiciária no Cível e Comercial*(São Paulo: Max Limonad, 1970), 41.

[50] Baudrillard formulates the concept of simulacrum, that is, the simulation that no longer takes the actual as its basis; the actual is only a reference, a virtual reality. The *reality show* is a hyper-real model, of a simulacrum that emancipates and disconnects from the commitment to reality. The simulation – simulacrum – tends to precede the actual (the real). Cfr. Baudrillard (2003), 8.

In a different way, in electronic proceedings it is possible to soften – but never exclude – representation. It is possible to *present* the representation of witnesses and even a performance of the reality of the records through images and sounds. In summary, the principle of hyper-reality, different from orality, which was translated in the trinomial: *actual truth-representation-celerity*, seeks the virtual-actual truth through *presentation*, with the substrate of instantaneity, *in real time, on line – rectius: on network*.

Finally, electronically reconstructed hyper-reality exponentializes orality, not only in the hearing, but, above all, in judicial hermeneutics. The judge may decide orally with the parties, in a more direct and interactive way, immediately correcting any material imperfections and mistakes. The link between feeling and speech is greater than in writing. The spoken sentence is more concise and felt, more of a sentence.

6.4.5 The Principle of Interaction

In paper proceedings, one of the most classical principles raised to the level of the constitution in Brazil, is the principle of pleading and counter-pleading. Practice has shown, however, that this principle, in its classical form, has served much more to the lack of rights, to the procrastination of proceedings, than to the assurance of rights of citizens. The millions of paper proceedings running through the courts speak for themselves.

It is necessary that the principle of pleading and counter pleading should be updated so that it does not continue to be used abusively. The already quoted saying that 'paper accepts everything' indicates that the principle of pleading and counter-pleading can be distorted and lose its noble purpose. As with any other right, it is not absolute and should be confined to certain limits.

Electronic medium may offer the opportunity to move to a kind of upgrade of the principle of pleading and counter-pleading, maximizing it, making it more immediate, instantaneous, in real time, that is, making it *interactive*.

Virtual proceedings allow us to overcome the old linear, segmented and static pleading and counter-pleading, where the deadline becomes the main drive, transforming from time for defense into time to find and excuse and hide the truth.

We may imagine a more intense, extensible pleading and counter-pleading, in real time, making everything more credible, authentic and instantaneous. Through linear, sequential departmentalized stages, it becomes a mechanical, manicheistic and artificial pleading and counter-pleading, since the essence of the principle is not even to contradict, but the pure possibility of participation in the proceedings[51] with equal opportunities.[52]

[51] Cfr. Elio Fazzalari, *Instituições de directo processual* – trad. Português Elaine Nassif (Campinas: Bookseller, 2006), 119.

[52] Cfr. Aroldo Plínio Gonçalves, *Técnica Processual e Teoria do Processo* (Rio de Janeiro: Aide Editora, 1992), 127.

With the virtual world in the records, everything is more instantaneous, the possibility of truth is wider, defense is wider, that is, participation is much wider and exponential.

This hyper-textual, hyper-real, intermediatic, immediate, non-mediated and participatory pleading and counter-pleading becomes much more *interaction* than mere *contradiction.* Interaction means a change of scale, a qualitative transformation in relation to linear and segmented contradiction. Interacting is contradicting and participating in real time, with greater synergy and authenticity.

Contradiction satisfies itself with mere equal participation and is reduced to a mere procedure, without any true commitment to reality or the truth; it is pure form. The principle of interaction is, therefore, a plus in comparison to traditional pleading and counter-pleading, since it also incorporates a substantial aspect of commitment to truth and virtual reality.

Finally, the principle of pleading and counter-pleading is more linked to procedural, competitive democracy, while the principle of interaction results from a new political, participatory and collaborative view.

6.4.6 The Principle of Instantaneity

In proceedings time is a crucial issue. The principle of celerity appears in all the handbooks. Constitutional Amendment 45 raised the issue of the reasonable duration of proceedings to constitutional status, but the hard reality in the courts is the direct opposite.

Obviously, electronic media makes everything much faster. Connection draws near,[53] interaction, hyper-reality and intermediality dynamize, immateriality flexibilizes, that is, everything in the electronic proceedings conspires to maximize celerity.[54]

Through virtual proceedings, mediation is drastically reduced. The lawyer, or the interested party, can take the pieces and evidence directly from the records. There is no request for individual examination of the proceedings, because they are available to the parties 24 h a day. There is not necessarily a conclusion for the judge, since the judge has immediate and non-mediated contact with the records in real time with the parties.

Agility is such that there is already jurisprudence dealing with the early opposition of appeals requesting clarification of a court decision, which is published on the Internet, but which takes time to appear on paper.

Another interesting aspect is that electronic proceedings break with the linearity of page numbering. There is no numbered sequence of pages, but a flow – workflow – of the proceedings, which is not necessarily linear, but conducted from procedural events.

[53] Cfr. Marshall McLuhan, *O meio são as massa-gens* – tradução de Ivan Pedro de Martins ; coord. por Jerome Agel (Rio de Janeiro: Record, 1969), 91.

[54] Cfr. Marshall McLuhan, *O meio são as massa-gens* – tradução de Ivan Pedro de Martins ; coord. por Jerome Agel (Rio de Janeiro: Record, 1969), 91.

In summary, virtual proceedings are online and on network proceedings, which leads to instantaneity, even before mere celerity. Instantaneity is more alive and interactive than the worn and inefficient principle of celerity of the paper proceedings.

6.4.7 *The Principle of Deterritorialization*

The dematerialization of the proceedings also dematerializes the idea of forum and judicial circumscription. The BACENJUD system in Brazil is proof of this. Before this, in order to seize funds in a bank account through a court action outside the territorial limits of the judge's district, it was necessary to issue a rogatory letter. Now all he needs is to log-in and to have a password in order to freeze accounts and financial investments anywhere in the country. The new INFOJUD and RENAJUD agreements open the same possibilities.

Non-penal summons is another example, since it is already possible to proceed to the electronic notification of an individual and of a company, even if they are outside Brazil, it being necessary only that they are enrolled and the proceedings are accessible over the Internet.

The concept of deterritorialization was formulated by philosopher Gilles Deleuze and psychoanalyst Felix Guattari and is a bit more complex than it might appear. The philosophy of Deleuze and Guattari is described by the authors in *O que é a filosofia?,*[55] as a geo-philosophy. According to them, the subject and the object are never the originators, founders, but they are derived, resulting and they do not provide a good approximation to reasoning, since reasoning is not a thread between to concepts. Reasoning is a double articulation[56] between 'earth' and 'territory'.[57]

The earth is the *plateaux* – the plane of consistency and immanence that does not presuppose any transcendence. But on this plateau there are phenomena of 'stratification', which are beneficial in one aspect, but unfortunate in many others. The territorial strata block the overland escape routes, imprisoning their intensities and virtualities to established territories. Territory is a trap, it works like a black hole that seeks to detain everything within its reach.[58] Territory is the demarcation of land, of power, a limit, a border on thinking.

Double articulation of thinking occurs through 'deterritorialization' and 'reterritorialization', that may be relative or absolute, because Deleuzean thinking and the thinking of pragmatic prudence, of reason that recognizes its limits, and which can

---

[55] G. Deleuze and F. Guattari, *Mil Platôs – capitalismo e esquizofrenia* – Vol. 1 – trad. Aurélio Guerra y Célia Pinto Costa (Rio de Janeiro: Editora 34, 1995(a)), 111.

[56] The 'double articulation', the 'double clawed lobster', the '*double bind*' Cfr. G. Deleuze and F. Guattari, *Mil Platôs – capitalismo e esquizofrenia* – Vol. 4 – trad. Suely Rolnik (Rio de Janeiro: Editora 34, 1997), 54, are the terms from and typical of Deleuze and Guattari.

[57] Cfr. G. Deleuze and F. Guattari, *Mil Platôs – capitalismo e esquizofrenia* – Vol. 1 – trad. Aurélio Guerra y Célia Pinto Costa (Rio de Janeiro: Editora 34, 1995(a)), 113.

[58] Cfr. G. Deleuze and F. Guattari, *Mil Platôs – capitalismo e esquizofrenia* – Vol. 1 – trad. Aurélio Guerra y Célia Pinto Costa (Rio de Janeiro: Editora 34, 1995(a)), 54.

no longer be naïve. Deterritorialization includes a movement of territory towards earth. Deterritorialization *in loco* exceeds the territory and becomes the movement of the lines of escape that coincide with nomadic movements. Reterritorialization is the other face, that is, the movement toward the territory, demarcation of power.

On the other hand, the doctrine indicates the idea of an internationalization of virtual material law, as happen in outer space or on the sea floor. There are those who say that maritime law would be the ideal dogmatics to serve as the basis for electronic law. It is worth saying that the idea of *navigating* on the Internet reinforces the linguistic image of this theory.[59]

The principle of deterritorialization[60] in the sphere of electronic proceedings[61] means, therefore, much more than the mere transposition of territories and jurisdictional districts and even jurisdictions; it means the fluidity of the effectiveness of rights that can no longer be simply contained by the material limits of physical space. The *longa manus* of the judge dematerialized becomes more extended, connected.

6.5 Conclusion

Therefore, it is expected that the operators of proceedings are able to learn with the mistakes and the inefficiency of traditional proceedings and do not lose the phenomenal opportunity catalyzing the so-called network externalities to benefit the social effectiveness of people's rights.

The electronic proceedings move in another order, different from the written tradition, since they translate the combination of the *immaterial* of the electronic, with the *reticular* and telematic aspect of the new technologies of communication, information and combination – *rectius*: connection.

Electronic proceedings have the potential to be much more than the mere IT infrastructure for traditional proceedings. Also, it is not the mere digital judicial *procedure*, nor the mere court records on digital paper. The new technologies of information and communication radically transform the nature of traditional

---

[59] Cfr. Carlos Alberto Rohrmann, *Curso de Direito Virtual* (Belo Horizonte: Del Rey, 2005), 27–33.

[60] The following decision by the Brazilian Supreme Court is relevant, still based on paper proceedings:

PROCEEDINGS : CC 66981 UF: RJ – STJ – TRANSMISSION OVER THE INTERNET OF PORNOGRAPHIC IMAGES INVOLVING CHILDREN AND ADOLESCENTS. COMPETENCY ESTABLISHED BY THE SITE OF ILICIT PUBLICATION. 1. As understood by this Court, the crime foreseen in article 241 of Law 8.069/90 occurs at the moment of publication of the images, that is, when there is the issuing on the Internet of photographs of pornographic content. It is irrelevant, for the purposes of establishing competence, the site where the person responsible for the server of access to the virtual environment is to be found.

[61] Chief Judge Fernando Neto Botelho, on this issue, calls this the principle of '*judicial ubiquity*. Cfr. Botelho, 2009, available at http://www.amatra18.org.br/site/Index.do [01.04.2011].

proceedings, which are marked basically by the separation of the court records from the world. Electronic proceedings are, above all, a network process which enable it to benefit, at the same time, from collective intelligence, the law of abundance, growing income and the synergy of interaction in real time.

We cannot follow the path of mere digitalization of the records, in the logic of the scanner, but we must begin a new process and not only a new *procedure*. To digitalize means to relieve the electronic procedure from the distorted logic of the paper, written proceedings.

The fear is that we should fall into the mere *computerization of inefficiency.*[62] We cannot miss the opportunity to take advantage of the arrival of electronic procedures to bring about a revolution in proceedings, which so far is no more than an unfulfilled promise. In other words, it is important to take advantage of the dematerialization of the court records to try to dematerialize the entrenched biases of the culture of writing in proceedings.

The old saying that 'paper accepts everything' brought the transformation of legal security as dogma, in this way losing the actual truth and, consequently, the material justice of decisions. In the immaterial world, the monitor will *accept* even more, therefore, it is necessary to check paranoiac raptures for virtual security.

The concern should move from security, thought of as mere stability, to the idea of preservation of intimacy and privacy in the electronic world, that is, it is more important to assure these constitutional guarantees to citizens than an excessive concern with technological security, since the possibility of redundancy is the great key to security and integrity of electronic files.

On the other hand, not only the technologies already available should not be ignored, but we should be alert to those which are in the pipeline so that electronic proceedings do not start their lives obsolete.

We must insist: electronic proceedings cannot be considered mere standard proceedings – which, in the final analysis, is a mere migration (including the vices) of writing to the new virtual proceedings. It is essential that electronic proceedings should be a relational, semantically user friendly database, with a "referential integrity", not a database of segmented documents.

On the other hand, it is necessary to avoid an obscurantist stance, a sentimental appeal to the specificities of human dignity. The essential in electronic proceedings, the potential of emancipation that it carries, is precisely the fact that it is a network proceedings. Not a network of wires and circuits, but a network that connects people, human beings: the judge, the parties and human society. It is not awe of technology, but the political, cultural, economic and sociological potential of the network.

Paper proceedings are the very incarnation of the separation between actors in the proceedings and in the world. What is not in the court records is not in the world.

---

[62] The expression was used by Magistrate and Law Doctor Antônio Gomes de Vasconcelos, in the Workshops of the *1st Minas Gerais Conference – Digital Justice and Labor Law*, held in Caxambu, State of Minas Gerais, Brazil, August 21–23, 2008.

It is an individual process, isolated from the world. Therefore, what is necessary is the development of a *legal technology* to deal with new proceedings that will connect the court records to the world. To use the same procedural principles of the paper procedures would be the same as to operate the computer using a native club. If we believe that a sheet of paper has the same political and social effect as a monitor on the network, a computer interface, we are losing the historical opportunity to make the long overdue revolution in the judicial proceedings.

Chapter 7
Digital Technology and the Character of Civil Procedure

Viktória Harsági *

Information technology (IT) does not have the same impact on all basic principles; problems of conflict may arise to a greater extent with respect to some principles and to a lesser extent regarding other ones. At the same time, in recent years as a result of modern technology, several possibilities have become available that could – in case of the incorporation of the appropriate legal guarantees – serve the better realization of some principles. This may lead us to the relative reconstruction of some traditional principles. Problems present themselves more intensively concerning litigious proceedings and to a lesser degree in case of non-litigious or, at least, some types of non-litigious proceedings. However, both experts of procedural law encouraging change and those holding more conservative views are likely to agree that the new IT may generate positive changes concerning both the procedural input and output (the forwarding of petitions, service of documents, issue of official copies, administration). The issue characterised by stronger differences in opinion is the trial itself accompanied by questions of procedural law relating to it, the problem of the virtualisation of the trial.

In case of both litigious proceedings and the various types of non-litigious proceedings, basic principles prevailing in the given proceedings may play a decisive role with regard to the introduction of electronic proceedings. This essay endeavours to find an answer to the question in which area the basic principles of procedural law, the majority of which originated in the nineteenth century, obstruct the application of IT and whether it may be justified to review or even redefine these

* The essay was written within the framework of OTKA (Hungarian Scientific Research Fund) research (registration number K-67951).

V. Harsági (✉)
Department of Civil Procedure Law, Faculty of Law, Pázmány Péter Catholic University, Budapest, Hungary
e-mail: harsagi@jak.ppke.hu

M. Kengyel and Z. Nemessányi (eds.), *Electronic Technology and Civil Procedure*, Ius Gentium: Comparative Perspectives on Law and Justice 15,
DOI 10.1007/978-94-007-4072-3_7, © Springer Science+Business Media B.V. 2012

principles in the twenty-first century.[1] When looking for an answer, the following two basic possibilities must be examined: either one may adjust the limits of electronification to the standard of basic principles or, going beyond this, one may reflect on whether the *traditional principles still stand the test of modern times*[2] *and fulfil the original function* they were meant to fulfil *in litigation* in the era when they were created and incorporated into the systems of procedural law.

During the analysis, attention should be paid to the fact that *the complex structure of litigious proceedings* impedes, to some extent, the application of new information technologies – if only for technical reasons. If in case of specific proceedings *automated processing* is made possible, the "decision-making" programme must contain examinations equivalent to those forming part of the traditional proceedings.[3] This is further complicated by the fact that for private individuals, a court proceeding – as opposed to electronic commerce and even some areas of electronic public administration – cannot be considered an everyday routine process but rather an exceptional phenomenon.

7.1 Access to "Electronic Justice"

In this respect, the main problems are centred round the question *of making the electronic form of proceedings obligatory*. If the electronic form were made obligatory, this would be irreconcilable with the principle of access to justice. Thus the electronification of the judicial administration would repeatedly lead to "tense relations" between accelerated and the so-called "social" civil proceedings.[4] It may infringe the right of access to justice and the right to be heard of the party lacking a computer or internet access unless a form of solution is found within the framework of legal aid[5] that could effectively eliminate this disadvantage. It should also be taken into consideration that making the electronic form of proceedings obligatory could, in the extreme case, de facto and indirectly result in mandatory representation by lawyers for a party inexperienced in informatics or lacking the required technical background.

---

[1] Cf. P. Gilles, "Civil Justice Systems in East and West 2007 Plus – Fundamental Current Reform Movements and Some Speculations About Civil Conflict Resolution Systems of the Future," in *The Recent Tendencies of Development in Civil Procedure Law – Between East and West*, ed. M. Storme, et al., 31–2 (Vilnius: Justitia, 2007).

[2] N. Fischer, *Justiz-Kommunikation* (Berlin, 2004), 59.

[3] U. Salten and K. Gräve, *Gerichtliches Mahnverfahren und Zwangsvollstreckung* (Köln: O. Schmidt, 2005), 34.

[4] Fischer: op. cit. (see Fn. 2), 65.

[5] J. Walker and G. D. Watson, "New Technologies and the Civil Litigation Process. Common Law – General Report," in *Direito Processual Comparado*, ed. A. Pellegrini Grinover and P. Calmon, 123 (Rio de Janeiro, 2007).

The parties' right to disposition – in the litigious proceedings at least – must extend to the *choice between the electronic and the traditional forms*. In this respect, for example, the new system of regulation of administrative proceedings in Hungary seems reasonable where, at any stage of the proceedings, the client is allowed to transfer from the electronic administration of affairs to the traditional form and the authorities are only allowed to maintain contact electronically with clients who have provided their e-mails to them for this purpose. Similarly, in the field of civil proceedings, it also seems reasonable to leave open for the parties the possibility of transferring or retransferring from electronically instituted proceedings to the traditional form.

In non-litigious proceedings where legal representation by lawyers is mandatory, no problem can be caused by the one-way solution, i.e. by directing the proceedings into the electronic communication channel. On the other hand, non-litigious proceedings where legal representation is optional should be individually examined. There are non-litigious proceedings (e.g. order for payment proceedings) where exceptions could be made to the voluntary principle with reference to the nature of proceedings and the requirement to increase effectiveness. Here unequal opportunities for parties may be eliminated easily if, by the court or within the framework of organizations providing legal aid, offices are set up where the paper-based documents of the parties would be transformed into electronic ones (through scanning or otherwise) and where the petitions of parties acting without a legal representative or lacking the technical background required would be prepared in electronic form and immediately forwarded to the competent court of jurisdiction. Then, duplicates of the documents in the electronic file would be issued in paper form for the party. This method of solution functions effectively in several parts of the world already (e.g. in Brazil). However, the question arises how the party would be informed of the availability of the paper-based version of the electronic document. Traditional service may result in "hybrid proceedings". There is a solution (in Austria) where the party is not delivered the document itself but a notification (e.g. by fax or phone) about the possibility of collecting the document.[6]

In consideration of the above, one must examine mixed proceedings, which are often referred to as *hybrid proceedings* in legal literature. In these situations the court is forced to keep contact with one of the parties in the traditional way while it is already conducting electronic communication with the other (others). "Two-track"[7] proceedings are closely connected with the principle of party control, the requirement of effective proceedings and the equal opportunity. Unequal opportunities may be mostly eliminated by the method outlined above. With regard to the other two principles, one may say the following. It is an interesting question whether the hybrid (asymmetric) solution resulting from the mixing of the old paper-based and the "purely" electronic models of proceedings could be justified despite the fact

[6]M. Glanemann, *Vergleich des deutschen und östereichischen Zustellsystems* (Hamburg, 2006), 117–8.

[7]Gilles: op. cit. (see Fn. 1), 31–2.

that it is probably neither less costly nor faster. Taking into consideration the fact that this model could thwart the very purposes for which electronic proceedings have been created, it would not be unreasonable to eliminate two-way proceedings or restrict their use narrowly. This could be realized by setting the parties' joint disposition as the condition for diverting proceedings into the electronic channel. Then during the proceedings diverted legally into the electronic channel, the authorities in charge would assist the party by providing the legal aid outlined above, thus opening up the possibility of "electronic access to justice". The other solution would be to continue to use the rather ineffective hybrid model for a transitory period as, following a change in generations, a decrease may be expected in the number of potential parties who would not take advantage – for other than financial reasons – of the possibilities provided by modern information technologies.

IT also offers possibilities by which access to justice may definitely be improved and proceedings without a legal representative may become better available. Let us think of *e.g. the introduction of electronic forms and their publication on the internet.* Through them a higher level of automatism may be achieved, especially if their actual *filling in online* is alleviated by *supporting programmes.* Such solutions also render the court's work less complicated from several aspects, e.g. the later preparation of statistics; thus it may, to some extent, reduce the workload on courts in the long run. In addition, the supporting programme could reduce mistakes made during the filling-in process, e.g. through the examination of jurisdiction and competence and the automated calculation of duties, costs and deadlines.[8]

7.2 Increasing Efficiency of the Procedures

During the electronic administration of affairs, the obvious advantage of electronic documents is that they are not dependent on place, they do not have to be moved around and safe-kept physically, they can be easily found and processed automatically.[9] With the help of document management systems, the systematic binding together of the electronic file is rendered possible, as a matter of fact, this will replace folders and file numbers.[10] Electronic court files save space, they are accessible all

[8]Cf. R. H. Selbmann, *Das Mahnverfahren. Außergerichtlicher und gerichtlicher Forderungseinzug und konventionelles und automatisiertes Mahnverfahren* (Freiburg, 2005), 283; See in more detail: B. Sujecki, *Mahnverfahren* (Heidelberg, 2007), 186–91.

[9]A. Roßnagel and D. Wilke, "Die rechtliche Bedeutung gescannter Dokumente," in *Neue Juristische Wochenschrift* (2006): 2145; A. Landoni Sosa, "New Information Technologies in Civil Procedure. Synthesis Report," in *Direito Processual Comparado*, ed. A. Pellegrini Grinover and P. Calmon, 191 (Rio de Janeiro, 2007); M. Taruffo, "Orality and Writing as Factors of Efficiency in Civil Litigation," in *Oralidad y escritura en un proceso civil eficiente*, ed. F. Carpi and M. Ortells Ramos, 202 (Valencia, 2008); R. L. Marcus, "The Impact of Computers on the Legal Profession: Evolution or Revolution?" in *Northwestern University Law Review* 102, no. 4 (2008): 1836–37.

[10]Wolfram Viefhues, "Das Gesetz über die Verwendung elektronischer Kommunikationsformen in der Justiz," in *Neue Juristische Wochenschrift* (2005): 1013.

the time and "from anywhere"; the judge may download them on his laptop in the courtroom, during the spot inspection or at his home desk. Through the electronic index system and the search function, information is accessible more easily and the proceedings are more transparent. Inspection of documents is less complicated and may be possible for several persons simultaneously, regardless of their whereabouts. During the electronic administration of affairs, as a result of the automated processes, electronic files may be better used for statistical purposes.[11]

The rationalizing and accelerating potential resulting from the application of modern IT in court proceedings is dependent on the structure of the given proceeding to some extent. It offers more possibilities for proceedings with a simple structure and a rather routine-like course, where the decision-making process is schematic and more standardizable.[12] It is no accident that in numerous countries order for payment proceedings have been considered to fall within this category. Automated processing is typically characterised by centralisation,[13] as the building of infrastructure requires fewer financial resources this way. Moreover, in the case of automated proceedings, the geographic location of the processing court is insignificant, as communication takes place through the internet anyway. The whole proceeding must possibly – not least for reasons of cost-efficiency – be supplied with few but specialized staff luckily classified both structurally and infrastructurally.

7.3 Communication Between the Court and the Parties

Electronic communication between the court and the parties – also producing legal effect – is the field most affected by the use of IT. As a starting point, it must be noted that this has a two-way impact. In some fields the introduction of IT may restrict the realization of the mentioned basic principles, as automated proceedings presuppose the written form of communication.[14] In other cases technical possibilities may give new impetus to the realization of orality[15] and they may also contribute to the realization of immediacy. Orality may be implemented without physical presence via a video conference as well. Or let us consider the situation when the request concerning the taking of evidence is not complied with traditionally but with the help of a video conference. The weight of both orality and immediacy could be enhanced if the visual recording (video recording) of trials and spot inspections could become widespread and the recording stored digitally as a part of the

[11] A. Stadler, "Der Zivilprozeβ und neue Formen der Informationstechnik," in *Zeitschrift für Zivilprozeß* 4 (2002): 423.

[12] G. E. Kodek, "Der Zivilprozeβ und neue Formen der Informationstechnik," in *Zeitschrift für Zivilprozeß* 4 (2002): 481.

[13] Salten and Gräve: op. cit. (see Fn. 3), 14–6.

[14] Cf. Taruffo: op. cit. (see Fn. 9), 202.

[15] S. Amrani-Mekki, "El impacto de las nuevas technologías sobre la forma del proceso civil," in *Oralidad y escritura en un proceso civil eficiente*. ed. F. Carpi and M. Ortells Ramos, 96 (Valencia: Universitat de València, 2008).

electronic file could be consulted by the judge before the decision-making several months (or maybe years) later.[16] The advantages of such a system are obvious, this way the transformation of oral evidence into written evidence could be avoided,[17] as through the judicial intranet the recording would be available to the appeal court, the court could have access and replay it easily. However, the retention of video recordings and access to them may pose a problem from the aspect of data protection.

Orality plays an important role mainly at the trial, where it also ensures the implementation of the principle of immediacy. In the traditional model of litigation, public control is also mainly guaranteed by the trial. The trial (oral proceeding) and the evidentiary proceeding have always been considered the central element, the core of civil litigation.[18] Consequently, they may also be regarded the "most sensitive" points of proceedings. Any effect on them may concern questions of basic principle directly or indirectly. As a result of technological progress, the culture of court speeches and trials will also undergo changes.[19] With regard to the evidentiary proceedings, electronic communication will widen the pool of the means of evidence. In addition to the electronic documents the legal background of which has become stable by today, other documents with various file formats[20] (video and music files) may be presented to the court and attached to the electronic file easily. Moreover, a Power Point presentation (simulation) may render the reconstruction of facts less complicated. However, besides constituting evidence, the presentation may also be used as a means of reasoning and persuasion by the lawyer during advocacy work.[21]

The question of orality and immediacy arises primarily concerning the taking of evidence, where during the process of establishing the facts a decisive role is played by personal conviction, directly and personally perceived impressions concerning people or objects, which serve as the guarantee for the legitimacy of the decision. The situation is different concerning the use of video conferences. Video conferences only serve as substitutes for personal and direct contact.[22] While in most

[16] Cf. Landoni Sosa: op. cit. (see Fn. 9), 191; J. W. Peyrano, "La prueba entre la oralidad y la escritura," in *Oralidad y escritura en un proceso civil eficiente.* ed. F. Carpi and M. Ortells Ramos, 168–9 (2008).

[17] Taruffo: op. cit. (see Fn. 9), 202.

[18] Cf. H. Nagel, *Die Grundzüge des Beweisrechts im europäischen Zivilprozeß* (Baden-Baden, 1967), 20; D. Coester-Waltjen, *Internationales Beweisrecht* (Ebelsbach am Main 1983), 1; W. H. Rechberger and D.-A. Simotta, *Grundriß des österreichischen Zivilprozeßrechts* (Wien, 2000), 363.

[19] Stadler: op. cit. (see Fn. 11), 443.

[20] S. C. Benneth, "Records Management: The Next Frontier in E-Discovery?," in *Texas Tech Law Review* (2009): 528; cf. R. K. Sherwin, N. Feigenson and C. Spiesel, "Law in the Digital Age: How Visual Communication Technologies are Transforming the Practice, Theory and Teaching of Law," in *Boston University Journal of Science and Technology Law* (2006): 233–5.

[21] Landoni Sosa: op. cit. (see Fn. 9), 189.

[22] Kodek: op. cit. (see Fn. 12), 485–6.

countries the use of video conferences in criminal proceedings has been regulated primarily with regard to the protection of witnesses, in civil procedure reasons of cost efficiency play the dominant role. From the aspect of the implementation of immediacy, video conferences may be described as being somewhere "halfway". The trial and taking of evidence via a video conference ensures the realization of immediacy to a greater extent than the records prepared and submitted following a request by the court, on the other hand, its possibilities are limited compared to direct perception. With a paraphrase, the use of video conferences will increase the "distance" without doubt, but it may also enhance the role of orality at the expense of the written form.[23] The interposing of a technical device restricts direct perception (technical limit), but it does not mean the interposing of another means of evidence or person or the substitution of another means of evidence for the witness. The witness whose testimony is broadcast in the courtroom via picture and sound will remain the same witness; similarly, the judge hearing him is not being substituted by an intermediary.[24]

As the principle of immediacy cannot be a purpose in itself, lawyers specializing in the field of litigation must also find an answer to the question as to what extent the discovery of truth should be sought at the expense of other aims of the litigation and whether the principle of immediacy could be interpreted to mean that the technically mediated (but otherwise direct) impression is to prevail over other solutions.[25]

When examining the relation between the use of video conferences and basic principles, it is worth distinguishing between the personal presence of parties and that of witnesses. With respect to statements by parties or their representatives, technology may be used more widely. Regarding witnesses, however, the use of modern technology would primarily be justified in special cases where the applicable rules of civil procedure allow the taking of evidence by request.

7.4 Transparency

In the era of modern IT, electronic publicity has a rather different meaning from the first definition of the principle of publicity. The use of the internet – as a result of its nature – may lead to wider publicity in any case; it may ensure access to the data of the lawsuit and the litigation materials for a substantially wider group of people than publicity taken in its traditional sense would imply.

The question arises as to whether the requirement of public trials refers only to the courtroom. Is it possible or required to go beyond the traditional interpretation

---

[23] Cf. Stadler: op. cit. (see Fn. 11), 443.

[24] Cf. G. Bachmann, "'Allgemeines Prozessrecht' – Eine kritische Untersuchung am Beispiel von Videovernehmung und Unmittelbarkeitsgrundsatz," in *Zeitschrift für Zivilprozeß* (2005): 139–40.

[25] Cf. ibid 148.

according to which the principle of publicity would mean that apart from the participants of the lawsuit, third parties or outsiders could also be (physically) present at the trial? Could it be the aim of litigation that, by broadcasting the trial via the internet, anybody could keep track of it? Could the principle of publicity really mean this in our era? In all circumstances, the answer is no, as the over-extension of the principle of publicity in this direction could cause conflict with other constitutional rights (e.g. data protection, protection of secrets, personal rights). For this very reason, it is also true about this principle that technology may offer more possibilities than one should necessarily incorporate into the "texture" of the rules of civil procedure. It is not easy to strike the right balance in this field though.

Concerning the harmonization of basic principles and the possibilities offered by IT in the service by publication, the balance is definitely positive. Summons by a notice posted on the court's bulletin board as ultima ratio – considered rather out-of-date in our days and, it must be noted, of little effect in practice – must be re-evaluated in any case. Neither could publication on the internet mean absolute solution, but – considering that it may reach a wider circle of persons in an incommensurably larger geographical area – it could mitigate the problem.

Substantially more theoretical and practical problems are generated by the question of possible public access to files and that of data protection. Providing access to electronic court files may be easily implemented technically and it may have numerous practical advantages. Making the whole material of files accessible for anybody is obviously undesirable (from the aspect of data protection). Electronic files may contain some parts (such as judgments) which may be of "public importance". Within this circle the primary question presenting itself could be the need for anonymizing, which, in the majority of cases, may lead to the over-restriction of publicity. On the other hand – despite the main rule of public trials – in legal literature, for example, one may encounter positions objecting to the electronic publication of trial records.[26] This seems closely connected with the question of whether the webcasting of trials should be allowed; therefore the counter-arguments are also similar.

However, a solution must be found to ensure the security of the system from the aspect of data protection and to prevent unauthorised access to electronic court files. During the inspection of documents, access should be recorded by the programme and a solution must be found to preserve the intactness and integrity of content of e-files. Another important data protection problem is constituted by files sent by electronic delivery to incorrect (e-mail) addresses, which may occur more easily than in case of delivery by mail.

There are several technical possibilities for "inspecting" electronic court files, e.g. a court office may ensure the public access to these documents by printing them; or by forwarding them as electronic documents; or by the installation of a court terminal for the inspection of documents; or perhaps by providing online access to the files so that they could be viewed on the monitor of a distant computer. Online access to court files would have numerous advantages for both the court and

[26] Kodek: op. cit. (see Fn. 12), 486.

those entitled to inspection. It is possible for several persons to have simultaneous access to the online file for the purpose of reading. The inspection of files would be possible 24 h a day; it would not depend on court time table. Consequently, the protraction of proceedings partly caused by the transportation of files could be reduced to some extent. As a result of the lists of contents and the search functions provided by the software and applicable to the whole text, some files would become more transparent and easier to handle. Access to information about ongoing proceedings and their different stages would be improved substantially both for the court and for the parties.[27] In Germany in the case of the electronic processing of order for payment proceedings, apart from the forwarding of petitions to the court electronically, it is also possible for parties to receive electronic messages about the developments in the proceedings (notification is made this way about the actual service of documents, the issue of the decision, the statement of opposition, the submission of petitions and costs.)[28]

[27] S. Hähnchen, "Elektronische Akten bei Gericht – Chancen und Hindernisse," in *Neue Juristische Wochenschrift* (2005): 2258; Viefhues: op. cit. (see Fn. 10), 1014.

[28] Salten and Gräve: op. cit. (see Fn. 3), 34.

Part II
Specific Applications of Electronic Technology Applications to Court Proceedings: Service and Other Applications of Communications Technology Issues

Chapter 8
Electronic Service of Documents National and International Aspects

Fernando Gascón Inchausti

8.1 E-Justice and E-Service: Previous Considerations

The implementation of so-called 'electronic justice' or 'e-Justice' is one of the major challenges facing the Justice system at present. In broad terms, e-Justice is understood to mean the use of 'new technologies' in the field of justice,[1] i.e. in the performance of the activity of the courts. The term 'new technologies' means information and communication technologies (ICT),[2] which are, by definition, technologies of an electronic nature, which depend upon the use of computers, and more specifically, the internet. E-Justice therefore means the use of electronic systems in order to carry out activities that had hitherto been carried out in some other way, or in a way that was much less reliant on the said systems than is envisaged for the future.

It is important to understand that the use of electronic systems affects *how* an activity or institution functions, but not what it *does*. Thus, the term e-Justice does not refer to a different kind of Justice, or to a different Judicial Power, or to any

---

[1] *Communication from the Commission to the Council, the European Parliament and the European Economic and Social Committee*, "Towards a European e-Justice Strategy", 30 May 2008, COM(2008) 329 final (pp. 4–5).

[2] According to the definition given by the European Commission in its Communication: "The e-Justice approach uses ICT to improve citizens' access to justice and to make legal action more effective, the latter being understood as any type of activity involving the resolution of a dispute or the punishment of criminal behaviour" (p. 3). The European Parliament, in its *Resolution of 18 December 2008 with recommendations to the Commission on e-Justice* [2008/2125(INI)], states that "e-Justice has a broad definition including, in general, the use of electronic technologies in the field of justice". The Council, in its *Multi-annual European e-Justice action plan 2009–2013* (OJ C 75, 31 March 2009), talks about "the use of information and communication technologies (ICT) in the field of justice".

F. Gascón Inchausti (✉)
Departamento de Derecho Procesal, Facultad de Derecho, Universidad Complutense de Madrid, Ciudad Universitaria s/n, Madrid 28040, Spain
e-mail: fgascon@der.ucm.es

M. Kengyel and Z. Nemessányi (eds.), *Electronic Technology and Civil Procedure*, Ius Gentium: Comparative Perspectives on Law and Justice 15,
DOI 10.1007/978-94-007-4072-3_8, © Springer Science+Business Media B.V. 2012

difference in jurisdictional functioning: e-Justice is ordinary Justice, but making use of the tools that ICT provides in the organization and performance of the tasks of judicial bodies.

It is important to emphasize that e-Justice is not a field of Justice that is based on electronics and on the operation of machines running software programs, presented as an *alternative* to a field of Justice carried out by human beings: anything of this sort seems inconceivable to me. To be precise, it is the 'tools of e-Justice' or the 'instruments of e-Justice' that have their own separate existence, and these are or may be used by judicial bodies in the performance of their activities. But the said new technologies continue to be applied by human beings, and this fact should never be overlooked.

Therefore, the fact that judicial bodies use new tools does not mean that their nature or functions have changed, and it is clear that the use of new tools should not serve as a pretext for the introduction of any changes in their nature or functions.[3] The only difference is in the external configuration of the procedural acts, which are carried out electronically, but service continues to be service, even if it is carried out by e-mail, just as the evidence of a witness does not lose its essence because it is given by video link. The changes that e-Justice entails should therefore only be external and should only affect the form of the procedural acts when carried out using ICT tools. But the use of electronic systems should under no circumstances result in any of the safeguards applicable to judicial activities being prejudiced.

One of the sectors where the use of ICT is envisaged as part of proceedings is that of service of notice in its widest sense, which includes the filing of pleadings, the service of documents, and the notification of judicial decisions. It is well known that from a certain point of view, proceedings are just a system of notices exchanged between different persons all playing different roles (the court and the parties in dispute): sometimes notice is served orally, in hearings, but often what the parties and the courts have to say to each other has been drafted in writing and compiled into a document, which then needs to be delivered to its addressees in accordance with legally-defined forms and procedures. The correct service of notice is necessary during the proceedings, from at least two different perspectives:

– On the one hand, the fairness of the proceedings depends on notices being served correctly: where a document served by one of the parties is not correctly received by the court, the court will deem that service has not taken place, to the (illegitimate) prejudice of the litigant, and the same unfair result can arise the other way around, where a litigant does not correctly receive a document that has been served by the

[3] See also M. Taruffo, "Orality and Writing as Factors of Efficiency in Civil Litigation," vol. 1, in *Oralidad y escritura en un proceso civil eficiente – Oral and written proceedings: Efficiency in Civil Procedure,* ed. F. Carpi and M. Ortells, 202 (Valencia: Universitat de València, 2008).

court or the other party. However, experience has shown us how the right to due process of law is infringed on a daily basis as a result of the way in which notices are served.[4]

- On the other hand, notice must be served properly in order to avoid unnecessary delays in proceedings: inadequate internal organization within many courts, together with the difficulties and obstacles that are often put up by the addressee of the notice, mean that the service of notice is one of the elements that give rise to litigation being unduly prolonged.

The use of ICT in this field consists in taking advantage of electronic communications systems as a means to file documents and serve notices: computer programs for sending messages based on the internet or on similar telecommunications networks are used, allowing documents to be sent in electronic format from the electronic address of the sender to the electronic address of the addressee.

The main benefit of the use of electronic notices is to speed up proceedings. In addition, there can also be the advantage of reducing costs (especially postage costs, the cost of court officials, and the cost of having to rely on certain professionals whose business is to serve notices).[5] Likewise, electronic service can be seen as giving rise to and being part of what is known as *paperless procedure*: in order for electronic service to take place, it is necessary for the documents to be available in electronic format (either because they have been created from scratch in such a format, or because they have been subsequently digitalized), and in turn *paperless procedure* can only function if notices are served electronically.

However, it would appear that when the use of ICT is encouraged in this area, not enough thought is given to overcoming the problems caused by the right to a fair trial. Electronic service is not put forward as a remedy to the problems inherent to ordinary service, given that it does not, of itself, escape them. In fact, it often seems that electronic service attracts certain safeguards that are not attached to notice served in the ordinary way. In my opinion, however, the dangers and difficulties are

[4] A quick overview on the case-law of the European Court of Human Rights offers many examples in the judgments delivered in cases like Goddi v. Italy (9 April 1984); Brozicek v. Italy (19 December 1989); Hennings v. Germany (16 December 1992); Pérez de Rada v. Spain (28 October 1998); Tsironis v. Greece (6 December 2001); Somogyi v. Italy (18 May 2004); Hermi v. Italy (28 June 2005); Strizhak v. Ukraine (8 November 2005); Examiliotis (no. 2) v. Greece (4 May 2006); Díaz Ochoa v. Spain (22 June 2006); Gregorio de Andrade v. Portugal (14 November 2006); Popova v. Russia (21 December 2006); Ern Marina Sanayi ve Ticaret A.S. v. Turkey (3 May 2007); Da Luz Domingues Ferreira v. Belgium (24 May 2007); Blandeau v. France (10 July 2008); Almeida Santos v. Portugal (6 October 2009); Société Anonyme Thaleia Karydi Axte v. Greece (5 November 2009); or Popovitsi v. Greece (14 January 2010).

[5] In Austria, which is without doubt the leading country in this area, it is estimated that the use of their system for the electronic filing of pleadings (*ERV*) results in a saving equivalent to 133 jobs. Furthermore, the use of the *ERV-Rückverkehr* – which allows the courts to serve documents on the parties – allowed a saving of 3.6 million euros in postage in 2008, according to information supplied by the Ministry of Justice at http://www.justiz.gv.at/internet/html/default/2c9484852308c2a60123708554d203e7.de.html;jsessionid=20BD802BA25B15B66F6DD7433A013172 [03.02.2011].

the same as arise for ordinary service, apparently amplified by the fact of being electronic: the truth, however, is that this factor only shifts the dangers and difficulties to another plane.

In the following pages it has been considered convenient to start with a review of how electronic service is being regulated in internal legislation and at an international level. This initial overview will allow us to identify the foreseeable problems and dangers, and how these can be averted, and we will also be able to detect the main challenges that need to be faced in the future by regulations that seek to make widespread use of e-filing and e-service of notices and documents in judicial proceedings.

8.2 National Regulations Concerning E-Filing and E-Service of Documents: An Overview

In a significant number of countries rules have already been approved that provide a legal framework for e-filing and/or e-service of judicial documents, although these are not being applied systematically and in full in all of these countries: electronic notification is still at an experimental and introductory stage, but general implementation is gradually approaching in a fair number of the said countries.

What is stated below is not intended to be exhaustive,[6] given that it is based, above all, on such normative and caselaw references as it has been possible to find, with regard to a limited number of States, belonging to a variety of legal traditions: in the Germanic tradition, we will examine the situation in Austria and in Germany; in the Roman-Law tradition, France, Italy, Spain, and Brazil; in the Anglo-American tradition, the United States and England. In all cases, I set out

[6]The workflow of the Conference was envisaged to encourage the participation of all interested academics, who were invited to contribute to the drafting of a national report in response to the questionnaire I prepared at the time. Unfortunately no national report has been received in relation to this subtopic. Although it was not my intention to provide a systematic and exhaustive description of the electronic communication systems existing in each State, this situation has conditioned the structure and content of this general report: I shall restrict myself, initially, to describing the normative development that has taken place in certain States, in order to then go on to identify the main problems which, in view of the rules that have been examined, seem to arise in respect of electronic notification when it comes to the right to a fair trial.

For a more thorough and still recent (2007) overview, see J. Walker and G. D Watson, New technologies and the civil litigation process. Common Law – General Report; E. Jeuland, Nouvelles technologies et procès civil. Rapport général pour les pays de droit civil; A. Landoni Sosa, "New information technologies in civil procedure. Synthesis report", in the context of the XIII World Congress of Procedural Law and published in Direito Processual Comparado (orgs. A. Pellegrini Grinover and P. Calmon) (Rio de Janeiro, 2007). See also S. Amrani-Mekki, "El impacto de las nuevas tecnologías sobre la forma del proceso civil", Vol. 1, in *Oralidad y escritura en un proceso civil eficiente – Oral and written proceedings: Efficiency in Civil Procedure*, ed. F. Carpi and M. Ortells (Valencia: Universitat de València, 2008).

the information that I have been able to obtain despite the limitation of not being able to have access to national reports, and with the assumption that the normative changes that occur so frequently in this area might have rendered some of the information stated out of date.

8.2.1 Austria

Austria is one of the pioneering countries in matters of e-Justice and the inclusion of ICT in the legal world[7]: it made the leap towards this kind of working practice in the 1990s, and nowadays its use would appear to be completely standard.[8]

The system used for this purpose is the so-called *Elektronischer Rechtsverkehr* (ERV), which was set up in 1990,[9] and it is currently governed by the *Verordnung des Bundesministers für Justiz über den Elektronischen Rechtsverkehr*[10] (abbreviated to *Elektronischer Rechtsverkehr*, ERV).[11] The ERV is not an e-mail program, but rather a special system that allows data to be transmitted securely, without any need for paper, between the parties to proceedings and the court. It was initially envisaged for the parties to file documents with the court, and since 1999 it has also allowed the court to serve documents on the parties (ERV-*Rückverkehr*).

In order to serve notices using the ERV it is necessary to be registered as part of a special system with a closed number of users (§ 7 ERV), through which service

---

[7] See the information provided by the Austrian Ministry of Justice at http://www.justiz.gv.at/internet/file/2c9484852308c2a601230eeed6f60127.de.0/folder_justiz-online_mai2008_+deutsch+v+1.0_neu.pdf [03.02.2011].

[8] According to the Austrian Ministry of Justice, in 2008 over 85% of initial applications for orders for payment (*Mahnklagen*) and over 65% of enforcement applications (*Exekutionsanträge*) were filed electronically: in total (together with the rest), 3.1 million applications were filed using this system. Also in 2008, the courts served 3.9 million notices on the parties using electronic means. See http://www.justiz.gv.at/internet/html/default/2c9484852308c2a60123708554d203e7.de.html;jsessionid=20BD802BA25B15B66F6DD7433A013172 [03.02.2011].

[9] The ERV was created in collaboration by the Ministry of Justice, the *Bundesrechenzentrum*, the *Österreichische Rechtsanwaltskammer* and *Telekom Austria*.

[10] Text available at http://www.jusline.at/Elektronischer_Rechtsverkehr_(ERV).html

[11] The Austrian regulation of notification is scattered. General directions and some specific rules are to be found at the *Zivilprozessordnung* (ZPO, §§ 87 a 121). However, § 87 (1) ZPO refers in general terms to the Notification Act (*Zustellgesetz*), applying to service done by public bodies and courts, http://www.bmvit.gv.at/telekommunikation/post/recht/downloads/zustellg_konsolidiert.pdf). Section 3 (§§ 28-37a) of the *Zustellgesetz* deals with electronic notification when made by public administrations, not by courts [§ 28 (2)]. For these, reference is made to the *Gerichtsorganisationsgesetz*, http://www.ris.bka.gv.at/GeltendeFassung.wxe?Abfrage=Bundesnormen&Gesetzesnummer=10000009, §§ 89a–89l. They don't include, however, a detailed regulation of the electronic service procedure, but only a general provision allowing any judicial notification to be made by electronic means and a reference to lower administrative rules [§ 89b (2) GOG], among which the one regulating ERV.

will be effected: indeed, e-mail is expressly excluded as a forum for the exchange of data [§ 5 (1a) ERV]. Although in principle the system was restricted to lawyers, notaries, banks, and insurance companies, from 2000 it has been open to any interested party.

Aside from the content of the ERV Regulations, a further two provisions deserve to be highlighted:

- On the one hand, § 89e GOG, which expressly governs liability for losses caused as a result of failings in the e-filing and e-service systems, and which is attributed as a general rule to the Federation, except in cases of inevitable or unforeseeable events that are not caused by poor design or malfunction of the computerized system.
- On the other hand, § 112 ZPO, which allows e-filing and e-service as a possible mechanism for the lawyers of the parties to transfer documents between themselves.

8.2.2 Germany

In Germany, the regulation of electronic procedure and electronic service has been carried out by basically three Laws, which have reformed the *Zivilprozessordnung* (ZPO)[12]: the *Zustellungsreformgesetz* of 25 June 2001 (which came into force on 1 July 2002); the *Gesetz zur Anpassung der Formvorschriften des Privatrechts und anderer Vorschriften an den modernen Rechtsgeschäftsverkehr*, of 13 July 2001 (which came into force on 1 August 2002), and the *Justizkommunikationsgesetz* of 22 March 2005 (which came into force on 1 April 2005).

The basis for electronic service is established by §§ 130a and 130b ZPO. The former provides for all pleadings that the parties must file being drawn up in electronic format, with the sole condition that the said format must be suitable for the court to work with the said document and that it must feature a qualified electronic signature.[13] The latter (§ 130b ZPO) recognizes the full validity of electronic documents drawn up by the court and by any other civil servants, provided that they feature a qualified electronic signature.

On the basis of these premises, the electronic notification is governed as a possible form for some communication mechanisms. Thus, the basic precept on this matter is § 174 ZPO, which governs personal notification carried out by the court, which must, as a rule, be carried out with acknowledgement of receipt (*Zustellung gegen Empfangsbekenntnis*) and which, with a view to the use of electronic means, must distinguish as follows in accordance with the identity of the addressee:

[12] http://www.gesetze-im-internet.de/zpo/

[13] However, the regional Governments will decide at what courts and from what moment it will be possible to use it.

- Where the notification is addressed to a lawyer, a notary, a *Gerichtsvollzieher*, a prosecutor, or to any other person who, by merit of his/her profession, ought to be considered reliable, and also when addressed to an authority, a corporation, or a public-law body: it may be served electronically, with acknowledgement of receipt.
- Where the notification is addressed to anyone else involved in the proceedings: service may be electronic provided that the addressee has expressly consented to being served with documents in this way.

Generically, the only requirement of the rule, in both cases, is that any document served in this way should feature an electronic signature and that the dispatch should be protected against third parties. In order to prove that notification has taken place, it is necessary to have acknowledgement of receipt with the date and the signature of the addressee, which must be returned to the court: said acknowledgement of receipt may be contained within an electronic document, but it must have a qualified electronic signature.

The electronic communication system in Germany is based on the Electronic Court and Administration Mailbox (*Elektronisches Gerichts- und Verwaltungspostfach*: EGVP)[14]: for virtually every aspect of electronic communication between the judicial system and citizens seeking justice, it is the EGVP that is used at national level and in the *Länder*.

The EGVP makes possible particularly secure transmission of communications to courts, authorities and parties to proceedings. This is done via "OSCI-Transport", which corresponds to the German e-government standard and is based on the XML and SOAP standards recognised worldwide. The transmission of data on content and use is done separately, thus making the delivery of communications unequivocally transparent. Moreover, particularly secure and reliable transmission is ensured with end-to-end encryption including electronic signature of the content. The EGVP combines the advantages of e-mail (simple to operate, free formulation of communications) with the requirements of secure judicial communication (integrity of the documents transmitted, identity of the parties to the proceedings, use of qualified electronic signatures and verifiability of document servicing).

For security reasons, data transmission by EGVP functions through a sort of "buffer"; the user does not communicate directly with the court, but via an "intermediary". This is a central server, on which all mailboxes are managed. No special court infrastructure is therefore required for the EGVP to operate. All the transmission technology is basically made available at a single point, which considerably reduces the technical cost for the courts. Users can obtain the necessary software

[14] See "Member States' national projects relevant for the e-Justice portal – Reply from the German delegation", Council of the European Union – Working Party on Legal Data Processing (e-Justice), Doc. 7628/09 Add 11 JURINFO 26.

free of charge and without a licence via the homepage www.egvp.de. Immediately on installation the user has his own electronic mailbox for receiving electronic documents.[15]

The EGVP provides access to courts "round-the-clock", as well as an immediate signed confirmation of receipt by the receiving facility of the court or authority and on-request automatic e-mail notification of receipt of communications.

Aside from what has been stated above, where personal notification should be impossible (cf. § 185 ZPO), the court may order, as a last resort, that notification be served by way of an edict (*öffentliche Zustellung*). Pursuant to § 186 ZPO, such a notification may be carried out by way of publishing the edicts on the court notice board or by way of including the edicts in an electronic court information system that is open to the public. Furthermore, the edict may also be published in the electronic version of the *Bundesanzeiger* (§ 187 ZPO).

Finally, the service of documents between lawyers is also possible (*Zustellung von Anwalt zu Anwalt*) within the terms laid down at § 195 ZPO: in these cases, notification may be in electronic form and may also be proven electronically, in accordance with the provisions of § 174 ZPO.

8.2.3 France

In France, the regulation of electronic service was introduced into the *Code de Procédure Civile* by way of a Decree in 2005,[16] which created new Title XXI (*La communication par voie électronique*) within Book I (which contains the common provisions applicable to all courts and to all proceedings), articles 748-1 to 748-7.[17]

The scope allowed to the system is potentially very broad: it may be used for filing and service of any kind (article 748-1). Proof of service is provided by way of electronic acknowledgement of receipt, sent by the addressee, which must state the date, and where appropriate, the time (article 748-3 I). Furthermore, when the electronic

---

[15] The EGVP has proved its worth in judicial practice and is used in supreme federal courts and in all *Länder* – particularly in high-frequency procedures (court registers, commercial registers and order-for-payment procedures). Since 1 January 2007, all registration processes relating to the commercial register, which has a good one million businesses on record, have been conducted via EGVP. In 2008, more than 772,000 incoming and 205,000 outgoing communications were transmitted to and from the participating courts and authorities. By May 2009, use had increased sharply; in April 2010, the system was transmitting over 280,000 communications. There were then more than 30,000 registered users (courts, authorities, citizens, above all lawyers and notaries).

[16] Through the *Décret* n° 2005–1678, 28 December 2005 (*Journal Officiel de la République Française*, 29 December 2005), in force on 1 July 2008. For a recent overview on e-Justice in France, see "Dossier – Le procès civil à l'épreuve des nouvelles technologies", *Procédures*, n° 4, avril 2010, pp. 6–45.

[17] http://www.legifrance.gouv.fr/affichCode.do?cidTexte=LEGITEXT000006070716&dateTexte=20100519

system is used, the service of multiple copies or the return of originals ceases to be necessary (article 748-3 III).

The technical requirements that must be met by the system used are also established generically (article 748-6): it must ensure (i) reliability in the identification of the parties, (ii) the integrity of the documents sent, (iii) the security and confidentiality of all communications, (iv) the preservation of all transmissions made, and (v) it must allow the date of sending and the date of receipt by the addressee to be established with certainty.

However, the system is optional: the addressee must expressly consent to its use (article 748-2). But as from 1 January 2011, it has become compulsory in all appeals where representation is necessary, pursuant to the provisions of article 930-1.[18]

Furthermore, a series of additional safeguards have been built in: (i) where the original of a document was created in paper format, the court may order this to be produced (article 748-4); (ii) the interested party may ask to be served with any enforceable judicial decision (*décision juridictionnelle revêtue de la formule exécutoire*) in paper format (article 748-5); (iii) there are minimum requirements for what must be done in the event of certain 'system failures': where on the last day within a time limit for performing a certain act it is not possible to carry out e-service or e-filing for reasons beyond the control of the serving or filing party, then the time limit is extended to the following business day (article 748-7).

In order to set up the system at the level of the *Tribunaux de Grande Instance*, the creation of a fairly complex system has been envisaged, which involves two different computer programs –'WinCi TGI' and 'ComCi TGI'– and two private communications networks, one for the courts –*Réseau privé virtuel justice*, RPVJ– and another for the bar –*Réseau privé virtuel avocats*, RPVA– and the interaction between the two is governed by a Convention between the Ministry of Justice and the *Conseil National des Barreaux*, executed on 28 September 2007,[19] which provides for subsequent development by way of further conventions with a more reduced scope between each TGI and the specific *Barreau d'avocats* to which it is linked.

8.2.4 Italy

In Italy, Law 2005/80 (which ratified Legislative Decree 2005/35 of 14 March) amended various precepts of the *Codice di Procedura Civile* in order to allow notifications by e-mail (and also by fax). Since then, articles 133 and 134 CPC allow the electronic notification of judgements and other judicial decisions, and article

[18] *Décret* nº 2009–1524, 9 December 2009: the court will only accept documents sent to it by electronic means, except when it is impossible for fair reasons; and any communication or summons from the court will also be served on the parties' representatives by electronic means (Art. 930-1 I CPC).

[19] http://www.heftmanavocat.com/rpva%20doc/convention-nouvelles-technologies%20RPVA.pdf [03.02.2011].

136 CPC allows, in general terms, the performance of any kind of electronic notification.[20] All of these rules contain a very wide-ranging reference back to any implementing regulations that may be enacted, although the CPC itself establishes two general safeguards:

- Notification by electronic means is only possible if the defence lawyer has stated in the first document registered for the purposes of the case at what e-mail address he wishes to receive notification (arts. 133–134 CPC).
- In the event that a decision, an act, or a document in electronic format is to be notified, and the addressee does not have a certified e-mail address, the notification shall be made in normal documentary format, and the court officials must preserve the electronic format for a minimum of 2 years. Where requested, notification shall also be carried out electronically to the address indicated by the addressee or his/her representative, or a copy of the document will be supplied in an unalterable electronic format.

The implementing provisions, paradoxically, precede the latest reform of the CPC, given that they are contained in Presidential Decree no. 123 of 13 February 2001.[21] This Decree envisages the creation of a system known as '*dominio giustizia*', made up of the collection of hardware and software resources which serve as the basis for electronic activity in the field of Justice, and within it, the '*sistema informatico civile*' is envisaged, as a part of the aforementioned system, focussing on the electronic process of civil proceedings [article 1, sections e) and f)]. The civil computerized system must be structured in such a way as to ensure the identification of the judicial office and the case, the identification of the party introducing, amending, or notifying an action, the effective receipt of the notice of the action, and the automatic authorization of the lawyer and the court officials. In fact, access to the civil computerized system is only allowed to the lawyers of the parties and the court officials (article 3).

As a general rule, all notices and service of documents must be carried out using the civil computerized system (article 2.2). However, certain notices addressed to the parties may also be carried out electronically and without using the civil computerized system (article 6): specifically, where the addressee is a lawyer, these may be addressed to the electronic address that the lawyer has notified to the professional College; in other cases, notice may be sent to the electronic address notified to the certifier of the digital signature (article 7).

Any communications and notices carried out using these channels shall be deemed to be made on the date assigned to the acknowledgement of receipt by the person responsible for serving the notice (article 8).

---

[20] http://www.altalex.com/index.php?idnot=33723

[21] http://www.cnipa.gov.it/site/_files/cd_Decreto%20del%20Presidente%20della%20Repubblica%2013%20febbraio%202001%20n%20123_c.pdf

8.2.5 Spain

The Spanish procedural system is open to electronic communication with great clarity following the approval in 2000 of the *Ley de Enjuiciamiento Civil* (LEC).[22] Articles 152.2.2 and 162 of the said Act allow notices to be served by electronic means, in accordance with the following provisions:

- Court offices and the parties or addressees of the items notified (lawyers and *procuradores*) must be in possession of the necessary electronic means. In this regard, the parties and the legal professionals participating in the process must notify the court that they are in possession of the means stated above, and of their address. Furthermore, a register (which shall be accessible electronically) is to be set up at the Ministry of Justice of the said means and of the addresses corresponding to the public authorities.
- The electronic communication systems must allow the sending and receipt of pleadings and documents in which the authenticity of the communication and its content are safeguarded,[23] with official record of the sending and of receipt in full, and of the exact times when this happened.

In order to implement the content of the *Ley de Enjuiciamiento Civil*, from 2004 work began on an experimental basis on a special system of notifications, known as 'Lexnet'. Subsequently, in 2007, the *Ley de Enjuiciamiento Civil* was amended to provide Lexnet with a legal basis and a specific decree was approved detailing its functioning.[24] The Lexnet system is based on the use of an advanced e-mail program, which is only available to those who have subscribed. It ensures the authenticity of the communications, their integrity, their confidentiality, prevents them from being repudiated, and ensures they are time-stamped, thereby allowing the precise moment the message was sent and received to be recorded. In fact, all transactions using Lexnet generate an electronic receipt; an error receipt also exists which provides proof of an attempted filing or service that has failed due to a system malfunction. Currently, it is reserved for court officials, lawyers, and *procuradores*, but it is not yet open to private individuals. The system allows three different types of communication to be managed:

---

[22] http://noticias.juridicas.com/base_datos/Privado/l1-2000.html

[23] Where the authenticity of decisions, documents, orders, or reports filed or served by electronic means can only be acknowledged or verified by way of direct examination, they may nonetheless be filed in electronic format by way of digitalized images; but they must be filed in their original paper format should any party, or the court, or the State Prosecutor's Office so request.

[24] Act 41/2007 (7 December 2007) and *Real Decreto* 84/2007, de 26 de enero, *sobre implantación en la Administración de Justicia del sistema informático de telecomunicaciones Lexnet para la presentación de escritos y documentos, el traslado de copias y la realización de actos de comunicación procesal por medios telemáticos*, http://www.boe.es/boe/dias/2007/02/13/pdfs/A06239-06244.pdf.

- Communication from the court to the parties: when the court needs to serve notice of a decision on a party, service is made to the Lexnet mailbox of the party's *procurador* or to the corresponding College of *Procuradores*, and service is deemed to take place at the time the *procurador* opens the mailbox. For this reason, Lexnet cannot be used to serve the originating writ on the defendant, given that no court advocate has been appointed at this time.
- Communication from the parties to the court: the parties may file their pleadings with the court using Lexnet, provided that they have been electronically signed, with such attached documents as may be required. Lexnet is operational 24 h a day, every day: if filing takes place on a business day during working hours, it is deemed to take place at the time stated on the receipt generated by the system; at other times, it is deemed to take place on the first following business day at the start of the working day.
- Communication between the parties, who have to serve on one another all documents and applications filed with the court: with Lexnet, this service occurs simultaneously, with joint dispatch to the court and to the other parties. Documents sent must be in 'rtf' format or some other standard form.

However, Lexnet is not the only valid system for electronic notifications. Other tools may also be used, provided that they comply with the general conditions laid down in the LEC. Thus, where a court addresses a notification to a litigant and there is proof of the correct dispatch of the communication using technical means, it shall be deemed that the communication has been lawfully carried out and that it has been fully effective once 3 days have elapsed without the addressee accessing its content. An exception to this rule is allowed where the addressee provides justification for not accessing the notifications system during this period. Where failure to access is due to technical reasons and these persist at the time they are to be served, the act of communication shall be carried out by way of the service of a copy of the decision. In any event, the notification shall be deemed to be validly received at the time when it is established that access to the system was possible.

Aside from all the foregoing, where other methods of communication should fail or be impossible, notice may be served by way of edicts, which must be published on the court notice board: article 164 LEC allows this form of publication to be replaced by the use of electronic formats, similar to as allowed in Germany.

8.2.6 Brazil

In Brazil, the regulation of the use of ICT in procedural law was contained in Law no. 11.419 of 19 December 2006.[25] This Law amends the *Código de Processo Civil*, article 154 of which now provides in general terms that all procedural acts may be carried out, transmitted, and stored electronically. On this basis, Law no. 11.419

[25] Available at http://www.planalto.gov.br/ccivil_03/_Ato2004-2006/2006/Lei/L11419.htm

regulates two different questions: the manner in which electronic communications are to be carried out,[26] and the performance of the proceedings in electronic format.

In general terms, the Law makes the validity of electronic procedural acts subject to the use of electronic signature and the prior accreditation of the user before the Judicial Authorities (always done in person). Once accreditation is obtained, the user will have a password to access the system that ensures the confidentiality, identification, and authenticity of all communications (article 2). Procedural acts shall be deemed to be carried out at the time they are sent to the Judicial Authority system: in the event they are subject to a time limit, they shall be deemed to be carried out in good time where the dispatch takes place prior to 24:00 h on the day the time limit expires (article 3). Furthermore, it is declared that communications between judicial bodies ought preferentially to be carried out by electronic means (article 7).

More specifically, and in respect of electronic communications, the law envisages two different forms:

- General communication (article 4): the courts may set up an electronic '*Diário da Justiça*', available on the internet, in order to publish their own judicial and administrative acts, as well as communications in general. Electronic publication in this way replaces any other notifications, except in those cases where the law requires service in person. The publication date shall be deemed to be the business day following the day on which the information is available in the Journal; and procedural time limits shall start to run from the first business day following the day on which publication was deemed to have taken place.
- Personal service (article 5): notifications where there is a specific addressee must be carried out electronically, using their own portal, to those persons who have registered on the system (thus service by ordinary e-mail shall not be sufficient). Therefore the acceptance of the addressee is required. Service is deemed to take place on the day on which the addressee accesses the electronic message: where this occurs on a non-business day, it shall be deemed to occur on the following business day; under all circumstances, it must be carried out within 10 days of the date on which the notification was sent, otherwise the notification shall be deemed to take place at that moment. In urgent cases where electronic service may prejudice the parties, or in those cases where there is evidence of an attempt to abuse the system, the procedural act must be carried out in some other way that is suitable in order for the objective to be attained.

8.2.7 United States of America

The United States of America has been one of the first countries to introduce the use of e-Justice tools. In the scope of electronic service and filing of documents, and in

[26] A general explanation at Demócrito Reinaldo Filho, *Comunicação eletrônica de atos processuais na Lei nº 11.419/06* (2007), en, http://jus2.uol.com.br/doutrina/texto.asp?id=9750 [03.02.2011].

the context of the federal civil procedure, it must be retained that the *Federal Rules of Civil Procedure*[27] regulate two different kinds of communication: on one hand, the first notification addressed to the defendant in order to inform him of the commencement of the proceedings (*summons*); on the other hand, any other service of documents that has to be made during the procedure.

Regarding the service of the originating summons to the defendant, there are no specific provisions allowing it to be made by electronic means: rule 4 (e)(2) FRCP states that the first service to the defendant may be done by delivering a copy of the summons and of the complaint to the individual personally, or leaving a copy of each at the individual's dwelling or usual place of abode with someone of suitable age and discretion who resides there, or delivering a copy of each to an agent authorized by appointment or by law to receive service of process.

By contrast, when it comes to the service of other documents, rule 5 FRCP allows electronic service, although the regulation is not very thorough. Apart from handing it to the addressee, leaving it at the person's office or at his dwelling or usual place of abode, mailing it to the person's last known address or leaving it with the court clerk if the person has no known address, rule 5 (b)(2)(E) allows service of a document to be made by "*sending it by electronic means if the person consented in writing – in which event service is complete upon transmission, but is not effective if the serving party learns that it did not reach the person to be served*".

In addition, *rule* 5 (d)(3) FRCP also allows the filing of documents in the court using electronic means of communication: *"A court may, by local rule, allow papers to be filed, signed, or verified by electronic means that are consistent with any technical standards established by the Judicial Conference of the United States. A local rule may require electronic filing only if reasonable exceptions are allowed. A paper filed electronically in compliance with a local rule is a written paper for purposes of these rules."*

There is, in a word, the possibility of e-filing and e-service of documents, with the exception of the originating summons.

8.2.8 England

The English regulation on e-service and e-filing of documents disserves a special attention, due to its thoroughness and pragmatism. The *Civil Procedure Rules* and their *Practice Directions*[28] deal with electronic communication when regulating the way the parties can file documents in the courts and, also, the different ways that can be used to serve documents to the parties during the procedure.

---

[27] http://www.law.cornell.edu/rules/frcp/

[28] http://www.justice.gov.uk/civil/procrules_fin/

8.2.8.1 Filing Documents by Electronic Means

According to rule 5.5 CPR a practice direction may make provision for documents to be filed or sent to the court by electronic means. This general rule finds its development in *Practice Direction* 5B, which deals with *Electronic Communication and Filing of Documents*, and in *Practice Direction* 5C, regulating the *Electronic Working Scheme*. Both practice directions leave room for three options: filing documents by e-mail or using two special electronic systems, the *online forms service* and the *Electronic Working Scheme*.

Filing of Documents by E-Mail

If the Court has published an e-mail address for the filing of documents on the Court Service website (www.courtservice.gov.uk) and the document is listed on the Court Service website as a document that may be sent to or filed in that court by e-mail, then a party to a claim may send a document to the court by e-mail.[29] However, e-mail can not be used to take any step in a claim for which a fee is payable.

Filing of Documents Using the Online Forms Service

As an alternative to e-mail, the online forms service is a service available at the forms website (www.courtservice.gov.uk): this website contains certain documents which a user may complete online and then submit electronically to a specified court. In order to be used, the court must be listed on the Court Service website as able to receive documents filed electronically via the online forms service; and the document must be available for completion on the forms website.[30] The online forms service can be used to take a step in a claim for which a fee is payable: the fee must be paid, using the facilities available at the online forms service.

Both the e-mail and the online forms service are ways of filing and sending a document 'electronically'. The Practice Directions have set some general provisions relating to the filing of documents electronically:

---

[29]Regarding the technical specifications, the e-mail message must contain the name, telephone number and e-mail address of the sender and should be in plain text or rich text format rather than HTML. Correspondence and documents may be sent as either text in the body of the e-mail, or as attachments (however, documents required to be in a practice form must be sent in that form as attachments). Attachments must be sent in a format supported by the software used by the specified court to which it is sent. The length of attachments and total size of e-mail must not exceed the maximum which a particular specified court has indicated that it can accept. This information is listed on the Court Service website.

[30]The online forms service will assist the user in completing a document accurately but the user is responsible for ensuring that the rules and practice directions relating to the document have been complied with. Transmission by the service does not guarantee that the document will be accepted by the specified court.

- Where a party files a document electronically, he must not send a hard copy of that document to the court.
- A document is not filed until the transmission is received by the court, whatever time it is shown to have been sent. The time of receipt of a transmission will be recorded electronically on the transmission as it is received. However, if a transmission is received after 4 pm the transmission will be treated as received and any document attached to the transmission will be treated as filed on the next day the court office is open. The party is responsible for ensuring that the transmission or any document attached to it is filed within any relevant time limits.
- The court will normally reply by e-mail where the response is to a message transmitted electronically and the sender has provided an e-mail address.
- Parties are advised not to transmit electronically any correspondence or documents of a confidential or sensitive nature, as security cannot be guaranteed.
- If a document transmitted electronically requires urgent attention, the sender should contact the court by telephone.

Using the Electronic Working Scheme (Practice Direction 5C)

The Electronic Working scheme operates from 1st April 2010 (there was an Electronic Working Pilot Scheme between 1st April 2009 and 31st March 2010) in certain jurisdictions of England and Wales: the Admiralty, Commercial and London Mercantile Courts, the Technology and Construction Court, and the Chancery Division of the High Court at the Royal Courts of Justice, including in the case of the Chancery Division the Patents Court and the Bankruptcy and Companies courts.

The EWS makes possible that proceedings be started and all subsequent steps be taken electronically. It also permits that proceedings which have not been started electronically be continued electronically after documents in paper format have been converted to an electronic format. The EWS will operate 24 h a day all year round, including weekends and bank holidays. This will enable claim forms to be issued and documents to be filed in electronic format out of normal court office opening hours (except for "down-time", planned or unplanned). It will receive an automated response to acknowledge receipt. Where a fee is to be paid, it may be paid by any method which Her Majesty's Courts Service may permit including any online or offline payment facility.

Persons wishing to use the EWS are required, wherever possible, to communicate with the court by means of e-mail.[31] When a claim form is received electronically at

[31] EWS users must ensure that all forms, documents, schedules and other attachments filed at court are in PDF format. They also must use the PDF forms which have been created by Her Majesty's Courts Service specifically for Electronic Working. If they wish to file any document which has not been created specifically for Electronic Working, before filing that document they must convert the document to PDF format and attach the document to the multi purpose form for that case which has been created specifically for Electronic Working.

the address provided by the court, the claim form will be issued, sealed and returned to the claimant for service: it is then a party's responsibility to print and serve the claim form, unless the party or parties to be served have agreed to accept service by email or other electronic means. The court will accept receipt of claim forms filed through EWS out of normal court office opening hours: claim forms received by the court up to midnight will bear the date they are received as the issue date.

In order to grant security, Her Majesty's Courts Service will take such measures as it thinks fit to ensure the security of steps taken or information communicated or stored electronically. It may provide such method of encryption to promote security of e-mail communications as may be deemed appropriate.

8.2.8.2 Serving Documents by Electronic Means

Rule 6.3 CPR establishes the methods that can be used to serve a claim form: among them, rule 6.3 (1)(d) allows the use of "other means of electronic communication in accordance with Practice Direction 6A". For the service of documents other than the claim form rule 6.20 (1)(d) allows also the use of "other means of electronic communication in accordance with Practice Direction 6A".

In both situations –service of claim form, service of other documents–, rule 6.23 (6) states that where a party indicates that she will accept service by electronic means other than fax, the e-mail address or electronic identification given by that party will be deemed to be at the address for service.

Practice Direction 6A (paras 4.1 and 4.2) develops the general provisions of rule 6 CPR. As a general condition, where a document is to be served by electronic means, the party who is to be served or the solicitor acting for that party must previously have indicated in writing to the party serving that the party to be served or the solicitor is willing to accept service by electronic means and the e-mail address or other electronic identification to which it must be sent. For those purposes, the following are to be taken as sufficient written indications: an e-mail address set out on the writing paper of the solicitor acting for the party to be served but only where it is stated that the e-mail address may be used for service; or an e-mail address or electronic identification set out on a statement of case or a response to a claim filed with the court.

In addition, where a party intends to serve a document by electronic means that party must first ask the party who is to be served whether there are any limitations to the recipient's agreement to accept service by such means (for example, the format in which documents are to be sent and the maximum size of attachments that may be received). Where a document is served by electronic means, the party serving the document need not in addition send or deliver a hard copy.

The use of electronic means of communication should also be possible, according to rule 6.11 CPR, if it is the contractually agreed method: a contract contains a term providing that, in the event of a claim being started in relation to the contract, the claim form may be served by a method or at a place specified in the contract.

8.2.9 Other European Countries

In May 2007 the Council of the European Union published the "Study of the situation regarding the use of information and communications technology (ICT) in Member States' judicial systems with particular attention to eJustice",[32] carried out by the Europäischen EDV-Akademie des Rechts GmbH (EEAR) [European IT Academy of Law], Merzig, Germany. As a general description of the situation, following is stated: "In recent years many Member States have made it possible under their court rules of procedure for documents to be transmitted in electronic form to courts and public prosecutors. Other Member States plan to introduce such a system in the near future. Electronic communication already applies, in particular in civil proceedings and summary proceedings. Overall, however, the level of use of electronic communication is not very high, although the limited experience gained thus far is described as positive across the board." A few months later (July–September 2007) all Member States sent to the Council a description of the use of information and communication technology in its judicial system. Although the information might not be as fresh as it would be desirable, it can help to get an idea of the situation in European countries other than those studied with further attention in this report. Those are the general answers to the topic "Electronic communication with parties involved in proceedings" furnished by each State:

Denmark: "In Danish judicial proceedings lawyers and other parties involved cannot yet send documents to courts and public prosecutor's offices electronically. Appropriate legislative rules are planned for all types of proceedings, however. The courts and public prosecutor's offices are set to be able to send documents back electronically to parties involved in proceedings under 2004 rules already established but not yet in force. There are no technical standards for electronic communication with parties involved in proceedings. The authenticity and integrity of documents transmitted electronically are guaranteed only in data communication with public prosecutor's offices using advanced electronic signature procedures." (Doc. 9083/07 Add 3 JURINFO 13) http://register.consilium.europa.eu/pdf/en/07/st09/st09083-ad03.en07.pdf

Bulgaria: "Parties involved in proceedings before Bulgarian courts have hitherto been unable to submit documents electronically, although there are plans to put the corresponding arrangements in place for civil and enforcement proceedings and the relevant technical systems have already been partially realised. The same applies to the dispatch of electronic documents by Bulgarian courts and judicial authorities. Here, too, technical systems have already been put in place for civil proceedings in general and for debt enforcement and enforcement proceedings in particular, but there is still no legal basis for their use. Consequently, nor have any

[32] Doc. 9083/07 JURINFO 13 and Doc. 9573/07 JURINFO 17). The contents of the study can be read at http://register.consilium.europa.eu/pdf/en/07/st09/st09083.en07.pdf [03.02.2011] (main study) and at http://register.consilium.europa.eu/pdf/en/07/st09/st09573.en07.pdf [03.02.2011] (Study Part II: Comparative Analysis).

technical standards been laid down by law for electronic communication between the courts and outsiders, although there are plans to establish such standards. The advanced electronic signature within the meaning of Article 2(2) of the Directive on electronic signatures may be used in Bulgaria as a means of authentication." (Doc. 9083/07 Add 2 JURINFO 13) http://register.consilium.europa.eu/pdf/en/07/st09/st09083-ad02.en07.pdf

Belgium: "In some Belgian court procedures, lawyers and other participants in proceedings can send documents electronically to courts and public prosecutors' offices. The legal and technical conditions for this have existed since 2006, but have not yet been implemented. Courts and public prosecutors will also be able to return documents electronically to persons involved in proceedings under legislation already framed but still to be implemented. There are no technical standards for electronic communication with persons involved in proceedings. The authenticity and integrity of electronically transmitted documents is ensured by simple and advanced electronic signatures." (Doc. 9083/07 Add 1 JURINFO 13) http://register.consilium.europa.eu/pdf/en/07/st09/st09083-ad01.en07.pdf

Estonia: "Since recently, the filing of documents in electronic form has been permissible for all types of proceedings in Estonia and the corresponding systems have been implemented. Although it is not yet used in the majority of cases, in some types of proceedings it is in widespread use. A similar situation applies to the transmission of documents by courts, although this is less widely used than electronic filing. Technical standards exist for that purpose and binding rules are planned. At present, structured documents are communicated via the Internet. Where authentication is required, this is achieved by means of the qualified signature within the meaning of Article 5(1) of the Signatures Directive." (Doc. 9083/07 Add 5 JURINFO 13) http://register.consilium.europa.eu/pdf/en/07/st09/st09083-ad05.en07.pdf

Ireland: "For a short time now, the electronic filing of documents by outsiders via an internet portal has been possible in some cases in respect of Irish social court proceedings, and this system is already in frequent use. Where documents are filed electronically, the data are transmitted in structured form. In other proceedings, electronic filing of documents is not yet possible, though there are plans to introduce such a system for general civil-law actions and debt enforcement and enforcement proceedings. There is no provision for the electronic transmission of documents by the courts. Consequently, no standards are laid down in that connection in Ireland." (Doc. 9083/07 Add 9 JURINFO 13) http://register.consilium.europa.eu/pdf/en/07/st09/st09083-ad09.en07.pdf

Finland: "Since the middle of the 1990s it has been possible to file documents in electronic form in Finnish civil law proceedings and many are indeed filed in this way. Since 2003 this has also been possible for administrative and other types of proceedings, but less use has been made of this option in certain cases. The situation regarding the transmission of documents by the courts is very similar; applications initiating proceedings may not, however, be transmitted electronically. The primary technical tool is secure e-mail. In addition, Finland has standards on data file formats and the structure of documents for transmission to the courts, although these standards are not mandatory. If an authentification is required, a simple electronic signature within the meaning of Article 2, point 1 of the Signatures Directive is provided." (Doc. 9083/07 Add 6 JURINFO 13) http://register.consilium.europa.eu/pdf/en/07/st09/st09083-ad06.en07.pdf

Greece: "In some court proceedings in Greece lawyers and other parties involved can send documents electronically to courts and public prosecutors' offices. In some proceedings courts and public prosecutors' offices can also send back documents electronically to the parties involved. However, these possibilities are used only infrequently. There are technical standards for electronic communication with parties involved in the proceedings but there is no legal obligation to observe them. An electronic signature to protect the authenticity and integrity of documents sent electronically must first be introduced". (Doc. 9083/07 Add 8 JURINFO 13) http://register.consilium.europa.eu/pdf/en/07/st09/st09083-ad08.en07.pdf

Lithuania: "In Lithuanian legal proceedings, lawyers and other parties involved in proceedings may not file documents with courts and public prosecutors in electronic form, nor are any corresponding legislative rules planned. Electronic signatures are not used." (Doc. 9083/07 Add 12 JURINFO 13) http://register.consilium.europa.eu/pdf/en/07/st09/st09083-ad12.en07.pdf

Netherlands: "At present there are no legal proceedings in the Netherlands in respect of which it is legally permissible for lawyers and other parties involved in proceedings to file documents with courts and public prosecutors in electronic form. Rules which would make this possible in the future are planned for only a few types of proceedings. In that connection, experiments are being conducted to test the use of technical solutions. As a rule, courts and public prosecutors may not transmit documents to persons involved in proceedings by electronic means. No legal amendments are planned in that respect. For certain proceedings, however, judgments are sent automatically and electronically to various bodies outside the judicial system. Electronic standards are applied to the electronic communication of judgments." (Doc. 9083/07 Add 14 JURINFO 13) http://register.consilium.europa.eu/pdf/en/07/st09/st09083-ad14.en07.pdf

Latvia: "The electronic filing of documents by public prosecutors and other parties involved in the proceedings is currently legally permissible and technically feasible in criminal proceedings only. Very little use is made of this possibility. The same applies to the submission of documents by courts and public prosecutors. This is also for the moment only possible in criminal proceedings. Basically only those offices in the Latvian judicial system with sufficient electronic signatures may be involved in electronic communication. Family and inheritance matters are excluded by law from the electronic transmission of documents. Certain document containing personal data may only be communicated in electronic format using certain procedures. In Latvia there are technical standards for electronic communication with courts, but these have not yet been established by law. However, a uniform court information system is currently being developed on the basis of international standards. Authentification is provided for using a qualified electronic signature within the meaning of Article 5(1) of the Signatures Directive." (Doc. 9083/07 Add 11 JURINFO 13) http://register.consilium.europa.eu/pdf/en/07/st09/st09083-ad11.en07.pdf

Luxembourg: "At present there are no legal proceedings in Luxembourg in respect of which the electronic filing or electronic dispatch of documents by the judicial system is legally permissible or implemented, nor do any technical standards exist yet in that connection." (Doc. 9083/07 Add 13 JURINFO 13) http://register.consilium.europa.eu/pdf/en/07/st09/st09083-ad13.en07.pdf

Sweden: "In some Swedish legal proceedings, lawyers and other parties involved in proceedings may file documents with courts and public prosecutors in electronic form; however, this does not apply to documents which must be signed. Courts and public prosecutors may transmit documents to persons involved in proceedings by electronic means, but as a complement to and never in place of physical documents. No electronic standards exist for electronic communication with persons involved in proceedings. No electronic signatures exist to protect the authenticity and integrity of electronically transmitted documents." (Doc. 9083/07 Add 19 JURINFO 13) http://register.consilium.europa.eu/pdf/en/07/st09/st09083-ad19.en07.pdf

Slovakia: "In many Slovakian legal proceedings, lawyers and other parties involved in proceedings may file documents with courts and public prosecutors' offices in electronic form. Courts and public prosecutors' offices may send documents back to persons involved in proceedings in electronic form. However, they make very little use of either option, despite the fact that they have been technically implemented, at least in part. There are mandatory technical standards for electronic communication with persons involved in proceedings. No electronic signatures exist as yet to protect the authenticity and integrity of electronically transmitted documents. However, the law on a certified electronic signature is to enter into force in mid-2007." (Doc. 9083/07 Add 20 JURINFO 13) http://register.consilium.europa.eu/pdf/en/07/st09/st09083-ad20.en07.pdf

Cyprus: "In Cypriot legal proceedings, lawyers and other persons involved may not file documents electronically with courts or public prosecutor's offices. Nor may courts or public prosecutors' offices send back documents electronically to persons involved in proceedings." (Doc. 9083/07 Add 26 JURINFO 13) http://register.consilium.europa.eu/pdf/en/07/st09/st09083-ad26.en07.pdf

Poland: "In some Polish legal proceedings, lawyers and other parties involved in proceedings may file documents with courts and public prosecutors in electronic form. In some proceedings courts and public prosecutors can also return documents electronically to persons involved in proceedings. No technical standards exist for electronic communication with persons involved in proceedings. The authenticity and integrity of electronically transmitted documents are ensured by advanced electronic signature." (Doc. 9083/07 Add 16 JURINFO 13) http://register.consilium.europa.eu/pdf/en/07/st09/st09083-ad16.en07.pdf

Romania: "In Romanian legal proceedings, lawyers and other parties involved in proceedings are not yet able to file documents with courts and public prosecutors in electronic form, nor does the judicial system currently transmit documents by electronic means." (Doc. 9083/07 Add 18 JURINFO 13) http://register.consilium.europa.eu/pdf/en/07/st09/st09083-ad18.en07.pdf

Czech Republic: "In some Czech legal proceedings, lawyers and other parties may file documents with courts and public prosecutors in electronic form. Courts and public prosecutors may return documents to persons involved in proceedings by electronic means. However, neither option is used very much, despite being technically feasible. No technical standards exist for electronic communication with persons involved in proceedings. The authenticity and integrity of electronically transmitted documents are guaranteed by simple electronic signature." (Doc. 9083/07 Add 23 JURINFO 13) http://register.consilium.europa.eu/pdf/en/07/st09/st09083-ad23.en07.pdf

Hungary: "There are currently no Hungarian court proceedings which allow lawyers or other parties involved in the proceedings to file documents electronically with courts and public prosecutors' offices. Likewise, courts and public prosecutors' offices for their part have no legal possibility of sending documents back electronically to parties involved in the proceedings. Both options are planned for most types of proceedings, and appropriate laws and rules are under preparation. In order to protect the authenticity and integrity of electronically transmitted documents, Hungary uses both advanced and qualified electronic signature within the meaning of the Signatures Directive." (Doc. 9083/07 Add 24 JURINFO 13) http://register.consilium.europa.eu/pdf/en/07/st09/st09083-ad24.en07.pdf

Slovenia: "Lawyers and other parties involved in proceedings may file documents with courts and public prosecutors in electronic form in debt enforcement and other enforcement proceedings under the Slovenian systems of civil and administrative law; rules are planned for all other types of proceedings. A similar situation obtains as regards judicial authorities sending out procedural documents electronically; in administrative court proceedings this is already permitted and feasible; for all other proceedings it is planned. Technical standards exist for the transfer and structuring of electronic data, but are not specifically laid down. XML and secure e-mail are relied upon in the main. Where authentication is required in connection with electronic communication, a qualified electronic signature within the meaning of Article 5(1) of the Signatures Directive is available, as well as the Justice Ministry's own certificates. PGP encryption is also used." (Doc 9083/07 Add 21 JURINFO 13) http://register.consilium.europa.eu/pdf/en/07/st09/st09083-ad21.en07.pdf

Portugal: "In many Portuguese legal proceedings, lawyers and other parties involved in proceedings may file documents with courts and public prosecutors in electronic form. The legal and technical framework conditions have been in place since 2001 and this type of communication is now well-accepted. Courts and public prosecutors can also return documents electronically to persons involved in proceedings and this option is widely used. No technical standards exist for electronic communication with persons involved in proceedings. The authenticity and integrity of electronically transmitted documents are ensured by advanced electronic signature." (Doc. 9083/07 Add 17 JURINFO 13) http://register.consilium.europa.eu/pdf/en/07/st09/st09083-ad17.en07.pdf

A short description of the situation in ***Norway*** (non EU-Member State) can be found at the Presentation on "The Use of Information Technology in the Norwegian Justice Sector", directed to the Working Party on Legal Data Processing (e-Justice) of the Council (Doc. 14485/09 JURINFO 120).

More recently, the Working Party on Legal Data Processing (e-Justice) of the Council of the European Union has decided to examine the Member States's national projects on e-Justice[33] and the topic of electronic service of documents in greater detail[34]: it is clear that it will be an area in which further developments are to be awaited.

The main general conclusions on our topic are the following:

c) Electronic submission of writs of summons and other documents

11 Electronic submission of writs of summons and other documents concerns the transmission of legally relevant documents from the parties to the courts by electronic means. Authentication of the sender is therefore an important issue in order to avoid abuse.

12 In some Member States such electronic submission of documents is already possible. However, in some cases this applies only to lawyers, notaries and other persons especially registered for this purpose (DE, AT). While some systems use qualified electronic signatures for authentication (DE, EE), in some systems other methods such as a simple username/password combination apply (AT, UK).

13 In some Member States the electronic transmission of written submissions is obligatory for lawyers, subject in some cases to technical constraints on the lawyer's side (AT).

14 Subject to the requirement of electronic signature, online application open to the general public is available in some Member States for certain special procedures (DE: payment orders, forms for other applications under construction; EE: payment orders, online applications are possible to most registers and administrative authorities for all citizens and a special e-Notary environment has been created for notaries; FR payment orders, administrative applications and guardians, all under construction; PL: pledge register, obligatory; UK: county court money claims and possession claims), and under construction for all cases in other Member States (EE).

15 In other Member States there are projects regarding the electronic transmission of applications (DK, HU, MT) or electronic communication between lawyers and the registries of the courts (FR).

d) Electronic service of documents

16 Electronic service of documents concerns judicial and extrajudicial documents sent from the courts or the bailiffs to the parties. It should be borne in mind that in some Member States service of documents is effected by a specific profession, the bailiffs, whereas in other Member States documents are notified to the parties directly by the courts. Nevertheless, some practical problems encountered with efforts to digitalise service of documents remain similar, regardless of whether the service of documents is effected by the courts or by the

[33] See Doc. 13759/09 JURINFO 112 (13 October 2009).

[34] See Doc. 7355/10 EJUSTICE 28 JUSTCIV 43 (12 March 2010), including a list of questions prepared by the Presidency to structure the discussions, http://register.consilium.europa.eu/pdf/en/10/st07/st07355.en10.pdf. Some countries have already submitted their answers (Spain, Italy and Poland).

bailiffs and regardless of whether the document to be served is e.g. a writ of summons to be served on the defendant, or a judgment: In order to take legal effect, service will in most cases require that the addressee has at least the opportunity to receive the document and that the court is informed about this fact by an acknowledgment of receipt.

17 In some Member States documents can already be served electronically with the electronic service having the same legal effect as service of documents in paper form (AT; IT: only some courts so far). Other Member States have projects running to this effect (HU). In most cases this is, however, restricted to lawyers, notaries or other persons who have expressly agreed to it, or to persons who have made use of the possibility of electronic transmission of applications (in the case concerned or in other cases). Sometimes, certain types of documents are excluded from the option of electronic service. From these restrictions it appears that in most Member States it is possible only in exceptional cases to serve documents instituting proceedings electronically.

18 Most systems create a return receipt automatically, some as soon as the document has reached the mailbox (IT, AT), some only when the document is actually downloaded/opened actively by the recipient (HU), the latter providing for a fiction of receipt some days after the document has reached the mailbox. An electronically signed acknowledgment of receipt is possible in some systems (HU).

19 Several other Member States have projects on the electronic service of documents. In some Member States (IT, HU) it will be obligatory in certain cases to sign up to the system for electronic service, so that as a result any service in these cases can be effected electronically.

8.3 E-Service at International Courts

Just like at a national level, the tools of e-Justice are gradually finding their place in international courts. There are indeed some rules allowing the use of electronic means to file documents and to make service at least in procedures before the *Court of Justice of the European Union*, the *European Court of Human Rights*, the *Inter-American Court of Human Rights* and the *International Criminal Court*. Proceedings before those courts are quite special and the service of documents is rarely threatened by the same dangers existing in national procedures. That is probably the reason why the rules dealing with this matter are little thorough. However, international regulations show a general concern with the same questions that are source of worries at a national level.

8.3.1 The Court of Justice of the European Union

The Rules of Procedure of the Court of Justice of the European Union[35] include some provisions regarding e-filing of documents addressed to the Court and e-service of documents addressed to the parties.

[35] http://curia.europa.eu/jcms/upload/docs/application/pdf/2010-04/rp.en.pdf [03.02.2011].

8.3.1.1 E-Filing

According to Article 37 (6) of the Rules, pleadings to the Court can be filed by "other technical means of communication available to the Court", which includes e-mail.[36] Where transmission is by electronic mail, only a scanned copy of the signed original will be accepted[37]: an ordinary electronic file or one bearing an electronic signature or a computer-generated facsimile signature will not be treated as complying with the rules.

In such case, the date on which the copy of the signed original of the pleading is received at the Registry by telefax or email shall be deemed to be the date of lodgment for the purposes of compliance with the time-limits for taking steps in proceedings, provided that the signed original of the pleading, accompanied by the annexes and copies, is lodged at the Registry no later than 10 days thereafter.[38] E-filing or e-lodgment of documents, hence, is only a way to accelerate the delays and to comply with time-limits, but doesn't substitute traditional filing of the original paper documents.

8.3.1.2 E-Service

Every lawyer or agent acting before the Court shall state an address for service in the place where the Court has its seat and the name of the person who is authorised and has expressed willingness to accept service. However, Article 38 (2) allows the lawyer or agent to state that they agree that service is to be effected on him by telefax or other technical means of communication: the statement must specify the electronic mail address to which the Registry may send that party documents to be served. The recipient's computer must be equipped with suitable software (for example, Acrobat or Readiris 7 Pro) for reception and display of communications from the Registry, which will be transmitted in PDF format.

In such cases, any procedural document other than a judgment or order of the Court may be served to the addressee by the transmission of a copy of the document by such means and shall be deemed to have been duly served when such means are

---

[36] According to the *Practice Directions relating to direct actions and appeals – Use of technical means of communication*, a copy of the signed original of a procedural document may be transmitted to the Registry as an attachment to an electronic mail to the email address ecj.registry@curia.europa.eu, http://curia.europa.eu/jcms/upload/docs/application/pdf/2009-02/ins_prat2_2009-02-09_16-15-31_502.pdf.

[37] According to the same *Practice Directions*, documents should be scanned at a resolution of 300 DPI and, wherever possible, in PDF format (images plus text), using Acrobat or Readiris 7 Pro software.

[38] The Practice Directions also state that the signed original must be sent without delay, immediately after the despatch of the copy by e-mail, without any corrections or amendments, even of a minor nature. In the event of any discrepancy between the signed original and the copy previously lodged, only the date of lodgment of the signed original will be taken into consideration.

used [Article 79 (2)]. However, where for technical reasons or on account of the nature or length of the document transmission by e-mail is not feasible, the document shall be served by traditional means.[39]

8.3.2 The European Court of Human Rights

For procedures held before the European Court of Human Rights there is a Practice Direction concerning "Secured electronic filing".[40] In fact, the ECHR has set out a system of secured electronic filing, which can be used voluntarily by the Governments of the Contracting States. The Governments which have opted for that system shall send all their written communications with the Court by uploading them on the secured Internet site set up for that purpose and shall accept written communications sent to them by the Registry of the Court by downloading them from that site.[41] In order to do so, the Government shall possess the necessary technical equipment and follow the user manual sent to them by the Court's Registry.

Regarding format, it is foreseen that documents filed electronically shall be in PDF format, preferably in searchable PDF. Signed documents to be filed electronically shall be generated by scanning the original paper copy.

As to the relevant date with regard to time limits, the date on which the Government has successfully uploaded a document on the secured site shall be considered as the date of dispatch. To facilitate keeping track of the correspondence exchanged, every day shortly before midnight the secured server generates automatically an electronic mail message listing the documents that have been filed electronically within the past 24 h.

The secured electronic site shall not permit the modification, replacement or deletion of an uploaded document. If the need arises for the Government to modify a document they have uploaded, they shall create a new document named differently (for example, by adding the word "modified" in the document name). Where the Government has filed more than one version of the same document, only the document filed in time shall be taken into consideration. Where more than one version

---

[39] See also the *Notes for the guidance of Counsel in written and oral proceedings before the Court of Justice of the European Communities* (February 2009: http://curia.europa.eu/jcms/upload/docs/application/pdf/2008-09/txt9_2008-09-25_17-37-52_275.pdf) [03.02.2011] and the *Practice Directions relating to direct actions and appeals – Use of technical means of communication.*

[40] Issued by the President of the Court on 22 September 2008, http://www.echr.coe.int/NR/rdonlyres/C3F78149-F39D-48E9-B348-99C86FFDD273/0/SecuredDocumentsDecember2008.pdf [03.02.2011].

[41] There are only a few exceptions: (a) All written communications in relation to a request for interim measures shall be sent simultaneously through the secured site and by fax; (b) Attachments, such as plans, manuals, etc. which may not be comprehensively viewed in an electronic format may be filed by post; (c) The Court's Registry may request that a paper document or attachment be submitted by post.

has been filed in time, the latest version shall be taken into consideration, unless the President of the Chamber decides otherwise.

8.3.3 The Inter-American Court of Human Rights

The Rules of Procedure of the Inter-American Court of Human Rights[42] contain also some general provisions regarding electronic filing and transmission of documents.

8.3.3.1 E-Filing of Documents

According to Article 28, all briefs addressed to the Court may be presented by electronic mail. Every brief must be signed, in order to ensure authenticity. If it is transmitted to the Court by electronic means and has not been subscribed, the original documents must be received by the Court within a term of 21 days from the expiration of the deadline established for the submission of that brief.

8.3.3.2 E-Service of Documents

E-service is foreseen by Article 33 in very general terms: it allows the Court to transmit briefs, annexes, orders, judgments, advisory opinions, and other communications submitted to it by electronic means with adequate guarantees of security.

8.3.4 The International Criminal Court

According to the Regulations of the Court[43] the procedure before the International Criminal Court should be managed taking the most possible advantage of electronic means. Regulation 26 states that the Court shall establish a reliable, secure, efficient electronic system to support its daily judicial and operational management and its proceedings. The system will ensure authenticity, accuracy, confidentiality and preservation of judicial records and material.

---

[42] Approved by the Court during its LXXXV Regular Period of Sessions in November 2009, http://www.corteidh.or.cr/reglamento.cfm.

[43] Adopted by the judges of the Court on 26 May 2004, Fifth Plenary Session, The Hague, 17–28 May 2004 [Official documents of the International Criminal Court, ICC-BD/01-01-04], http://www.icc-cpi.int/NR/rdonlyres/B920AD62-DF49-4010-8907-E0D8CC61EBA4/277527/Regulations_of_the_Court_170604EN.pdf [03.02.2011].

8.3.4.1 E-Filing

All documents, decisions and orders shall, whenever possible, be submitted in electronic version for registration by the Registry. The electronic version of filings shall be authoritative [Regulation 26 (3)].

8.3.4.2 E-Service

In order to be served of any document registered by the Registry or any decision of the court, all participants shall provide to the Registry an electronic, facsimile or postal contact address for notification of documents (preferably in The Hague). If the participant provides an e-mail address, he will be notified by this means[44] and he will be deemed notified, informed of or to have had communicated to him, a document, decision or order on the day it is effectively sent from the Court by the Registry. Such date shall be written on the notification form to be appended to all copies of the document, decision or order, as relevant [Regulation 31 (1) and (2)].

As an alternative, a participant represented by counsel shall be deemed notified when his counsel has been notified of a document, decision or order at the electronic, facsimile or postal address which that counsel has indicated to the Registry [Regulation 32 (3)].

8.4 Main Challenges for Electronic Judicial Communication

Any regulation of the electronic service of documents benefits from the fact that electronic communications have become standard practice in most business and contractual relationships: for this reason it is being introduced quite naturally into legal proceedings, with little scope for resistance from members of the legal profession. It is not area in which one can say that there is any 'fear of the unknown'.

The legal overview, however, is very heterogeneous, as shown in the precedent pages. Most of the States have made legal reforms in a pretty short period of time in order to introduce new rules to make possible the e-filing and the e-service of documents. In addition, the reforms have been quite frequently carried out in two steps: legal norms with general contents, followed by much more detailed regulations of a lower rank. It is also quite striking to realize how, despite the diversity, there is a common core of questions which constantly catch the attention of lawmakers – even if with different degrees of concern.

---

[44] E-mail notification is not possible when personal service is required, and this happens with the following documents: (a) Warrants of arrest; (b) Summonses to appear; (c) Documents containing the charges; and (d) Such other documents, decisions or orders ordered by the Chamber to be notified by way of personal service.

8.4.1 The Importance of Technical Aspects

Legislators are clearly concerned about the technical aspects associated with electronic service. Some legislators, for example, are careful to clarify whether or not ordinary systems such as e-mail may be used, or whether, in contrast, it is preferable to use special systems, and on the same subject, whether the electronic communication systems are to be part of the internet, or whether, in contrast, different communications networks are to be used (Austria with its *Elektronischer Rechtsverkehr*, England with its *Electronic Working Scheme*, Spain with its *Lexnet*, Italy with its *sistema informatico civile*).

These appear to be decisions of a strictly technical nature, given that they are driven by efficiency, costs, and security criteria, which are regarded differently in each country. However, these may also have repercussions on the right to a fair trial: an electronic service system can only be implemented and made widespread if it is ensured that all interested parties or all parties under a legal duty to use it can have access to it, such that any technical options that do not have the potential to become universal ought to be rejected.

Some jurisdictions also contain much more detailed requirements, relating to the format or the size of the files that may be sent electronically, which once again may have an effect on the right to a fair trial (this may be clearly seen in English and in Spanish legislation, as well as in the rules of the European Court of Justice and the European Court of Human Rights). There is no doubt that the correct functioning of the computer systems may require all files to have specific standardized formats (such as RTF in Spain or PDF in England and in the European Court of Justice): however, these must be formats that are easy to use and that are universally accessible, in order not to prejudice access to justice. With regard to the requirements on file sizes, these may be technically justifiable, in the sense that the systems will not handle files that are too large. However, on this point the dangers to the right to a fair trial are more significant, because the size and weight of files do not always depend on the will of the parties: therefore any technical restrictions on capacity within the systems should not lead to restrictions on the rights of the parties to file pleadings and adduce evidence. For this reason, technical formulas ought to be found that allow large files to be sent electronically, or an alternative to electronic filing and service ought to be allowed, despite being defined as compulsory.

8.4.2 Reliability and Security of the System

A system of electronic legal notification must be designed taking into account the tension that exists between competing interests. On the one hand, the efficiency and speed of the system must be considered: this is something which is expected, by definition, of all computerized systems. Now then, there are certain requirements regarding security and reliability of the system which must also be respected and which are in relation to the generic notion of 'procedural safeguard'. In this regard, the system must be capable of ensuring the following points:

- Authenticity, by way of mechanisms that verify the identity of the issuer.
- Integrity, such that it is possible to prove what has been sent and that this has not been altered after being sent.
- Non repudiation, i.e. proof that the notice has been served, such that the issuer cannot deny that it has been sent, and the addressee cannot deny that it has been received.
- Confidentiality, such that only the addressee may have access to the content of the information.
- Timestamping, so that it is possible to verify the exact time a message was sent and received.

The above conditions must be met if the system is to be reliable and if it is to respect the rights of all those involved.

The analyzed legal systems deal with all or most of those questions and state in general terms that the electronic systems of service need to meet the technical requirements assuring their effectiveness. It has to be noticed, anyway, that they are not facing really new requirements, since they are common to every system of service or communication: the mentioned conditions must also be met when serving documents by "traditional" means. The difference lays in the origin of the problems and of the possible obstacles to comply with those requirements. When using traditional means, the system lays on the reliability given to the work of certain persons: specially court officers, *huissiers de justice*, *procuradores* or postmen.

In electronic systems, instead, there must be a double source of trust: there is of course a need to have trust in the persons that operate the systems; but there is an additional need that the electronic system itself is reliable: in a word, within the scope of e-Justice, in order to avoid errors, and where appropriate, correct them, one must have faith not only un persons, but also in computers; the problem is that legal staff and legal professionals, as a rule, do not have advanced computer skills – nor is there any reason why they should – and they feel uncomfortable when they are obliged to trust computer specialists. This is the reason why, at this point, national procedural laws can only include general provisions and are based on the assumption that any system introduced according to their rules will comply with the requirements set in the law: this can be easily seen when law relates to electronic signature and requires from it a certain degree of security.

In any case, there is a clear paradox in the high level of suspicion that is aroused by electronic means of service when compared to personal or traditional means of service: situations in which a service carried out by a person is defective are quite frequent and many legal systems allow service to be carried out by certain non electronic means that are merely fictive and provide no proof that notice has arrived to the addressee (like service by publishing the document at an official journal or service by deposit of the document in the addressee's mailbox).

To put it briefly, when it comes to assessing the suitability of using electronic systems as a tool for serving notices in judicial proceedings, it is best to avoid extremes. It is obvious that electronic communications can be subject to poor performance, errors, and system breakdowns that are difficult to explain and fix, or even abuse, as may be seen in a wide variety of electronic interaction. It is also true that people make mistakes and the legal systems of all countries are full of examples of notices being served incorrectly by people using traditional methods.

For this reason, legislators must also foresee how people will react in situations where things go wrong for technical reasons, and take into account the peculiarities of electronic systems when it comes to regulating their consequences and allowing for possible remedies. In that sense, some legislations consider that system's failures can happen and have established some safeguards for the litigants (as seen in Spain, France, Brazil, England or before the Court of Justice of the European Union). Austria has gone a step beyond and has set the general rule of public responsibility in case of malfunction of those systems.

Besides, the need to strengthen the trust in electronic service systems entails the additional need to turn to companies that carry out the managing ant the compliance of the system's reliability and security requirements, and this will certainly involve additional costs: the set-up and maintenance of secure computerized systems through which to undertake the electronic service of documents has an economic cost.

8.4.3 The Proof That the Document Has Been Filed or Served and the Moment It Is Deemed to Have Been Filed or Served

Any system of electronic service shall be able to provide proof that service has been carried out properly and at what moment. National laws deal with the question of the proof of service in a very general manner: they assume that the technical systems will be equipped with proper tools to generate an acknowledgement of receipt, which can be automatic (like in England or in Spain), or which has to be sent back by the addressee (like in Germany), or they rely on the information provided by the system concerning the moment in which the addressee got access to the sent message (like in Brazil, in France or, in a certain way, in Italy). Confidence in technical elements is once again required.

In particular, it is important to be able to prove the time at which electronic filing or service is deemed to have taken place, because in respect of the filing of documents by the parties, this activity is normally subject to time limits, and in respect of the service of decisions on the parties, this normally marks the start of a fresh time limit within which to reply.

One might expect that the use of electronic systems, which avoid the need to attend in person at the court house, would eventually lead to the disappearance of the restrictions imposed by business days or working hours. However, the fact is that most legal systems have not yet made this leap and continue to attach importance to the requirement that any electronic filing or service carried out on non-business days or at non-working hours should be linked to business days and working hours, normally the next business day at the start of the working hours (as in the case of Spain, or in England when using the online forms service, or in Brazil, in the case of notifications addressed to the parties): if working hours end at 20:00 h, for example, a dispatch sent at 23:00 h on the day on which a time limit expires is deemed to have been made at 08:00 h on the next day, which would make it out of time.

However, looking to the future, it is not hard to envisage that the widespread implementation of electronic service will end up affecting the rules on calculating time limits in proceedings. It is possible that the notion of 'business day' and 'working hours' will continue to apply to all activities that take place physically in court, performed by officials subject to business days and working hours. But their significance will be eroded for litigants and for the independent professionals who work for them and who are free to define their own working hours: it does not seem logical to require litigants and their lawyers to abide by working hours when filing documents electronically, when reality shows that their work goes on beyond the hours at which the courts are open, and that the electronic systems can operate 24 h a day. As has been seen, the legislation in most countries has not yet caught up with this reality. Fortunately, some examples of common sense can be found in English law – but only when EWS is used – or in Brazilian law and in the rules of the ECHR, in the case of filing pleadings with the court. It is to be hoped that other legislators will follow the same path.

8.4.4 The Voluntary Nature of the Electronic-Notification Mechanisms

The majority of national legal systems emphasize the voluntary nature that the use of electronic means for the filing of documents and service of notices must have. This voluntary nature is even more pronounced when it comes to the receipt of notices: sometimes there is an express requirement (France, Italy, Germany, United States, England), and in other cases there is an indirect approach, where the use of the system is made conditional on the prior registration of the user, and such registration is not compulsory (Austria or Brazil). Spain would appear to be an exception to this rule, where the electronic system is designed to be compulsory as and when it becomes available to the courts, and also in France, where it has shortly become compulsory in certain types of litigation. When it comes to supra-national courts, its voluntary nature is an absolute rule.

The question arises of whether it is possible to make the leap towards making the use of these systems compulsory – or at least towards increasing the degree of compulsion. In general terms, such a change would only be possible if various circumstances arise:

- On the part of the States, it is necessary that they should design an e-filing and e-service system that is adequate and that is capable of handling the entire volume of communications that are carried out every day.
- On the part of the courts, it is necessary that they should all be equipped with computer systems that allow the use of the e-filing and e-service system designed in each country, and that court officials should all be trained to use it properly.

- On the part of litigants, it is necessary for one of the two following situations to arise: either there must be a duty that all persons or legal professionals should have computer equipment and sufficient computer skills to use them, or it should be verified that all persons or legal professionals are already *de facto* in possession of the said means and of the said computer skills.

It is obvious that currently not all private individuals – potential litigants – are in possession of computer equipment and computer skills, but at least in those countries that have acquired a certain level of economic and social development, their lawyers and legal representatives do have them – or can be required to have them.

Therefore, at the moment, it is only possible to envisage a gradual generalization of the mechanisms of e-service where these rely on legal professionals. It is clear that the use of these systems seeks to reduce the workload of court officials and the postal services. It is however indispensable that other professionals (such as *huissiers* in France or *procuradores* in Spain) can become a part of the system, as the reference point for electronic notices to or from their client.

Ultimately, in certain geographical contexts, there are not that many obstacles to advancing towards the generalization of electronic communication systems in court proceedings.

8.4.5 The Possibility of Using Electronic Means to Serve the Writ Commencing the Proceedings

One of the most difficult questions –at the present moment, but also when facing a future mandatory use of electronic service– is the possibility of validly using electronic means of communication to serve on the defendant the writ or any equivalent document commencing the proceedings, which is nowadays only allowed in some countries if the defendant has previously accepted it [England, with *rules* 6.3 (1)(d) and 6.23 (6) CPR and *Practice Direction* 6A].

Indeed, the electronic service of documents within the framework of legal proceedings ought to be simple and relatively problem-free, once the parties to the proceedings have been identified before the court and have supplied to the court or to the other parties their corresponding electronic addresses. Where appropriate, this will involve having a user ID on the communications system that has been designed.

In the case of the party commencing proceedings, the system is quite simple: whilst it is true that litigants cannot be required to be in possession of the means necessary for the receipt of documents served electronically, this can be required of their lawyers or representatives, at least in countries with a medium economic and social development. Therefore, difficulties would only arise within a small segment, that of litigation where legal representation by a lawyer is not required.

In the case of defendants, the same is true, but only once the addressee of the claim has already been reached by the originating writ or similar document: even if he does not possess electronic means of communication, his lawyer or representative can be required to possess such means for the remainder of the proceedings.

It is clear that this allows proceedings to be managed in a much more streamlined way. But any absolute generalization on the use of the electronic service of documents will require that it should also be possible to use it to serve the originating writ on the defendant: it is of the utmost usefulness that it should be possible to serve the writ commencing the proceedings validly by electronic means. The use of the electronic service of documents in proceedings entails significant benefits, specially the saving of time; however, experience has shown that the most time is wasted in the service of the originating writ on the defendant, and so any savings will only be clearly observable and significant once it is possible to serve the defendant with the originating writ electronically.

There are still a lot of obstacles to overcome on this point: specifically, we must make the leap that will allow the creation of the duty – *rectius*, the burden – on all litigants to be in possession of an 'official' active electronic address in which they could be served of judicial documents, even if they wouldn't want to.

Currently all litigants (both private individuals and bodies corporate) are already under a duty to have an official domicile, which serves –amongst other things– as a channel for our relationship with public bodies. When it comes to Justice, and with regard to procedural purposes, this duty of being physically located is the basis of the legitimacy of systems where notices are –falsely– presumed served if published in an official gazette or done in some "special" ways, admitted by the procedural law of many countries in order to make possible procedures against defenders that can not be located. Therefore it is necessary to consider the possibility of taking one further step and establishing a duty on citizens to be in possession of an 'electronic domicile' to which official notices may be served – including those pertaining to court proceedings – provided that the conditions laid down by law are met: for example, there may be a requirement for an electronic acknowledgement of receipt or the like before it may be deemed that service has been successful.

I think that the introduction of a generic duty of being in possession of an electronic domicile and of being 'electronically located' for official purposes – just as, whether we like it or not, we are physically located in our homes – is currently justifiable at certain levels: I refer to companies and to private individuals who work as professionals. In the internet era, the public authorities may legitimately require all those who seek to pursue an economic activity in the marketplace to be in possession of an electronic address, just as they are required to keep their books and to have a physical domicile. It would be easy to extend this duty to the possibility of using this address to serve an initial electronic notice of the existence of the proceedings, which would allow a significant saving in time and energy.

Extending this duty to private individuals remains a utopian idea, especially in certain geographical areas. Looking to the future, governments ought to make an effort towards achieving the goal of all citizens being in possession of an accessible internet address where documents may be served electronically. In any event, and whilst the level of development that allows the introduction of a general duty to be in possession of an electronic address is reached, legal systems should allow originating writs to be served electronically in those cases where the e-mail address of the addressee is already known and it is possible to prove that it is current and in use (and this would be quite easy, e.g. where this address has been used by the party in previous notices served on the other party). However, the success of this first notice, at least for the moment, should maybe remain conditioned to acknowledgement of receipt from the addressee.

8.4.6 International Electronic Service of Documents? The Example of the European Procedures

The electronic service of documents is also extremely useful at an international level: borders are no barrier to the internet and to secure systems of electronic communication, and this is something that should be taken advantage of for the benefit of procedural efficiency.

In particular, and in relation to the questions examined in the foregoing paragraphs, one can envisage the use of electronic mechanisms for the international service of notices. Given that the internet knows no geographical boundaries, the use of e-mail could allow e-service of notices abroad directly, without the need to resort to the standard mechanisms of international judicial co-operation in this matter.

It is evident, however, that there is a problem: normally, international co-operation is necessary in order to serve the originating writ on the defendant, which is precisely the service that is most complex and for which it is most difficult to use electronic service mechanisms from the start (as has just been seen).

This is maybe the reason why electronic means of communication are absent from all the conventions and legal instruments regulating judicial international cooperation in the area of the service of documents. For obvious chronological reasons there are no references to e-service in The Hague Convention of 1 March 1954 on Civil Procedure, in The Hague Convention of 15 November 1965 on the Service Abroad of Judicial and Extrajudicial Documents in Civil or Commercial Matters, nor in the Inter-American Convention on Letters Rogatory of 30 January 1975.

But there are also no direct references to e-service in the most modern rule, Regulation (EC) No 1393/2007 of 13 November 2007 on the service in the Member States of judicial and extrajudicial documents in civil or commercial matters (service of documents), and repealing Council Regulation (EC) No 1348/2000. It is indeed notable that such a recent text should not contain an express reference to electronic service. The only possibility – although only in accordance with national legislation – is that once the documents that are to be served have been received by the authority responsible for carrying this out, this can then be done electronically: but in this case it would not be the international notification that was carried out electronically, but the internal notification.

For this reason, it might be reasonable to transpose the English model (direct service of the claim by e-mail in those cases where the addressee has consented to this prior to the litigation) to the field of international notifications – at least in the context of closely-related regions, such as Europe. In this regard, it could also make sense to use certain presumptions (similar to those of English Law), based on the premise that a private individual who or body corporate which has provided an e-mail address as a valid address for commercial relations is deemed to have given such consent. It is obvious that the practical impact of a provision such as the foregoing would be limited, but it is in any event reasonable that it should be expressly envisaged. In fact, if one takes a closer look, it is something that is perfectly possible nowadays in accordance with internal regulations which –as in England– allow for this, without any need for it to be envisaged in supra-national regulations: where a party with residence in Spain, but with business interests in England, consents in writing to electronic service, an originating writ may be served directly by electronic means, without any need to rely on the instruments envisaged in the European Regulation.

However, general regulations would be desirable in order to provide legal certainty to a sector where the differences between States can have repercussions for the recognition and enforcement of foreign judgements. In effect, within the scope of the Brussels Convention (article 27.2) and Regulation No. 44/2001 (article 34.2), the recognition and enforcement of judgements given in the absence of the accused can be rejected if it is held that the originating writ was not properly served and with sufficient time to allow the defendant to prepare his defence.[45] But it is the court of the enforcement State that must decide whether or not the notification issued in the State of origin complies with the requirements, irrespective of whether or not service was carried out in due manner pursuant to the laws of the State of origin.[46] Therefore, a situation could arise where an electronic notification could be validly carried out in England, but subsequently recognition could be denied in Spain because a Spanish judge could decide that the notification was invalid. Therefore there is a clear need to have common standards with regard to the manner in which the originating writ ought to be served. The setting of these common European standards[47] would also serve to establish a certain level of harmonization between national laws and could also be the opportunity to resolve problems regarding the language of the documents being served or filed and the right of the addressee to validly reject the notification.[48]

[45] See, generally, the Judgment of the European Court of Justice, Lancray v Peters und Sickert (C-305/88) [1990] ECR I-02725.

[46] See the Judgments of the European Court of Justice Klomps v Michel (166/80) [1981] ECR 01593 and Mærsk Olie & Gas 14 (C-39/02) [2004] ECR I-09657.

[47] And I shall recall that Article 81.2 f) of the Treaty on the Functioning of the European Union allows the European Parliament and the Council to "adopt measures, particularly when necessary for the proper functioning of the internal market, aimed at ensuring: (f) the elimination of obstacles to the proper functioning of civil proceedings, if necessary by promoting the compatibility of the rules on civil procedure applicable in the Member States".

[48] There are two important decisions of the European Court of Justice interpreting Regulation No 1348/2000: Leffler (C-443/03) [2005] ECR I-09611, regarding the possibility of the sender to remedy a faulty service –on the ground that it is not in an official language of the Member State addressed or in a language of the Member State of transmission which the addressee understands–; and Weiss und Partner (C-14/07) [2008] ECR I-03367, clarifying that it is not necessary to provide a translation of documentary evidence attached to the claim, when those documents have a purely evidential function and are not necessary for understanding the subject matter of the claim and the cause of action. Besides, and this time concerning the interpretation of Council Directive 76/308/EEC of 15 March 1976 on mutual assistance for the recovery of claims relating to certain levies, duties, taxes and other measures, reference must be done to the recent Judgment Kyrian (C-233/08) [2010], stating that the addressee of an instrument permitting enforcement must receive the notification of that instrument in an official language of the Member State in which the requested authority is situated.

Clear examples of what has been said are supplied by the European Regulations on civil procedure ruling the European Enforcement Order, the European Order for Payment and the European Small Claims Procedure.

8.4.6.1 Electronic Service as a Condition for the Expedition of a European Enforcement Order

According to Regulation No 805/2004 creating a European Enforcement Order for uncontested claims, certain judgments may be certified as European Enforcement Orders if they have been rendered in procedures where the debtor did not contest a money claim: the judicial decision is certified as a EEO in the State of origin and it will be directly enforceable in any other Member State, without the need for a declaration of enforceability in the Member State of enforcement prior to recognition and enforcement.

A judgment on a claim that is uncontested can be certified as a European Enforcement Order only if the court proceedings in the Member State of origin met certain procedural requirements and among them the method followed to serve on the debtor the document instituting the proceedings. Art. 13.1 d) of EEO-Regulation accepts service by electronic means such as e-mail, if it is attested by an acknowledgement of receipt including the date of receipt, which is signed and returned by the debtor. Art. 14.1 f) EEO-Regulation goes further and considers valid service by electronics means without proof of receipt by the debtor if it is attested by an automatic confirmation of delivery, "provided that the debtor has expressly accepted this method of service in advance".

Since there is no requirement that the debtor is domiciled in the Member State of origin, this could be a way to carry out an "undercover" international electronic service of documents. And the debtor will not be allowed to apply afterwards for the withdrawal of the European Enforcement Order certificate adducing that service was made on him by electronic means.

8.4.6.2 Electronic Service on the Debtor in the European Order for Payment Procedure and in the European Small Claims Procedure

It is also necessary to take account of what is established in the rules regarding the European Order for Payment procedure (Regulation No 1896/2006 of 12 December 2006 creating a European order for payment procedure) and the European Small Claims procedure (Regulation No 861/2007 of 11 July 2007 establishing a European Small Claims Procedure). Both procedures apply only to cross-border cases, which means that at least one of the parties is domiciled or habitually resident in a Member State other than the Member State of the court seised. Therefore there will be the need to carry out cross-border service of documents and the European legislator has been aware of the possibility of using electronic means as a way to avoid the need to use traditional systems of international service of documents.

(i) The European Order for Payment procedure is based on an application lodged by the creditor, upon which, if certain requirements are met, the court will issue a European order for payment addressed to the debtor. If the debtor does not pay nor lodges a statement of opposition within 30 days, the court will declare the European order for payment enforceable in all Member States. Two communications are at least necessary and one of them may have a cross-border nature:

- First there is the lodging of the creditor's application. This application may be submitted in paper form or "by any other means of communication, including electronic, accepted by the Member State of origin and available to the court of origin" (Art. 7.5 EOP-Regulation). There is hence a place for an e-filing of the request, which is possible, according to the European Judicial Atlas in Civil Matters, in several countries.[49]
- Once the European order for payment has been issued, it must be served on the debtor: this service originates the debtor's burden to pay or to contest the claim. Service on the defendant may be effected by electronic means such as e-mail, attested by an acknowledgement of receipt, including the date of receipt, which is signed and returned by the defendant [Art. 13 d) EOP-Regulation]. But it may also be served without proof of receipt by the defendant using electronic means attested by an automatic confirmation of delivery, provided that the defendant has expressly accepted this method of service in advance [Art. 14.1 f) EOP-Regulation]. It is important, therefore, to notice that a simple automatic confirmation of delivery could be enough to consider that the European order for payment has been correctly served and, eventually, to render it enforceable if the debtor doesn't pay nor contests the claim.

  The Regulation, however, includes the general requirement that service has to be made on the defendant in accordance with the national law of the State in which service is to be effected: national law, thus, should admit this kind of service and this will reduce the practical effectiveness of the rule. But it reveals clearly where the European legislator has placed the minimum standards for electronic service of documents.

(ii) The European Small Claims procedure is a written procedure allowing the recovery of sums less than EUR 2000 and based on the use of standard forms. The key for a proper functioning of the procedure is the lodging of a standard claim form

[49] It is clearly permitted in France, Slovenia, Cyprus, Estonia and Finland. In Slovakia and in the Czech Republic, if the application is filed without advanced electronic signature, the original copies of the forms must be sent no later than 3 days after submission. Austria permits using the WebERV –open to any person and company–, but doesn't admit e-mails. In the Netherlands the electronic submission of applications for a European Order for Payment is permitted, as long as this is provided for in the court's procedural rules, but currently none of the courts provides for this possibility. Germany is currently working on the development of an IT system which should make it possible to submit European payment order applications and objections electronically, but it doesn't seem to be available so far. The situation is similar in England and Wales, in Northern Ireland and in Scotland.

with the competent court and notifying the defendant, so he can submit a standard answer form. Both notifications may be of a cross-border nature and for both of them the use of electronic means of communication is permitted.

- To file his claim the claimant may use any means of communication acceptable to the Member State in which the action is taken (Art. 4.1 ESCP-Regulation) and this includes e-mail.[50]
- The ordinary way to serve the claim on the defendant shall be postal service attested by an acknowledgement of receipt including the date of receipt (Art. 13 ESCP-Regulation). If service by this means in not possible, then it may be effected by any of the methods provided for in Articles 13 or 14 of EEO-Regulation: e-mail will be therefore admissible, even without proof of receipt by the debtor, if there is an automatic confirmation of delivery, provided that the defendant has expressly accepted this method of service in advance (and without the condition, set by the EOP-Regulation, that this method of service is in accordance with the national law of the State in which service is to be effected).

(iii) The ruling of this question made by EOP and ESCP-Regulations has a very strong significance, when compared to what is established in the EEO-Regulation. In fact EEO-Regulation merely states the minimum standards that must be met in order to give immediate cross-border enforceability to certain judicial decisions. EOP and ESCP-Regulations go a step further and rule directly the methods considered acceptable to serve on the defendant the document instituting the proceedings: the European legislator considers that electronic service without proof of receipt by the defendant, merely attested by an automatic confirmation of delivery, is sufficient to create the burdens associated to the defendant's position, provided that he has expressly accepted this method of service in advance. Therefore, the harmonizing effect associated with European procedural rules may lead to the generalization of this criterion across the national legal systems.

Given this eventual approach, it would be important to have a clear definition of what is meant by 'having expressly accepted in advance' a notification method such as e-service. The Court of Justice of the European Union, in its judgment of 8 May 2008 delivered in the case *Weiss und Partner* (C-14/07), held that "the fact that the addressee of a document served has agreed in a contract concluded with the applicant in the course of his business that correspondence is to be conducted in the

[50] According to the European Judicial Atlas in Civil Matters, this kind of electronic filing is clearly permitted in the following countries: Portugal, France, Cyprus, Slovenia, Estonia and Finland. In Austria only WebERV can be used, but not e-mail. In the Netherlands the electronic submission is legally possible, but in practice it is not available. In Germany an electronic submission of the claim is only possible in the regions of Brandenburg, Bremen and Hesse. In Slovakia and in the Czech Republic, if the claim is filed without advanced electronic signature, the original copies of the form must be sent no later than 3 days after submission. England and Wales, Northern Ireland and Scotland are currently working on the development of an IT system which should make it possible to submit the claim electronically, but it doesn't seem to be available so far.

language of the Member State of transmission does not give rise to a presumption of knowledge of that language, but is evidence which the court may take into account in determining whether that addressee understands the language of the Member State of transmission". Likewise, it could be argued that agreeing that any communications linked to the business or contractual relations are to be conducted by e-mail to a specific e-mail address is sufficient evidence that the service of any claims linked to the same business relations may also be served to the same e-mail address by electronic means.

And it would also be necessary to bear in mind – and to take measures to prevent – the danger that in certain contractual relations, *a priori* acceptance of this kind of service might be imposed on the weaker party, despite not being familiar with them or not having the tools to access their electronic mailboxes frequently.

In addition, we can not forget the difficulties raised by the cross-border nature of the service (for example, those relating to languages). It is clear that these problems are not new and that the guidelines for resolving them are already contained in the normative instruments that govern traditional cross-border service of documents: therefore it would only be necessary to adapt these tools to the electronic nature of the service.

8.4.7 Electronic Communication and International Judicial Cooperation

In general, international cooperation systems, both civil and criminal, require communication between parties located in different States: the parties that need to communicate and to file requests and documents *lato sensu* might be judicial authorities or 'central authorities' in charge of overseeing cooperation, in accordance with the subject matter and the applicable normative instruments.

It is clear, in this context, that it is convenient to encourage the use of electronic means as tools for transmitting any pertinent requests for cooperation or actions between the parties participating in international judicial cooperation, at least for two reasons. Firstly, because cooperation would function more swiftly, especially if one takes into account the need to complete or correct documents of all kinds. And secondly, because this would require that any request for judicial cooperation, as well as any complementary documentation, should be in electronic format, which would in turn simplify the subsequent treatment and circulation of the request at an internal level, both in the plaintiff and in the defendant States. As we shall see below, this is one of the objectives of the European institutions in their plans for the promotion of e-Justice.[51]

[51] See the *Communication from the Commission to the Council, the European Parliament and the European Economic and Social Committee*, "Towards a European e-Justice Strategy", 30 May 2008, COM(2008) 329 final; the *Resolution of the European Parliament of 18 December 2008 with recommendations to the Commission on e-Justice* [2008/2125(INI)]; and the Council's *Multi-annual European e-Justice action plan 2009–2013* (OJ C 75, 31 March 2009).

8.4.7.1 Electronic Communication and International Judicial Cooperation in Civil Matters

In civil and commercial matters the use of electronic mail as a mean to activate the judicial international cooperation can find a place, at least within the European Union, for both the service of documents and the taking of evidence abroad.

As to service abroad, according to Art. 4 of the Regulation nº 1393/2007 on the service in the Member States of judicial and extrajudicial documents in civil or commercial matters (service of documents), the agencies involved in cooperation shall transmit any document “directly and as soon as possible”, and this “by any appropriate means, provided that the content of the document received is true and faithful to that of the document forwarded and that all information in it is easily legible”. The Member States have communicated to the European Commission which are the means of communication they accept to receive requests of cooperation. Some of them – not many, though – accept electronic mail for this purpose.[52]

Regarding the taking of evidence abroad Art. 6 of the Regulation (EC) nº. 1206/2001 of 28 May 2001 on cooperation between the courts of the Member States in the taking of evidence in civil or commercial matters, states that “requests and communications shall be transmitted by the swiftest possible means, which the requested Member State has indicated it can accept. The transmission may be carried out by any appropriate means, provided that the document received accurately reflects the content of the document forwarded and that all information in it is legible.” As a result of this provision, many Member States have communicated to the Commission their willingness to admit requests transmitted per e-mail.[53]

As a matter of fact, the European Action Plans on e-Justice include the proposal to introduce the use of electronic means of communication into the Regulations on taking of evidence and service of documents, taking advantage of the construction of secure networks for the transmission of information between courts.[54]

8.4.7.2 Electronic Communication and International Judicial Cooperation in Criminal Matters

Developments in matters of criminal judicial co-operation are event greater, at least within the scope of the European Union, thanks for the most part to the existence of the European Judicial Network in criminal matters.[55] The Decision that

[52] That’s what happens in Belgium, the Netherlands, Hungary, Denmark, Finland, Estonia, the Czech Republic and Cyprus; Latvia and Malta accept requests sent by e-mail if they are also sent by ordinary mail; Germany accepts e-mail only for informal communications.

[53] This happens in Portugal, France, Malta, Greece, Slovakia, the Czech Republic, Latvia, Estonia, Finland, Scotland and Ireland. There is no match with the countries mentioned in the precedent footnote!

[54] See the Resolution of the European Parliament (Recommendation 2, section 2).

[55] Council Decision 2008/976/JHA of 16 December 2008 on the European Judicial Network (OJ L 348, 24 December 2008).

sets it up envisages the creation of a secure telecommunications connection for the operational tasks of the contact points on the Network, the cost of which is to be funded by the general budget of the European Union. This secure telecommunications connection will allow the flow of data and of applications for judicial cooperation between Member States. Therefore, it constitutes a tool at the service of the transmission of letters rogatory and applications for cooperation or recognition as provided for in the various European texts or agreements and which are applicable to criminal judicial cooperation between Member States.

In order to articulate the supply of information and the use of the communications network, there exists an internet Portal for the European Judicial Network in criminal matters.[56] The features of the portal serve to facilitate international judicial cooperation in criminal matters, whether pursuant to the normative texts approved by the European Union, or also on the basis of other normative texts: now then, the principal participants in the cooperation – both plaintiffs and defendants – must always be Member States. Amongst other tools, the portal features a judicial Atlas which allows the judicial authority of the other State to which the application for cooperation ought to be addressed to be identified, in accordance with the type of activity in question and the normative framework being used.[57] In addition, there is also the *Compendium Wizard*, which can be used to create an on-line letter rogatory or equivalent application, and where appropriate, to dispatch it securely to the addressee.[58] Furthermore, and more specifically, an alternative Atlas and *Compendium Wizard* have been designed in order to assist in implementing the European provisions on the European arrest warrant. In any event, the European institutions continue to encourage the creation of a secure network for the exchange of information between judicial authorities, progressing on the basis of the various versions of the E-POC project developed by Eurojust.

Electronic service may indeed become the standard form for the service of documents between judicial bodies at all levels of international judicial cooperation. A significant part of the normative instruments that govern international judicial cooperation encourage direct service between courts, or service through central authorities acting swiftly. Electronic service may be a suitable tool for the transfer of those documents through which international cooperation operates, both in civil and in criminal matters. For this purpose, it is necessary to set up mechanisms that will allow documents to be served in secure conditions which are also required at an internal level; and it is also necessary to set up information systems, such as the European Judicial Atlas, which will allow the use thereof to be optimized by way of compiling a directory or catalogue with the electronic addresses of all bodies and authorities involved. The best examples of how it is possible to progress in this path are – there is no doubt about it – the works of European institutions.

---

[56] http://www.ejn-crimjust.europa.eu/

[57] http://www.ejn-crimjust.europa.eu/atlas_advanced.aspx

[58] http://www.ejn-crimjust.europa.eu/compendium.aspx

8.5 Final Remarks

When they function properly, electronic communication mechanisms allow the court and the parties to save time and money, and contribute to more agile and efficient proceedings. For this reason, it is convenient to encourage their implementation and to make their use widespread. Indeed, they have significant potential in cross-border situations and in streamlining international judicial cooperation: their use should also be encouraged in these sectors.

It may be seen, however, that legal development is variable in each State and that similar problems are sometimes met with different solutions. For this reason, it would be recommendable to attain a certain level of harmonization on this point, at least within integrated regional environments, such as the European Union. This harmonization, in my opinion, should consist in the setting of minimum standards at various levels: the technical requirements that ensure the reliability of the electronic communications system and the conditions under which e-service and e-filing may be carried out – in particular the requirements that must be met in order for them to be deemed to be validly carried out.

In the field of electronic notification there are certain 'red lines' that cannot be crossed or that can only be crossed if certain safeguards are in place and there are clear rules that ensure that the safeguards are upheld: electronic service of the originating writ on the defendant, especially in cross-border cases, is the clearest example.

In general, one must be aware of the dangers that the automation of human activities might pose for procedural rights, and we must ensure that the development of judicial activities in electronic form does not covertly or unwittingly end up infringing the rights of litigants. Therefore, the roll-out of e-Justice should not be an end in itself, but rather a means of allowing the Justice system to better perform its function of safeguarding the legal system and protecting rights.[59]

For this reason, I would like to emphasize the idea, mentioned at the start of this work, that e-Justice is made up of *tools* at the service of Justice, amongst which are included electronic communications: but the tools should never dictate the development of the purpose they serve. The process of computerizing all kinds of activities, however, has made us accustomed to just the opposite: it is commonplace to come across obstacles to carrying out certain activities where the computer system controlling them does not envisage them. It is evident that this should not occur in the field of Justice: it would be intolerable that the legitimate exercise of rights, or that their proper legal protection, should be precluded or hindered by the deficiencies of the software tools available to the courts. Therefore it would be sensible for the phenomenon of the implementation of e-Justice to be accompanied by a general

[59] In that sense, see the *Opinion of the European Economic and Social Committee on the Communication from the Commission to the Council, the European Parliament and the European Economic and Social Committee: Towards a European e-Justice Strategy*, 30 September 2009 (Doc. INT/457, 1455/2009, section 4.3).

safeguard or get-out clause: litigants shall always have the right for judicial bodies to carry out all activities necessary for the protection of their rights by 'traditional' methods, where there should be a failure, deficiency, or design defect in a computerized system for court procedure: this maxim should be upheld at all levels, including e-service and e-filing.

Recently, the European Court of Human Rights has recognised that the decision of one national court not to allow a litigant to use the electronic systems envisaged by law – specifically, the electronic filing of originating writs – was contrary to the right to have access to the courts enshrined at article 6.1 ECHR.[60] In the same manner, a refusal by a court to allow the use of non-electronic mechanisms in the event of the failure or malfunction of the electronic systems, even if they have been defined by law as being compulsory for the performance of a particular step in proceedings, could also amount to a breach of the right to a fair trial.

---

[60]Lawyer Partners A.S. v. Slovakia (16 June 2009).

Chapter 9
Enforcement of a Claim with the Support of the New Information Technology. Protection of the Creditor and the Debtor*

Michele Angelo Lupoi

9.1 Introduction

This paper's topic is the enforcement of a claim with the support of the new information technology. In other words, my task is to try and describe (or maybe even imagine) ways in which the new information technology may be helpful in relation to the enforcement of a claim.

A lot has been written and actually practised in relation to the application of IT to ordinary civil proceedings but the area of enforcement of claims appears to be

*Finding materials for this paper has not been easy. Warm thanks are therefore due to all the colleagues and friends who were so kind as to reply to my queries, send feedback, give advice, present papers and ideas Credit and thanks are due, in alphabetical order, to N. Andrews (United Kingdom), for materials and encouragement; L. Cadiet (France) for sending me his paper "Le procès civil à l'epreuve des nouvelles technologies"; P. H. Dos Santos Lucon and G. Recena Costa (Brazil), for sending me their most complete report on the situation in Brazil, from which I widely "stole" in the preparation of this paper; A. Gidi (Brazil-Usa), for reviewing the original draft of this paper and friendly questioning my questions; P. Gilles (Germany) for sending me materials and giving moral support; G. C. Hazard (Usa), for hints on the American situation; A. W. Jongbloed (the Netherlands), for kindly providing me with a copy of the article he wrote with D. Struiksma, "Bailiffs on the e-Highway"; S. Mironova (Russia), for insights on the situation in the Russian Federation and reference to websites and online discussion groups; E. Oteiza (Argentina), for sending me a copy of the paper "Diagnostico sobre el proceso argentino de ejecucion de sentencias, en el marco de una nueva Mirada sobre el anteproyecto de codigo procesal civil modelo para iberocamérica", which he wrote with J. M. Salgado; A. Pellegrini Grinover (Brazil) for invaluable support; D. Reiling (the Netherlands) for precious hints and suggestions; J.J. Stettin (Usa), for feedback and remarks; M. Stuerner (Germany) for an insight on the most recent German reforms; S. Subrin (Usa) for reading my draft paper and sending enlightening feedback: R. van den Hoogen, for introducing me to Dr. Reiling's dissertation; V. Vebraite (Lithuania) for hints on the Estonian situation; F. Verbic (Argentina), for feedback and support. Of course, all the mistakes are mine. Thanks to Camilla Barker for her final linguistic revision.

M.A. Lupoi (✉)
Faculty of Law, University of Bologna, via Cesare Battisti 33, 40123 Bologna, Italy
e-mail: miclupoi@tin.it

M. Kengyel and Z. Nemessányi (eds.), *Electronic Technology and Civil Procedure*, Ius Gentium: Comparative Perspectives on Law and Justice 15,
DOI 10.1007/978-94-007-4072-3_9,

somehow less mapped out. In this field, therefore Prof. Cadiet's remark that "Les nouvelles technologies (...) ne sont pas si simples à identifier: c'est une nébuleuse aux contours incertains"[1] rings even truer.

This paper is not a "general report" on the assigned topic in the traditional sense of the expression. No regional reports were requested nor organized and most of the materials I collected were found with personal late-night "leg-work", or, to put it better, websurfing-work. The picture which has been drawn here is therefore incomplete, but still, hopefully, a good starting point for further research and discussion.

My report is mainly focused on the enforcement of pecuniary claims, the most common form of enforcement.

Proceedings for enforcing civil claims after judgment (or sometimes in the absence of a previous judicial decision) play a fundamental role in any system of civil justice. In many legal orders, however, they are often considered to be exceedingly lengthy, costly and leading to unsatisfactory results. In other words, enforcement proceedings are often ineffective as concerns the satisfaction of a creditor's claims: most often, this is the area where judicial implementation of rights is at its most ineffective and inefficient.

As Neil Andrews clearly stated, with reference to the English situation: "(...) money judgments, or other judgments concerning goods or land, should be speedily, economically, and effectively carried into effect so that the victorious claimant's quest for justice is not denied 'at the last hurdle'".[2]

On the other hand, one of the risks involved in enforcement proceedings is that the debtor's rights might be violated, so that the patrimonial effects he or she suffers are heavier than the amount of any original debt he or she was under the obligation to pay. An obvious example is when the debtor's assets are sold at a fraction of their market value.

The starting question, therefore, is whether IT may improve upon this quite dramatic situation. There is a wide spread notion that it indeed may. Action plans are being drawn worldwide and pilot projects are being experimented in several countries. In some jurisdictions, the digitization of enforcement proceedings is indeed already a successful reality.

As a matter of fact, the first pilot projects and implementation programs worldwide have mostly dealt with other areas of civil justice, namely service of documents, civil proceedings at large and the creation of e-justice networks. Only recently have enforcement proceedings come to the fore. Therefore, they may take advantage and exploit the positive results achieved in other areas of civil litigation, possibly posing fewer problems of adaptation. Enforcement presents, indeed, a more antiseptic environment than cognizance proceedings, where issues of fact have to be tried and evidence gathered and stronger hostility and perplexities against the idea of online-trial have been voiced. Enforcement of monetary claims requires taking control or seizing the debtor's assets and then sell them and distribute the sum made from that sale among all the participating creditors. The enforcement

[1] L. Cadiet. Le procès civil à l'épreuve des nouvelles technologies.

[2] N. Andrews, Enforcement of civil judgments in England.

of other claims requires even less articulated activities. The procedural position of the parties are more clear cut (sometimes inevitably so) and the court mostly plays a supervising role of activities performed by other enforcement agents. In some legal order, as a general rule, courts are not even involved in enforcement of claims and other (private or public) agents perform all of the necessary actions.

Enforcement may therefore be defined as an ideal training ground for the implementation of IT in civil litigation.

9.2 Some IT Background

This paper deals with IT from a layman perspective. I will not even try to pretend that I am able to understand, not to say explain, how IT systems applied to civil justice operate. A brief look at the technical side of the problem, however, needs to be taken. At this stage, two main models of network information technologies may play a role in the context examined in this paper[3]:

(*a*) On the one hand, one should mention network IT: according to Dr. Reiling, a Dutch judge, "the network group includes IT that provides a means by which people can communicate with one another".[4]

In our context, such network may be used, for example, to deliver documents and claims in electronic format to the court or to the enforcement agent which then stores then in an electronic file which may be accessed by any person involved in the procedure. Moreover, information technology might be used for any communication between the parties, the judge, the court appointed expert, the bailiff and\or any enforcing agency involved.

Allegedly, this network represents an update of the traditional "paper system", where e-files take the place of material "paper" files. Most of the actions of the procedure (e.g. the attachment and selling of the debtor's assets), however, continue to be performed in the traditional way, through material activities of the parties and of the enforcement agents involved.

In this context, the procedure is only partially digitized. Such network technology therefore requires a relatively low level of adaptation of the existing procedural rules. The main challenge in this context is the creation of a safe electronic two-way external communication system between the parties and the court. Many successful programs however have already achieved such results and this may be considered an established result. As Dr. Reiling remarks: "It looks as if electronic files have become a regular feature in a majority of courts", while at the same time noting that: "so far, most courts have largely used IT to assist their paper based processes"[5];

(*b*) Enterprise IT represents a development on network IT: it has been described as "work flow management systems, customers relations management systems and

[3] For further discussion on IT functionalities, see D. Reiling, *Technology for Justice. How Information Technology can Support Judicial Reform* (Amsterdam: Leiden University Press, 2009), 45.

[4] Reiling, op. cit., 45.

[5] Reiling, op. cit., 55 and 59.

electronic interaction with customers. The ideal type for this group is an entirely electronic management process (…): cases are filed electronically, they are managed as electronic files by electronic work flow systems, both individually and according to load, outputs are filed in an electronic archive".[6]

In these interactive systems, "costumers" and "agents" communicate and exchange information, not necessarily to perform strictly procedural functions and activities.

Automation has been defined as "creating a process that can be handled by a machine without human intervention (…). This means the process runs electronically, this does not mean that the entire process need to be automated. It is also possible to automate parts of a process (…) Automation means translating policies and routines into programs for electronic processing. (…) The more the input into the court can be done as structured data, the less human activity is needed for the product".[7]

In other words, with this technology, mostly or wholly automated enforcement proceedings may be conceived of, where all or most of the procedural activities are performed online, in a secured environment, with minimum paperwork or material performances. Here, not only an interactive network must be established and made operative but also the existing rules might require more or less extended adaptations and modifications.

As a matter of fact, since different types of enforcement proceedings exist, different levels of automation may be realized. Proceedings where the enforcement agent is required to physically accede the debtor's premises to seize goods or implement an eviction order by necessity may be only partially automated (e. g., as concerns the selling of the assets or the distribution of the money made from such sales).

Other forms of enforcement, on the other hand, lend themselves to almost complete automation, such as in the case of enforcement over (at least certain types of) credits or real property.

Arguably, the challenges for a network dedicated to enforcement proceedings are the same which over time were faced and dealt with in other areas of civil litigation. In other words, experience shows that such networks may indeed be created and grant the necessary standards of privacy and safety.

Since the implementation of IT in enforcement proceedings should not be an end in itself but a means to achieve specific goals, one should consider which benefits the application of IT may bring to enforcement activities.

In her PhD dissertation,[8] Dr. Reiling argues that the beneficial impact of IT should be tested in relation to the time necessary for the solution of judicial cases, access to justice and judicial corruption. While Dr. Reiling's dissertation does not tackle the issue of enforcement of claims in itself, such reference points appear to be relevant also in the context examined here. Moreover, it is worth considering also the possible positive impact of IT on the general efficiency, effectiveness and transparency of enforcement proceedings.

In the next paragraphs I will examine each of these points.

---

[6] Reiling, op. cit., 45 and 57.

[7] Reiling, op. cit., 126 ff.

[8] Reiling, op. cit.

9.3 IT and Access to Enforcement Proceedings

As concerns us here, access to justice primarily means enabling the creditor to gather the information required to put her in the position to start enforcement proceedings against the debtor.

Enforcement of claims usually is preceded by ordinary proceedings where the claimant's right is adjudicated upon. In most civil proceedings, a lawyer needs to be involved. This lawyer will also proceed with any necessary enforcement phase, thus reducing the need for the individual creditor to seek information about her enforcing prerogatives. Nonetheless such information may indeed be required, for example for occasional claimants, like maintenance creditors or even small sized commercial creditors.

Undoubtedly, IT may be used to provide easily accessible information for creditors. Many courts or governmental agencies worldwide have indeed established website for such purposes. Today, courts' websites provide different kind of services, from online information, to forms downloading, forms processing and case handling.[9] Also private enforcement agents and even lawyers have set up websites, providing a first introduction to the "secrets" of the enforcement of judgments and possibly make business with perspective claimants.

When claims have to be recovered across borders, even lawyers may put to good use specific public or private websites providing information for international claimants.

Just to give a few examples, I have copied some web pages which I have "stumbled upon" during my research for this paper.

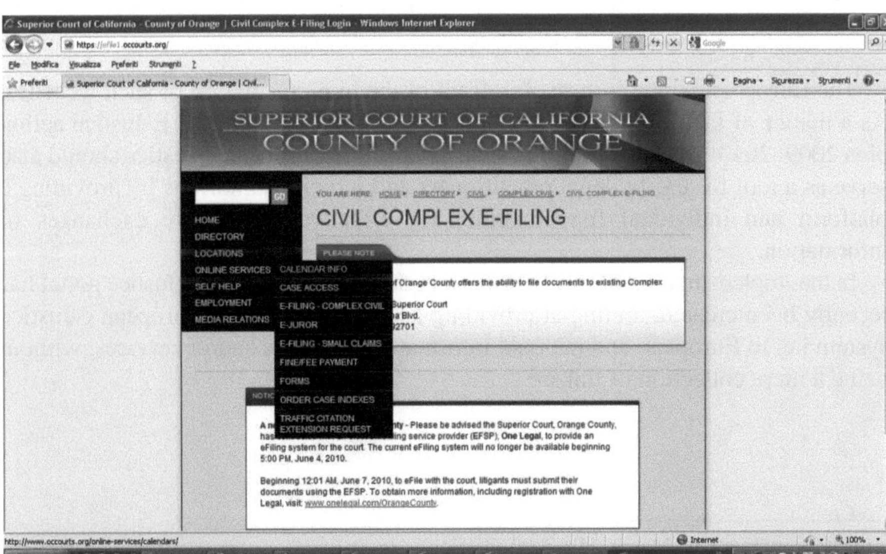

[9] More details in Reiling, op. cit., 58.

The above picture is taken from the website of the Superior Court of California, County of Orange, showing some of the services provided therein.

Certainly very impressive is the website created by the Justice and Attorney General of the Canadian province of Alberta in relation to the local Maintenance Enforcement Program, where an alimony creditor may find all the necessary information to enforce his or her claim against reluctant debtors:

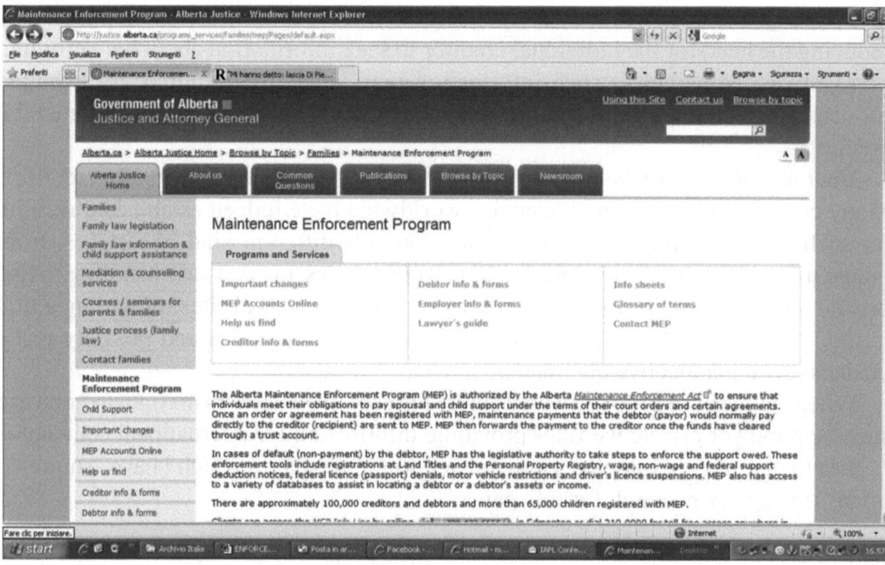

The European Union is at the forefront of the implementation of such websites. As a matter of fact, the European Council is actively pursuing its E-Justice action plan 2009–2013[10] which, *inter alia*, states that (# 22) European e-Justice should also serve as a tool for use by legal practitioners and judicial authorities by providing a platform and individual functionalities for effective and secure exchanges of information.

In the implementation of this E-Justice action plan, a European e-Justice portal has recently been created, aiming at providing access to the whole European e-Justice system i.e. to European and national information websites and/or services, without being a mere collection of links.[11]

[10] In Off. jour. Eur. Un., March, 31st 2009, C 75\1.

[11] European Union Council E-Justice action plan 2009–2013, § 31.

Below is a picture of the homepage of the European e-Justice Portal, referring to all of the information provided therein:

This is the page dedicated to the recovery of monetary claims in the cross-border European context:

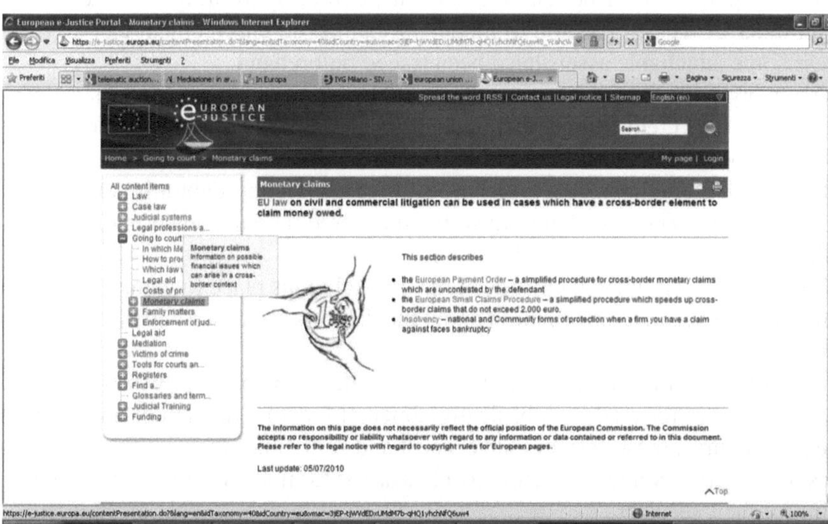

Finally, this a picture of the page of the portal dedicated to the enforcement of judgments:

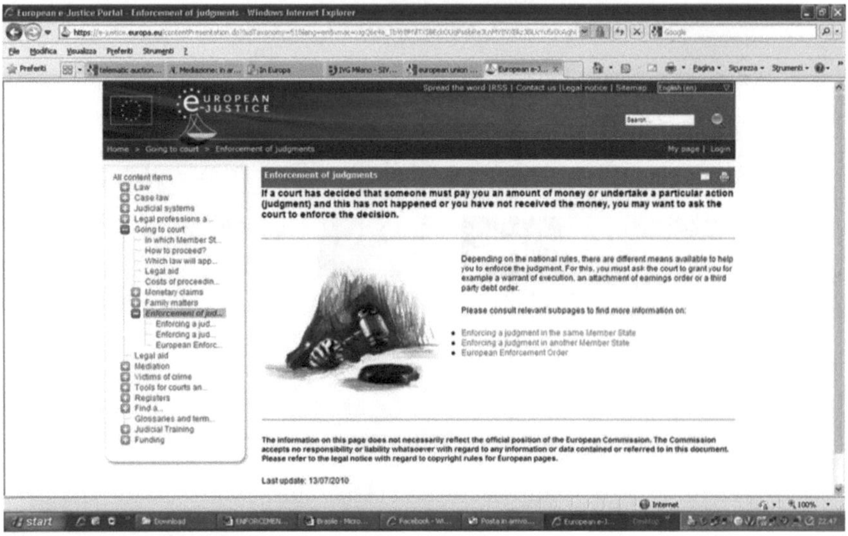

9.4 Locating the Debtor's Assets Through the Use of IT

Access to enforcement proceedings, moreover, implies enabling the creditor to effectively pursue his or her claim against the debtor by locating the latter's assets. As a matter of fact, in relation to monetary claims, no enforcement can take place unless some attachable assets have been traced beforehand.

Arguably, for a creditor, getting a favourable judgment is often the easiest part of the process. The creditor may then lack efficient devices or not have enough information to locate the debtor's assets so that his enforcement attempts are deemed to be frustrated. Sometimes, it is even hard to discover where a debtor is actually living (or hiding…). When a claim must be recovered abroad, even more practical difficulties may have to be faced.

Today, the worldwide net is indeed the most obvious place to gather information from. Finding accurate information about the debtor's residence and his assets is not however an easy task.

As a quick online search would show, private agencies exist worldwide which provide services for creditors, basically aimed at locating the debtor and his or her assets. Of course, these services have a cost and sometimes the necessary investigations are lengthy. Some enforcing agents ask for as much as 50% of the creditor's claim for their services.

Creditors (or their lawyers) may find it more convenient to pursue their own private investigations, through the available channels. IT has indeed made such investigations more accessible to a larger number of creditors. As a matter of fact, in a growing number of legal orders, many public registers (like land, automobile or company registers) are now easily accessible online.

Recourse to these registers by lawyers and even individual creditor may indeed reduce time and costs for the search of the debtor's assets.

Just to give a few examples, here is the picture of the website of the German Commercial register (*Handelsregister*)[12]:

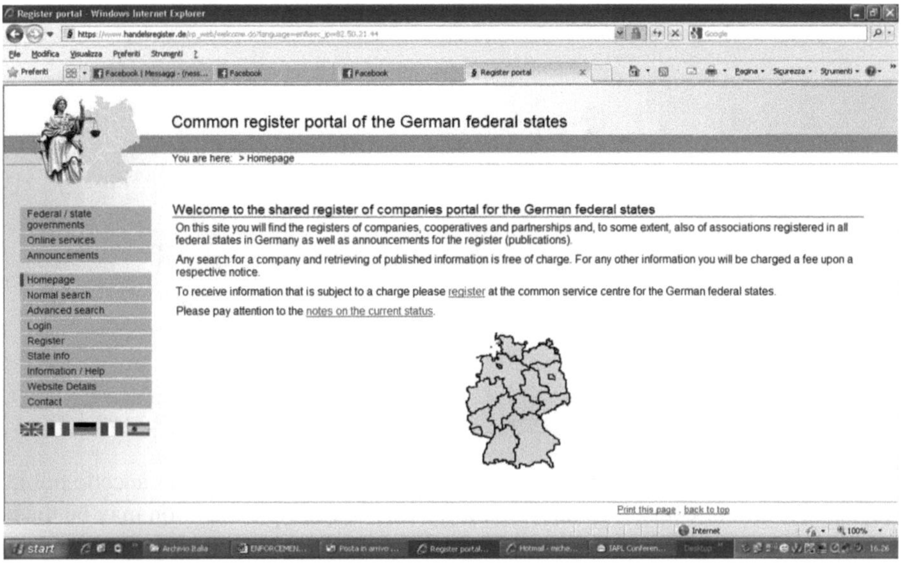

And here is the website of the German land register (*Grundbuch*):

In some legal orders, it is possible to access several different types of public register from the same website. As an example, I have chosen a picture of the website

[12] Thank you Michael Stuerner.

created by the Estonian Centre of Registers and Information Systems (RIK),[13] providings access to the commercial register, land register and e-Notary:

At the European level, the creation of comprehensive websites to accede public registers is still not very well developed. By way of example, mention may be made of the European Business Register (EBR), providing commercial information from all of the member States:

[13] RIK is a state agency working under the Ministry of Justice. It develops and administers about 50 registers and information services for law and criminal jurisdiction such as court information system, e-File, e-Law etc. There is also e-Business Register, e-Land Register and e-Notary.

In following picture the page listing the member countries of the EBR is shown:

Worth mentioning is also the European Land Information Service (EULIS) website, providing access to official land registers in a limited number of European countries:

E-Business Register makes it possible to register a new company over the Internet and also make detailed inquiries about other companies, look at annual reports and much more. E-Land Register is a register of ownership which has made electronic conveyance much easier and all the data is available over the internet. E-Notary is the environment that helps notaries in their work and organizes electronic communication between notary and state. Thank you Vigita Vebraite.

The E-Justice action plan 2009–2013 actually mentions, among the actions to be taken in the next few years by the European institutions, the interconnection of insolvency registers, land registers (integration of EULIS) and of commercial registers (integration of EBR).

Arguably, the creditor's search for the debtor and\or his assets would be more complete if the former would have the chance to have (electronic) access to registers or databanks which are not open to the public, such as tax or social security registers or to perform "aggressive" online investigations to track down the elusive debtor or the hidden bank account.

Such a course of action is indeed frowned upon in most jurisdictions, in the light of the need to preserve the debtor's privacy and safeguard personal data, as Prof. Hess' study on the transparency of the debtor's assets clearly shows.

The general perception is that investigations in these "confidential" databanks should be done either directly by the court or by the public enforcement agent or under strict public surveillance. Sanctions are provided for the unauthorized use of information gathered from the debtor or third parties.[14]

That this is arguably the most sensitive problem in relation to the use of IT in the context of enforcement proceedings is clearly shown by the reactions to the European Council proposal for a regulation on the transparency of the debtor assets.

As it is well known, in March 2008 a Green Paper on the "Effective enforcement of judgments in the European Union: the transparency of debtors' assets" was published, containing several proposals for discussion concerning common European devices to obtain information about a debtor's assets. As concerns us here, increasing the information available and improving access to registers and the exchange of information between enforcement authorities have been proposed. In particular, the Green Book remarks that work should be carried on in the area of e-Justice with a view to creating at the European level a technical platform giving access, in the sphere of justice, to existing or future electronic systems. The priorities for future work include creating the "conditions for networking of insolvency registers and of commercial and business registers". The Green Book notes that, at present, enforcement bodies are not able to access directly the (non-public) registers of other Member States which are open to the enforcement bodies of that State: if

[14] See e.g. Sec. 15B ("Offence of unauthorized use or disclosure") of the English Act of 2007:

(1) This section applies if the Commissioners make a disclosure of information ("debtor information") under section 15A(5).
(2) A person to whom the debtor information is disclosed commits an offence if—
    (a) he uses or discloses the debtor information, and
    (b) the use or disclosure is not authorised by subsection (3), (5), (6) or (7).
(3) The use or disclosure of the debtor information is authorised if it is—
    (a) for a purpose connected with the enforcement of the lapsed order (including the direction of the order to the debtor's current employer), and
    (b) with the consent of the Commissioners.

access to these registers by enforcement bodies exists at all, it is strictly limited to national enforcement bodies. There are currently no international instruments dealing with the exchange of information between national enforcement bodies. Consequently – in the absence of Europe-wide registers – one option could be to enhance cooperation between national enforcement authorities and introduce direct exchange of information between them. Enforcement bodies in one Member State would be able to request the assistance of the competent bodies in other Member States. In this regard, the existing Community instrument on the mutual assistance of tax authorities could serve as a model.

In the words of the Green Paper, the extent to which the Internal Market Information system (IMI) could be used for the exchange of information between national enforcement authorities could also be examined. IMI has been developed as an electronic tool for information exchange between Member States administrations working in all official languages, and it can potentially support any piece of Community legislation. Following this approach, a future Community instrument could provide a list of national enforcement authorities entitled to request information from registers in another Member State and could set time limits within which a request for information should be implemented. There could be standardized question and answer forms in all Community languages, and data could be exchanged electronically as far as possible.

The reactions to the Green Paper's proposals, however, have been lukewarm, if not openly negative. As the summary of the replies to the Green Paper makes clear, at the end of the consultations on the proposals contained therein, only very few respondents agreed with the idea that non-public organisations should participate in and have access to the sources of information needed for enforcing judgements. The need for non-public organisations to have a statutory framework and a legally defined status was voiced and concern was expressed regarding the compatibility of such measures with Directive 95/46/EC of the European Parliament and the Council of 24 October 1995 on the protection of individuals with regard to the processing of personal data and on the free movement of such data. The need to avoid misuse of incoming data through sanctions was also aired, noting the importance of the right balance between the interests of debtor and creditor. A large majority of replies expressed concerns regarding data protection and the protection of the debtors' privacy and agreed that public access should not be increased. A widely shared opinion was that the right to privacy, including data protection, should not be overridden by the necessities of law enforcement and must be upheld.

The European economic and social Committee, at the plenary session of 3\4 December 2008, voiced worries in relation to proportionality and the safeguard of fundamental rights and expressed open critical remarks on some of the proposals of the Green Book. In particular, as concerns us here, the Committee strongly opposed a general and indiscriminate access to tax and social security registers. It also expressed the opinion that no new register is required, existing ones being sufficient. The Committee showed appreciation for a more traditional "cooperative" approach, improving the exchange of information between national authorities. In the Committee's opinion, easier and faster access to existing data should be enough to grant equal opportunities and treatment in the locating of the debtor's assets. National enforcement authorities should be able to have easier access to existing population registers in order to obtain information about the debtor's place of residence, in particular through electronic means. Moreover, only judicial authorities should have access to tax\social security registers, ensuring the protection of the data contained therein. In cross-border enforcement, such access will have to require the cooperation of the local judicial authority. A system of direct information exchange between national enforcement authorities should be created, through electronic means in order to identify and locate the debtors and determine their patrimony, following the model of regulation 1206\2001. The use of electronic devices or the creation of an intranet system connecting the national authorities should be considered fundamental. The Committee also offered some proposals for discussion: *inter alia*, as concerns us here, the creation of a European automobile register; online registers of all pending enforcement proceedings in every Member State; access to land or real property registers.

Also the European Parliament, in its resolution of 22 April 2009, appears to prefer a not too aggressive approach, keeping electronic access to "sensitive" personal data under control of the judicial authorities of the Member States. In particular, the Parliament expressed the view that the need to protect sensitive data should be balanced with proportionate measures that provide the requisite transparency and significantly reduce processing and management costs. Besides publicly available information, the creditor should have access to the data required – subject to supervision by, or with the assistance of a competent authority – in order to initiate the enforcement procedure and recover the debt by procedures readily applicable throughout the internal market. The European Parliament "strongly believes that the publication of national directories of foreign lawyers exercising their internal market rights under Directives 77\249\EEC and 98\5\EC would be useful (..); such national directories could be linked to a Commission website and could be complementary to a proposed manual of national enforcement law and practices. Serious worries are however expressed about the misuse of confidential data if indiscriminately accessed to. The idea of improved cooperation between public enforcement bodies may be worth exploring further, even though such bodies do non exist in all Member States".

In light of these reactions to the Green Book, it appears unlikely that, in the near future, creditors will be granted an unsupervised electronic access to databanks and public records not generally open to the public. A more traditional network of cooperating between national judicial authorities will most likely be established. In order for such network to really be effective, its operations should be as inexpensive and as fast as possible. IT communication and electronic exchange of information will surely be the key for the eventual success of this network.

In order to solve the problem of finding the debtor's assets, one possible solution could be to enlarge the investigating powers of official enforcing authorities, especially as concerns online research.

A good example may be found in the Netherlands. Dutch bailiffs are crown appointed but they are not in the government service and they run their own office. They cooperate with the courts in a system where they can supply information electronically. A system exists which enables bailiffs to obtain information from a central agency: e. g., they can thus verify addresses. Dutch Bailiffs are also authorised to obtain info on debtors from third parties: no longer necessarily in writing but now also electronically: a "sofi" number must be used in communications with a certain number of bodies which are themselves authorised to use "sofi" numbers. According to a recent study, 300,000 applications were made in 2004 alone. Apparently, this is a cheap, fast and secure (only bailiffs can apply for data) method for obtaining sensitive and confidential information. Most notably, data may be obtained without human intervention. Unlawful use of data is prevented: in case of misuse, a bailiff may be excluded from this information network. Such power to obtain information has been extended to all the bodies and agencies responsible for paying out salaries or benefits in the Netherlands, enabling bailiffs to gather information on a debtor's source of income from a central database, ensuring more efficiency, less costs and less time.[15]

Information available online could also stimulate a recalcitrant debtor to voluntarily pay his creditors. As a matter of fact, websites may also be used as tools for indirect coercion. As an effective example, mention can be made of the site of the Attorney General of Alberta (Canada), which has a part dedicated to the enforcement of maintenance claims. Clicking on the "Help us find" section, a visitor is connected to a page where the names and photographs of defaulting debtors who owe child or spousal support are exposed, on a region by region basis within the province:

[15] D. Struiksma and A. W. Jongbloed, "Bailiffs on the e-Highway," in *Information & Communications Technology Law* 15, no. 2 (2006): 201–6.

Such websites may indeed prove very useful in locating debtors and\or their assets and possibly in inducing them to voluntarily perform the obligations. However, they appear to be more familiar to common law jurisdictions, as the reactions to the EU Green Paper clearly show.

9.5 e-Filing and e-Processing of Cases

IT may improve access to enforcement proceedings also on a more procedural level. As a matter of fact, in various legal orders, a network has been created enabling the creditor to file her claim in an electronic form, for example as a pdf file attached to an e-mail sent directly from the lawyer's office to the court or to the enforcing agent.

Also other actions in the proceedings may then be performed online, with no need to produce "paper" documents or briefs.

Such an electronic filing of attachment procedures is now possible, for instance, in some Italian courts, in relation to real property (*pignoramento immobiliare*).[16] In Italy, however, sending the electronic request to attach the debtor's real property to the court's offices does not (yet) completely substitute the required "paperwork", since the creditor is also required to deliver a "traditional" written request to the bailiff. At the moment, then, the only real advantage of this "*pignoramento elettronico*" is that, when the bailiff brings the relevant documents to the court, after service has been performed on the debtor, all of the case data will already be present in the court's file, thus reducing the processing time.

Also in Germany, under para. 130(a) *ZPO* it is possible to institute (enforcement) proceedings by means of an electronic document. However, not all the courts do have the necessary equipment to accept electronic documents and the whole thing is in a test phase. Electronic communication with the court is only possible providing parties use the so-called electronic post box. A web page[17] shows the standards as well as participating courts and administrative bodies.[18]

Also the American Federal Bankruptcy code enables electronic filing of claims and notice through electronic transmission.[19]

---

[16] See Salemme, "L'impiego dell'informatica nelle esecuzioni immobiliari," in *Rivista trimestrale di diritto e procedura civile* (2010): 1333.

[17] http://www.egvp.de/.

[18] Thank you M. Stuerner. See http://www.justiz.de/elektronischer_rechtsverkehr/index.php for details.

[19] See e.g.:

Rule 5005 (Filing and Transmittal of Papers):

(a) Filing. (…)

(2) Filing by electronic means.

A court may by local rule permit or require documents to be filed, signed, or verified by electronic means that are consistent with technical standards, if any, that the Judicial Conference of the United States establishes. A local rule may require filing by electronic means only if reasonable exceptions are allowed. A document filed by electronic means in compliance with a local rule constitutes a written paper for the purpose of applying these rules, the Federal Rules of Civil Procedure made applicable by these rules, and Â§ 107 of the Code.

At the European level, the implementation of a European network between National courts is actively pursued in the E-Justice action plan 2009–2013. According to the European Council, in Europe, "the highest possible degree of compatibility between the various technical and organisational measures selected for the judicial system applications must be ensured (...). It is necessary to reach agreement on standardised communications formats and protocols in line with European or international standards, allowing for interoperative, effective, secure and rapid exchange at the lowest possible cost".[20]

9.6 Making the Definition of Enforcement Proceedings Faster

Allegedly, there is no real evidence that the injection of IT in civil proceedings will shorten their length. According to a source: "experience-based research has not identified the implementation of IT as a very successful factor in reducing case

---

Rule 8008 (Filing and Service):

(a) Filing.

Papers required or permitted to be filed with the clerk of the district court or the clerk of the bankruptcy appellate panel may be filed by mail addressed to the clerk, but filing is not timely unless the papers are received by the clerk within the time fixed for filing, except that briefs are deemed filed on the day of mailing. An original and one copy of all papers shall be filed when an appeal is to the district court; an original and three copies shall be filed when an appeal is to a bankruptcy appellate panel. The district court or bankruptcy appellate panel may require that additional copies be furnished. Rule 5005(a)(2) applies to papers filed with the clerk of the district court or the clerk of the bankruptcy appellate panel if filing by electronic means is authorized by local rule promulgated pursuant to Rule 8018.

Rule 9036 (Notice by electronic transmission):

Whenever the clerk or some other person as directed by the court is required to send notice by mail and the entity entitled to receive the notice requests in writing that, instead of notice by mail, all or part of the information required to be contained in the notice be sent by a specified type of electronic transmission, the court may direct the clerk or other person to send the information by such electronic transmission. Notice by electronic means is complete on transmission.

Rule 9037 (Privacy Protection For Filings Made with the Court):

(a) Redacted filings.

Unless the court orders otherwise, in an electronic or paper filing made with the court that contains an individual's social-security number, taxpayer-identification number, or birth date, the name of an individual, other than the debtor, known to be and identified as a minor, or a financial-account number, a party or nonparty making the filing may include only:

1. the last four digits of the social-security number and taxpayer-identification number;
2. the year of the individual's birth;
3. the minor's initials; and
4. the last four digits of the financial-account number.

[20] European Union Council E-Justice action plan 2009–2013, § 37.

disposition times".[21] Also Peter Gilles remarked that there is no reason to believe that the use of IT will make proceedings as a whole "faster".[22]

When network IT is used, the possibility of creating e-files does not, *per se*, reduce the length of the proceedings as a whole. However, when the filing parts of the proceedings are digitized, the processing times of the file are greatly reduced, the court and enforcement offices need to perform less "manual" activities (like inserting the parties and claims details into the system) and therefore the cases may "move" along more swiftly.[23]

As concerns the general public (and lawyers in particular), the possibility to interact with the court and\or the enforcement agents through the exchange of e-information eliminates the need to travel to the courts or to the agents themselves, making distances irrelevant and reducing the costs of the procedure for the creditor.[24]

Arguably, less paperwork will also reduce the general cost of the proceedings.

Moreover, as Prof. Cadiet remarked: "Grâce aux marques de révision et aux liens hypertexte, les nouvelles technologies permettent un accès instantané et dynamique à l'ensemble du dossier, l'archivage personnel des données s'en trouve simplifié, ainsi que la circulation des actes et des pièces du dossier, qui s'affranchit de difficultés de l'écrit sur support papier comme celles de la transmission en plusieurs exemplaires ou celle de la restitution des actes et pièces remis ou notifiés".[25] Undoubtedly, the formal simplification of procedures connected with the dematerialization of certain steps of the proceedings is "un gain de temps", as Prof. Cadiet writes.[26]

The time necessary for the disposition of a case can be greatly reduced when proceedings are wholly or greatly automated.

The electronic processing and enforcement of attachment requests and orders would surely contribute to the speeding up of the procedure. The creditor would no longer need to go to the enforcement agents to file her claim nor would the latter be burdened with material activities or paperwork. Surely, such electronic attachments and seizures would have to be performed in relation to assets which do not need to be physically apprehended by the enforcing agent. Bank accounts, credits in general, real property or other registered assets (cars, aeroplanes, company stocks and so forth) could be surely subject to such forms of attachment with a minimum adaptation of existing rules, after a safe and operative network has been created between the creditors, the enforcement agents and any interested third party.

---

[21] Cfr. (4) Reiling, op. cit., 108.

[22] P. Gilles, Oral, Written and Electronic Civil Court Procedures and Questions About Their Efficiency (Paper Presented at Tifilis, 2010).

[23] See also D. Reiling, op. cit., p. 155; L. Cadiet, op. cit.

[24] See also L. Cadiet, op. cit.

[25] L. Cadiet, op. cit.

[26] L. Cadiet, op. cit.

In various jurisdictions, bailiffs and enforcement agents already perform online or electronic attachment. In the Argentinean province of Buenos Aires, for example, an administrative procedure exists enabling the local tax agency to attach the debtors' assets through electronic devices.[27] In Estonia, attachment instruments may be sent electronically by the bailiff to a bank, where the debtor owns an account.[28] In the Netherlands, since 2005, bailiffs may execute electronic attachment on real property through the Land registry which records the title for real property: an official report is sent to such registry as a pdf file with an authorised electronic signature. Dutch bailiffs may also serve electronic attachments on third parties such as banks and social security agencies.[29]

At the European level, as it is well known, a project is being developed concerning the attachment of bank accounts in the European common space of justice, whereby a creditor would be allowed, in certain circumstances, to secure the payment of a sum of money by preventing the removal or transfer of funds held to the credit of his debtor in one or several bank accounts within the territory of the European Union.

It is beyond the scope of this paper to enter into the details of this project. As concerns us here, it is worth mentioning that in the Green Paper of the Commission of 24 October 2006, the question is posed of how to implement such an attachment order after if has been issued by a court in a Member State. Given the need to act swiftly and the purely protective nature of the instrument, the Green Paper suggests that an attachment should take effect directly throughout the European Union without any intermediary procedure (like a declaration of enforceability) in the Member State requested being required. The ways of transmitting the attachment from the issuing court to the bank holding the account to be seized will need to be considered. The procedure will have to balance the creditor's interest to effect a speedy transmission with the interests of the debtor and the bank to minimise unjustified seizures. The Green Paper, noting that the cross-border transmission of documents is governed by Regulation 1348/20009 which provides for the direct transmission of an attachment order from the court to the bank by postal services, notes that while this method already allows for a relatively rapid service of judicial decisions, additional consideration should be given to the question whether the use of electronic communication can be used to further speed up the transmission process. In order to achieve the policy objective to render the freezing of accounts more efficacious, the Green Paper suggests that a bank attachment should operate electronically at all or most of the stages of the procedure, i.e. from the court granting it to the bank holding the account. It will be necessary however to assess which mechanisms will have to be devised to provide for an appropriate degree of security in the transmission process and whether the use of an electronic signature would

---

[27] Thank you y Francisco Verbic.

[28] Thank you Vigita Vebraite.

[29] D. Struiksma, A. W. Jongbloed, op. cit.

suffice to certify the identity and competence of the issuing authority and guarantee the accuracy of the data transmitted.

These ideas are aired again in the Annex to the Green Paper, remarking (p. 18): "given the need for speed, it has to be ensured that once the attachment order is granted, it can be implemented as quickly as possible in the Member State where the bank account is situated". In order to pursue that aim, the Annex considers the possibility that the attachment order be "transmitted by electronic means (...). It is clear that the cross-border transmission of attachment orders by electronic means would require the creation of common interfaces between the respective judicial systems of the Member States and of common communication standards with regard to security of transmission and the electronic acknowledgment of receipt. At the same time, some Member States have already introduced means of electronic communication in their national procedure and have reported positive experiences with the use of new technologies".

The European economic and social Committee, at plenary session of 26\27 September 2007, however, expressed strong critical remarks on the Green Book and serious doubts as to the necessity of the proposed regulation in the light of the subsidiarity and proportionality principles. It appears therefore unlikely for it to be enacted in the near future.

9.7 Online Auctions

After assets have been seized, online advertisement of their judicial sale may be organized, in order to contact a wider audience of perspective buyers. A proposal for such forms of advertisement, for example is discussed in Argentina, in order to reduce costs and delays.[30]

Experiences show that, failing disputes between the creditors and the debtor, the time required for the definition of enforcement proceedings of monetary claims is connected with the formal requirements for selling any attached property. Such selling procedures are greatly speeded up when IT is used. Every step of the selling process is liable to be digitized. Specific websites may expose attached assets to the general public, auctions may take place online, payments may be performed electronically and programs may be devised to plan the distribution of the money made from such auction between the participating creditors, taking into consideration existing liens and privileges.

Today, online sale transactions are quite common. Millions of such operations and payments are performed every year on the internet. The "e-commerce" techniques

[30] E. Oteiza, Salgado, Diagnostico sobre el proceso argentino de ejecucion de sentencias, en el marco de una nueva Mirada sobre el anteproyecto de codigo procesal civil modelo para iberocamérica.

may be (and indeed are) applied to the liquidation of seized assets in the course of enforcement proceedings. In several jurisdictions special websites have been created to expose seized goods to a larger public, making any relevant document accessible for the consideration of would-be buyers (e.g. the expert description and evaluation of real estate).

Commercial online auctions and bidding systems have been adapted to the requirements of enforcement procedures, in particular as concerns the protection of the debtor's privacy and the identification of the bidding party. As an example, in Italy, art. 4 of law n. 24 of 22 February 2010 has introduced urgent measures for the digitization of judicial services. As concerns us here, such measures include the reform of arts. 530 and 569 *c.p.c.*, regulating the sale of seized property. The judge may now decide that the deposits, the bids, the auction and the payment of the price be performed though electronic means. Also art. 533 *c.p.c.* has been modified, making it possible, now, for the persons interested in buying any attached assets, to view them online before the actual sale takes place. More specific rules are handed down in new articles in the so called *Disposizioni di attuazione* of the *c.p.c.*: in particular, art. 162 *ter* requires that such auctions, to be further regulated by the Ministry of Justice, should respect and safeguard the principles of competition, transparency, simplification, effectiveness, safety, accuracy and regularity of the telematic procedures. Moreover, Art. 169 *quarter* allows the price to be paid with a credit, debit or prepaid card or other electronic means of payment.

Here is an example of an Italian website dedicated to judicial auctions online:

Such websites allow the interested visitor to choose among various categories of assets, clicking on a scroll window:

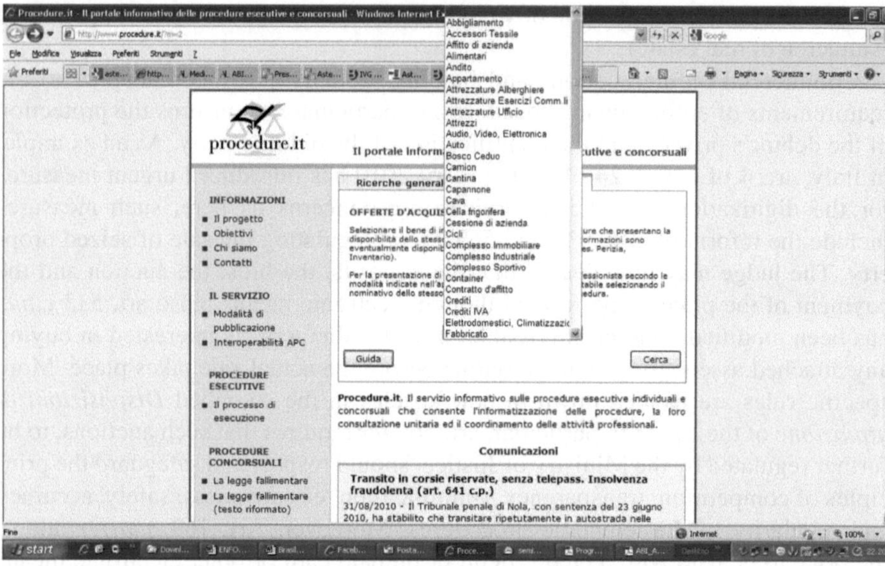

For example, in my search, I have chosen the "lorry" category and here are the details of lorries available for online purchase in the context of enforcement and insolvency proceedings:

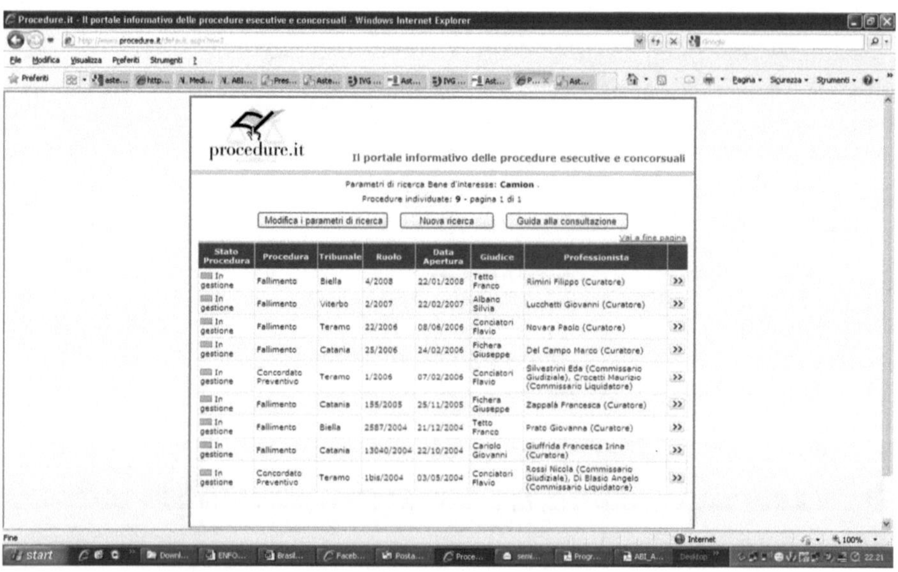

Here is just another example of a website showing pictures of the seized assets for online purchase:

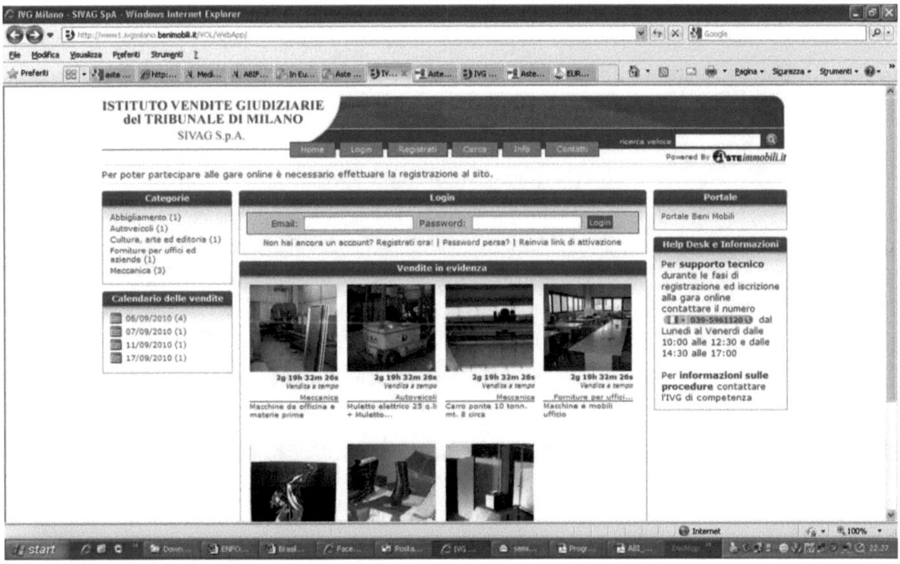

Some more examples from Italy, showing the wide variety of website design and services:

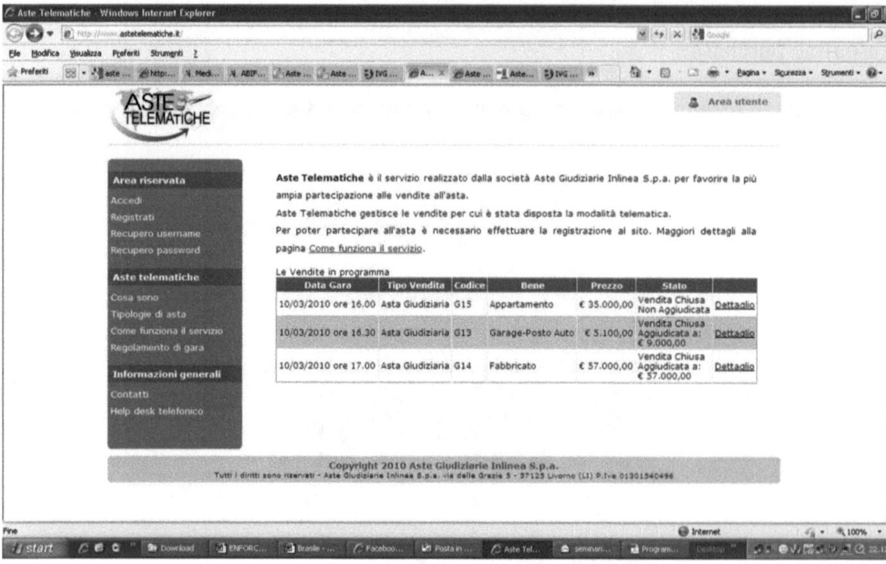

The next picture shows the details of some auctions of real property:

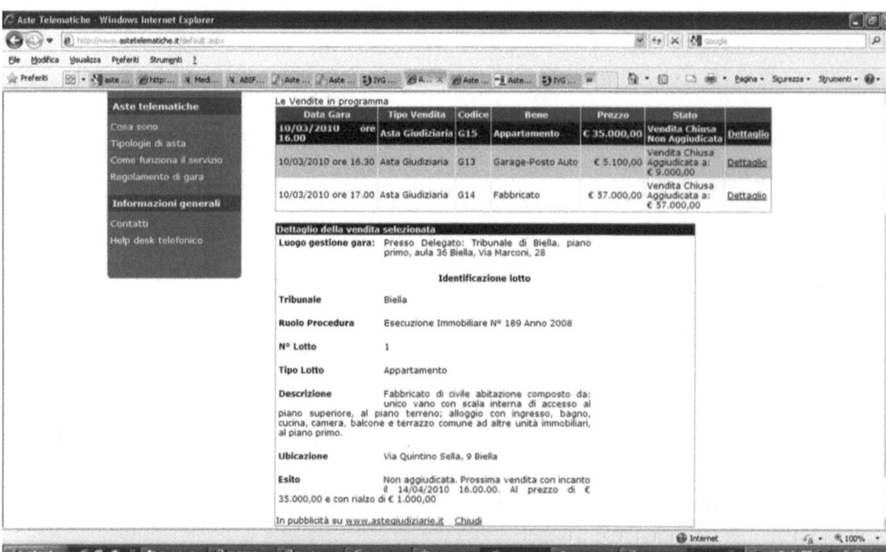

Also in Germany, internet auctions can take place in the course of enforcement proceedings: a recent piece of legislation (from July 2009) has introduced such auctions (para. 814 (2) *ZPO*). The bailiff can chose between the traditional auction and the internet auction[31] and official web page shows current auctions.[32]

It also seems possible to substitute "real life" court hearing with videoconferences, at least in proceedings involving a limited number of participants. Apparently, videoconferencing technology is already widely employed in American Federal Insolvency Courts.

At the European level, the use of videoconferences is indeed encouraged, even though so far, mostly in the context of the cross-border gathering of evidence.[33]

Arguably such techniques could be extended to the context that concerns us here.

9.8 Fighting Corruption and Improving Transparency in Enforcement Proceedings

Corruption in the context of enforcement proceedings is a documented problem worldwide. Most notably, in this context, the possibility to influence and corrupt the outcome of judicial auctions is a major source of concern.

The dematerialization (and therefore the reduction of the "human factor") of enforcement proceedings (and judicial auction in particular) may indeed prevent distortions and abuses.[34]

In general, IT may contribute making every step of enforcement procedures more transparent and reviewable by both the creditor and the debtor, since relevant data, facts and individual behaviours will be more publicly acknowledged.

As an example, it will prove impossible to block or interfere with bids during judicial online auctions. More in general, the use of IT might be instrumental in "piercing the veil" of judicial auctions in many jurisdictions.

Moreover, the integrity of the proceedings will arguably be increased by accuracy and experimentation capacity.[35]

---

[31] See the press release of the Federal Ministry of Justice at: http://www.bmj.de/enid/0,3d9d1570 6d635f6964092d0935383838093a0979656172092d0932303039093a096d6f6e7468092d0930 35093a095f7472636964092d0935383838/Pressestelle/Pressemitteilungen_58.html

[32] http://www.justiz-auktion.de/: thank you M. Stuerner. See also P. Gilles, Public and Private Justice: Dispute Resolution in Modern Societies (Paper Presented at Dubrovnik, 2009).

[33] For further reference, see: https://e-justice.europa.eu/contentPresentation.do?&lang-=en&idTaxonomy=71&idCountry=EU&vmac=C8KGtdqVdhjWb41xu5WQ4Xo-7HfUd9-3RgJbUr54zSaYIiDKIym0E4XVumfQDc7FgAYGZ23JLDrB1jmAtlIhAAAAGwUAAAFF and https://e-justice.europa.eu/contentPresentation.do?&lang=en&idTaxonomy=151&idCou-ntry=EU&vmac=HPv4XPhSBKzvDfkMlabYuGjpcqJI6ld1ctKZ9GV9kvDNY0Ds5iVDX4Xn G582enleKbVshNGSMVP8Ca8ftNxQ6AAADfQAAAMs

[34] See D. Reiling, op. cit., 210, stating: "Introducing and using information technology is often believed to reduce corruption in courts and judiciaries because it is expected to introduce more impersonal ways of working".

[35] D. Reiling, op. cit., 258.

9.9 Improving the Overall Efficiency of Enforcement Proceedings

When enforcement procedures are more easily accessible and transparent, they will also be more efficient and effective, in particular as concerns the discovery of the debtor's assets and the outcome of the auctions of any attached property. The traditional mechanisms to liquidate the debtor's assets are often lengthy, costly and may eventually award only a fraction of the real value of such assets. This appears to be detrimental for both the creditor and the debtor. Implementing a system of online advertisement and electronic auctions should lead to wider participation in the auctions and to the collection of larger sums of money in a more transparent and fast way. From wider chances of disclosure and better earnings from the auctions, the creditors will recover a larger fraction of their claims, therefore better pursuing the final function of enforcement. Deadlines and time limits will be automatically checked and verified, reducing errors and procedural defects. Formal requirements will also be automatically and instantly verified and any necessary correction will be made immediately.

9.10 Some Inconclusive Remarks

Experience shows that developing IT programs in relation to judicial activities is costly and time consuming.[36] Often, such projects even fail to deliver and vast quantities of public and private money are wasted.[37] Why bothering about IT and the enforcement of civil claims, then? As a matter of fact, the doubt might arise that this apparently unstoppable rush for the digitalization of society at large and civil justice[38] in particular is not really worth the effort and that our lives will not be improved by definition by such changes. As Peter Gilles has remarked[39] "the "irresistible" implementation of tele-courts will be achieved regardless to the questions how necessary and useful its outcomes will be". In other words, sometimes one cannot help but wonder whether court proceedings are receiving an IT "electroshock" simply because computers are everywhere, so why should we not bring them into the courts too?

People generally assume that IT will make civil litigation more efficient and effective. The problem then arises of how to define this improvement in efficiency

[36] See also L. Cadiet, op. cit., remarking: "Il a fallu sept ans à la Cour de Cassation pour parvenir au point où elle en est aujourd'hui".

[37] See D. Reiling, op. cit., 61 ss.

[38] Undoubtedly, IT "is attractive because it implies the promise of things becoming better and easier": D. Reiling, op. cit., 16. The a., however, also remarks, p. 273: "Claims that IT will improve things are taken as a certainty and not as mere claims".

[39] P. Gilles, Oral, written and electronic civil court procedures, cit.

and effectiveness in order to be able to verify whether any progress has been made after all the efforts and costs involved.

Arguably rendering enforcement of pecuniary claims more efficient and effective implies making it possible for more creditors to obtain more money out of them, in less time, with a proportionate cost for the civil justice system, while at the same time preserving the basic rights of the parties involved, in particular those of the debtor. I believe that the arguments I put forward in the previous pages show that IT may indeed make enforcement proceedings more efficient and effective in the rough meaning used above.

By varying degrees, both a simple system of e-filing of cases and a wholly digitized network may improve on the accessibility, swiftness, transparency, accuracy and, ultimately, profitability of enforcement proceedings. As a matter of fact, the enforcement of civil claims appears to be a very good operating ground for the implementation of digitized procedural mechanisms. For several years, proceduralists debated about the possibility and indeed the desirability of what may be called the "virtual trial", in the context of cognizance proceedings. Arguably, enforcement proceedings offer a more antiseptic context and therefore most of the objections raised against virtual trial (most notably, loss of direct contact between the court\jury and the parties\evidence) are not be applicable here. Enforcement proceedings are not concerned with the determination of issues of fact and no evidence needs to be collected or evaluated. Moreover, the activities of the subjects involved have a prominently "operative" character (in particular, seizing and selling assets). In some jurisdictions, the judge is not even involved in such activities; in others, the court's role is sometimes minimal (e.g., solving problems and difficulties arisen in the course of the enforcement operations).

The examples mentioned above show that IT may indeed support enforcement proceedings as concerns access to justice, reduction of delays, fight corruption, grant transparency and efficiency to the proceedings.

Today, it appears that in most jurisdictions traditional "paper" practices are being converted into digitized ones. In other words, with a minimum of adaptation, IT is being grafted on to existing rules. Most courts, so far, have used IT to support their paper processes. Using technology for radical innovation in courts is said to be still rare.[40] As a matter of fact, some are concerned that with the electronic form taking hold, the substantial contents will remain unchanged. Peter Gilles fears a reform of the form and not a reform of the content, pursued by a "conservative" movement: according to this eminent scholar, the modernization of justice by the new media does quite often mean not much more than the effort to replace or to augment the conventional form of actions and reactions.[41]

In the long run, adapting IT to existing procedures will prove unsatisfactory and insufficient. For an enhanced use of IT in civil litigation (and in enforcement proceedings in particular), the creation and implementation of new procedural

[40] D. Reiling, op. cit., 271.

[41] P. Gilles, Oral, written and electronic civil court procedures, cit.

models are required. For this, engineers will have to work alongside lawyers, judges and scholars: the former to explore and develop the technological side of the issue, the latter to set the standards and the functions which the new systems have to comply with. A strong interaction between IT experts and legal scholars will ensure that technology and procedural fundamental principles combine in an efficient and just way.

Contents need to be reformed. Allegedly, the procedural codes in most jurisdictions are outdated, retrospective and past-oriented. Future-oriented regulations are required.[42] Existing rules will have to thoroughly revised, introducing simplified and possibly more rational[43] procedures, possibly taking some of the relevant activities out of the court and into the hand of (either public or private) enforcing agents belonging to a safe and strictly supervised network. As Dr. Reiling wrote, the need arises to "mitigate complexity by reducing it and starting small and moving forward in small steps".[44]

For example, the notion of the "paper" title has become outdated. A new idea of e-enforcement title needs being conceived and implemented. Arguably, existing technology makes it possible to enforce e-judgments and e-notary documents, which would travel through the electronic highway from the court or the notary's offices to the creditor and from her to the enforcement agents.

As the first pilot experiences show, the creation is also feasible of networks between courts, bailiffs (or other enforcement agents), creditors and their lawyers, third parties and even debtors. As concerns the latter, the problem of their willingness to take part in such network will have to be faced. In certain cases, previous acceptance of this method of service\interrelation with the enforcement bodies may be obtained, for example through the use of *ad hoc* contractual clauses. But can a debtor be compelled to participate in an enforcement network? A positive answer may be given *vis à vis* institutional (like a public administration) or commercial debtors. For private and\or occasional debtors, devices will have to be provided in order to safeguard their basic rights. As the European experiences reported in this paper clearly show, the need is felt to safeguard the debtor's fundamental rights, avoiding the use of IT to implement some sort of Enforcement Big Brother. Third parties (e.g. Bank, employers and so forth), on the other hand, may benefit from registering in the network, being enabled to handle their role in enforcement procedures more swiftly and with less costs.

Enforcement network would be most useful at the transnational level, in order to enable direct contacts between the creditors and the enforcement agents of the State where the debtor's assets are located. The European space of justice appears an ideal practicing ground for such international cooperation. Just to make an example, a network could be conceived for the cross border enforcement of European enforcement orders, which do not require the grant of an *exequatur*.

---

[42] P. Gilles, Oral, written and electronic civil court procedures, cit. See also L. Cadiet, op. cit.

[43] Cfr. L. Cadiet, op. cit.

[44] Cfr. D. Reiling, op. cit., 77.

After all, in the jurisdictions where pilot projects and new programs have already been implemented, some beneficial outcomes are being experienced. There are good reasons therefore to be optimistic and give a positive answer to the question which I asked myself at the start of this (rather inconclusive) paragraph of my paper. (E-) writing the future is going to be a challenging and exciting endeavour.

Chapter 10
Remarks on the Polish Electronic Proceedings by Writ of Payment

Robert Kulski

10.1 General Information

The recent development of electronic technology of communication is reflected in and has been influencing the Polish civil justice system. The possibility to use sources and techniques of electronic communication in Polish civil procedure has been in force since January 1st, 2010, which is since the amendment to the Code of Civil Procedure (under the Act of January 9th, 2009 on amendment of the Act – Code of Civil Procedure and some other Acts, Official Journal of Laws of 2009, No. 26, item 156) was passed. The amendment involves the introduction of a new manner of conducting legal action which is electronic proceedings by writ of payment based on the plaintiff's statement (hereinafter: electronic proceedings by writ of payment)[1]. Electronic proceedings by writ of payment (*Polish abbreviation*: *Epu*) are of innovative character[2] in Polish civil law procedure. These proceedings are based on solutions from Germany and Great Britain. The goal of the following essay is to present the general characteristic of the course of Polish electronic proceedings by writ of payment.

[1] Electronic proceedings by writ of payment are basically regulated by the provisions of the Code of Civil Procedure (especially art. 505^{28}–505^{37}). Beside of the said regulation, executive orders are in force (resolutions of the Ministry of Justice) which refer to the manner of submitting statements of claims under electronic proceedings by writ of payment, manner of court fee payments, creating the account, manner of using the electronic signature and electronic deliveries in these proceedings.

[2] It is worth noticing that use of electronic communication sources is permitted in administrative proceedings, proceedings before administrative courts, criminal proceedings, as well as in other proceedings (e.g. application of the entry to the National Court Register may be filed online).

R. Kulski (✉)
Assistant Professor, Lodz University, Faculty of Law and Administration, Poland,
Lodz University, Kopciński Street 8/12, 90-232, Łódź, Poland
e-mail: rkulski@wpia.uni.lodz.pl

M. Kengyel and Z. Nemessányi (eds.), *Electronic Technology and Civil Procedure*, Ius Gentium: Comparative Perspectives on Law and Justice 15,
DOI 10.1007/978-94-007-4072-3_10, © Springer Science+Business Media B.V. 2012

Electronic proceedings by writ of payment make it possible to settle civil cases (through issuance of orders for payment) with the use of IT technologies and through the agency of the e-court. These proceedings introduce further improvements and facilitation to civil law procedure. The said is possible thanks to the development of modern IT technologies (communicative and IT services and computer technique) and importance of internet popularization.[3] According to T. Zembrzuski, the fact that procedural law has been opened to new technologies, as well as the fact that legal instruments have been adjusted to requirements and expectations of a growing information society will contribute to the increase of effectiveness of legal protection in civil court procedure.[4] Electronic proceedings by writ of payment apply a new method (or tool) of communication, i.e. electronic, whose benefits lie in its simplicity and in the fact that the use of this method leaves evidence (tracks). It must be noticed that under electronic proceedings by writ of payment the defendant also may react to the plaintiff's acts of legal procedure using the same tool as the plaintiff.

In general, electronic proceedings by writ of payment may be defined as civil procedure which is: (1) simple, which means accessible, (2) fast, and (3) cheap.[5]

Annotation to (1). Circumstances influencing the said quality of electronic proceedings by writ of payment are as follows. Electronic forms in this kind of proceedings make it easier to draft statements of claims, moreover danger of formal mistakes in such statements should decrease significantly. Acts of legal procedure are performed online and in simplified manner with decreased engagement of the parties (for example, it is enough to upkeep the ordinary electronic signature, there is no obligation to enclose the extract from the statement of claim, annexes and the power of attorney, there is no obligation to ground the objection towards the order for payment or refer to evidence in the objection; art. 89 § 1, 126 and 128 § 2, 505^{35} of the Code of Civil Procedure). Acts of procedural bodies, including both courts of 1st instance as courts of 2nd instance, are also recorded in the teleinformatic system only based on electronic court files. It must be also acknowledged that making it possible for the party to initiate civil procedure in long distance, enables easier access to legal protection before the court.

Annotation to (2). Electronic proceedings by writ of payment were designed as procedure conducted in a maximum integrated (consolidated) manner. For example, it is required to make a payment while filing the claim (the fee is made with the

[3] K. Weitz, in J. Jodłowski, Z. Resich, J. Lapierre, T. Misiuk-Jodłowska and K. Weitz, Postępowanie cywilne (*Civil Procedure*) (Warsaw, 2009), 415–20.

[4] T. Zembrzuski, "Elektroniczne postępowanie upominawcze a skuteczność ochrony prawnej w postępowaniu cywilnym (*Electronic Proceedings by Writ of Payment and the Effectiveness of Legal Protection in Civil Procedure*)", in Materials from 10th Department Conference on "Effectiveness of Law" organized by Law and Administration Department of the Warsaw University on February 27th, 2009, (Warsaw, 2009), 81.

[5] R. Kulski, in Kodeks postępowania cywilnego. Komentarz pod red. K. Piaseckiego (*Big Commentary to the Polish Code of Civil Procedure*, ed. K. Piasecki), vol. II, (Warsaw, 2010), 751–53.

use of the specific mechanism to which access is permitted through online system). In lack of such payment, the party is not called for payment, and submission of the claim does not bear legal and procedural effects (unless the plaintiff is free from court fees against the provisions of law, see art. 130 § 3 of the Code of Civil Procedure). The exclusion of the possibility to apply to the court for release from the obligation to make a payment, as well as appointment of the court-assigned council (legal adviser) also influences the speed of electronic proceedings by writ of payments. Cases in electronic proceedings by writ of payments are considered (at least in the first period of regulations existence) by one district court for the whole country, the so called "e-court", which is the District Court of Laws Lublin-Zachód in Lublin, 6th Civil Division; see www.e-sad.gov.pl. The competent "e-court" to consider means of appeal is only one court, which is the Regional Court of Laws in Lublin, 2nd Appeal Civil Division (see art. 17 clause 4 of the Code of Civil Procedure, art. 16 § 6, and art. 20 clause 9 of the Act of July 27th, 2001 – organization of common courts, Official Journal of Laws no. 98, item 1070 as amended). Furthermore, it must be stated that in electronic proceedings by writ of payment, communication between the party of the proceedings and the court is fast, which significantly lies in the fact of electronic deliveries (art. 131^1 § 1 and 2 of the Code of Civil Procedure).[6]

Annotation to (3). Expenses resulting from correspondence sent via post to the parties of legal proceedings (especially to the plaintiff) in electronic proceedings by writ of payment are excluded thanks to electronic communication. Electronic circulation (transfer) of data included in the statements of claims filed in electronic form makes it possible to keep electronic archives for court files and decrease costs. The said data (information) are copied without bearing any expenses. What is more, resulting from art. 19 sec. 2 clause 2 of the Act of July 28th, 2005 on court fees in civil cases (Official Journal of Laws no. 167, item 1398 as amended), lower fee is collected from the claim filed in electronic proceedings by writ of payment than in ordinary proceedings. Additionally, it must be noted that the appeal from the order for payment filed by the defendant in electronic proceedings by writ of payment is not subject to any fee.

10.2 Facultative Character of Electronic Proceedings by Writ of Payment

Electronic proceedings by writ of payment are of facultative character. The use of this kind of proceedings shall be the plaintiff's right, not obligation. The plaintiff's will to consider the case in electronic proceedings by writ of payment shall be expressed through filing the claim online.

[6] Electronic delivery shall be made to specific e-mail address (account) which is created for needs of electronic proceedings by writ of payment.

10.3 Electronic Proceedings by Writ of Payment Within the Polish Civil Law Procedure System

As results from the explanatory statement of the Act,[7] electronic proceedings by writ of payment constitute a type of "ordinary" proceedings by writ of payment only within the scope of the manner of performing acts of legal procedure. The explanatory statement includes however a note that electronic proceedings have been established as separate proceedings and are of calling for payment character. According to T. Zembrzuski electronic proceedings by writ of payment indeed are a type of another "summarized" special proceedings, however they are not of autonomous character.[8]

It is justified to acknowledge electronic proceedings by writ of payment as a specific mode of traditional proceedings by writ of payment which are separated according to specific criteria, meaning use of online manner to settle pecuniary claims and other cases.

Finally, it must be stated that in a case considered under electronic proceedings by writ of payment, beside regulations which directly refer to these proceedings, i.e. art. 505^{28}–505^{37} of the Code of Civil Procedure, and art. 17 clause 4, 46 § 1, 68, 89 § 1, 125 § 2^1, 126 § 3^1, 126 § 5 and § 6, 128 § 2, 130 § 6, 131^1, 353^1 § 2, 397^1 of the Code of Civil Procedure, additionally the application will (always) find some regulations of another special proceedings (i.e. traditional proceedings by writ of payment), and some regulations of ordinary proceedings.

10.4 Scope of Application of Electronic Proceedings by Writ of Payment

Electronic proceedings by writ of payment find application in any case in which the plaintiff files pecuniary claims, regardless of the amount.[9] Generally, it must be stated that electronic proceedings by writ of payment are designed to make pecuniary claims which are not of complex character, which are proved by documents (most often invoices or other receipts), and which are not disputed by debtors.

There are no limitations as to the parties of the proceedings. Electronic proceedings by writ of payment find application in litigations between entrepreneurs, also without their participation, as well as in cases between employees and employers.

---

[7] Sejm printout no. 859, p. 23; http://orka.sejm.gov.pl

[8] T. Zembrzuski, Elektroniczne postępowanie upominawcze..., (*Electronic Proceedings by Writ of Payment...*), 73. See also: J. Gołaczyński, Elektroniczne postępowanie upominawcze. Komentarz pod red. J. Gołaczyńskiego (*Electronic Proceedings by Writ of Payment. Commentary*, ed. J. Gołaczyński), (Warsaw, 2010), 161 and J. Widło, Elektroniczne postępowanie upominawcze (*Electronic Proceedings by Writ of Payment*) (Warsaw, 2010), 9.

[9] Electronic proceedings by writ of payment consider also other cases referred to in specific provisions.

Electronic proceedings by writ of payment should appear to be an attractive mode for making claims, especially by the so-called mass plaintiffs, meaning telecommunications companies, energy, gas and water providers, companies dealing with sewage and waste removal or companies rendering services of people and luggage transfer in mass transportation.[10]

Under electronic proceedings by writ of payment all acts of procedural bodies (of different character, not only decisive decisions) are made by the court, division official or the chairman.

10.5 Initiation and Course of Electronic Proceedings by Writ of Payment

Initiation of electronic proceedings by writ of payment depend of the plaintiff.[11] In the case, the plaintiff decides on electronic proceedings by writ of payment in order to assert a claim, it is required to file the claim (only) online (art. 125 § 2^1 of the Code of Civil Procedure). The statement of claim shall be considered electrically filed at the moment it is filed through the online system (art. 505^{31} § of the Code of Civil Procedure). It is unquestionable that online manner of filing have become new, and additional mode (manner) for filing statements of claims (next to filing the claim in the day book of the court or during the hearing, see also art. 165 § 2 and 3 of the Code of Civil Procedure). Online system is a specific tool to transfer electronic data of acts of legal procedure[12] performed in electronic proceedings by writ of payment (data included into the acts of legal procedure).

It results from performance (consolidation) of acts of legal procedure online that the manner of the form of the said acts is electronic. There is no situation where statement of claim which is filed online is drafted in traditional (paper) form. Online system shall find application in reference to acts of legal procedure of both parties of the case and bodies of civil proceedings. According to S. Cieślak, it is more justified to acknowledge electronic form of acts of legal procedure as one of the qualified types of written form.[13] Substantive basis of the statement in common and

---

[10] T. Zembrzuski, *op.cit,* 71.

[11] The plaintiff may select between filing the claim online (with the use of electronic devices) or initiate traditional civil procedure.

[12] J. Gołaczyński, "Elektroniczne czynności procesowe (Electronic Acts of Legal Procedure)," in *Law of Electronic Media,* vol. 1, (Addition to Monitor Prawniczy, 2004/4, 3–6).

[13] S. Cieślak, Formalizm postępowania cywilnego (*Formalism of Civil Law Procedure*) (Warsaw, 2008), 125, 211–3, 226. See also: E. Rudkowska-Ząbczyk, "Pisma procesowe wnoszone w postępowaniu cywilnym na elektronicznych nośnikach informatycznych (Statements of Claims Filed in Civil Law Procedure on Electronic Devices)," in *Law of Electronic Media*, (Addition to Monitor Prawniczy, 2006/16, 36–7).

qualified-electronic written form is the same, which means that it is a subject covered with writing, but the manner of recording and reading is different. It shall be also noted that new technologies which may be applied in civil procedure may also refer to oral elements (oral form of acts of legal procedure) e.g. hearing the evidence from the testimonies of witnesses through videoconference.[14] Effects of acts of legal procedure executed electrically are equal to effects of common written form of the same acts.

Basically, there is no differentiation in formal requirements towards the claim depending of the claim filing manner to the curt (electronically or traditionally).[15] Under electronic proceedings by writ of payment the suit must also additionally include data required for identification of the entity filing the proceeding writ, and these are: name and surname (name), address, seat, and individual identification number (for natural/private person – PESEL number [*Personal Identification Number*], and for non-natural person – NIP no. [*Tax Identification Number*] and number in the National Court Register, and in the case such number does not exist, number in any other competent registry or evidence (art. 505^{32} § 1 of the Code of Civil Procedure)). The said identification is connected with the login and the password.

The plaintiff does not attach evidence in electronic civil proceedings (art. 505^{32} §1 of the Code of Civil Procedure). It is sufficient to indicate evidence to support presented statements on existence or non-existence of specific facts. In practice it refers to different kinds of invoices, bills of sale, and printouts whose existence and content shall be known to the plaintiff. "Indication of the evidence" shall be understood only as description of the evidence, however the description may not be too general. Description of the evidence in the claim must allow the defendant for verification of grounds for making the claim. If the evidence constitutes the invoice, it will be required to place (fill in) the following data in the claim: date, issuing entity, item and the amount due. As result, it must be emphasizes that electronic proceedings by writ of payment shall mainly be used in cases which do not require hearing the evidence. It is assumed that the decision is made in trust as to the authenticity of presented claims by the parties and to actual statements.

It is also worth stating that in the light of art. 89 § 1 *in fine* of the Code of Civil Procedure, in the case where the claim under electronic proceedings by writ of payment is filed by the proxy, no power of attorney must be enclosed. It is sufficient to refer by the proxy of the plaintiff to the fact on granting the authorization indicating data of the power of attorney, its scope and circumstances which grounds acting as a proxy.

---

[14] See: R. Kulski, Sprawozdanie z Międzynarodowej Konferencji Naukowej pt. Elementy ustności i pisemności w postępowaniu sądowym: skuteczność w procesie cywilnym (Report on International Science Conference titled Oral and Written Elements in Court Proceedings: Effectiveness in Civil Law Procedure). The conference organized in Gandia-Valencia on 6–8.11.2008, Państwo i Prawo, 2009/8, 129–30.

[15] Elements (parts) of the claim are regulated in art. 187, and art. 126–128 of the Code of Civil Procedure.

Before the court initiates consideration of basis to issue order for payment, the following must be considered: formal circumstances of the claim, payment of the fee and prerequisite for court proceedings (admissibility of court proceedings, capacity to be a party in civil cases, and capacity to be a party in a given civil case, national jurisdiction, *res judicata*).[16] The court may not issue order for payment in electronic proceedings by writ of payment when the claim is clearly groundless, circumstances given are doubtful, satisfaction of claim depends of mutual performance, place of residence of the defendant is not known or if the order for payment cannot be served on the defendant within the territory of the country (art. 499 in connection with art. 505^{28} of the Code of Civil Procedure). As the consequence of lack of possibility to issue the order for payment because of the aforesaid circumstances the case is referred by the "e-court" to the court of general jurisdiction (art. 505^{33} § 1 of the Code of Civil Procedure.). The court to which the "e-court" refers the case, is the competent court as of material jurisdiction and as of general territorial jurisdiction if the plaintiff had not filed the claim in electronic proceedings by writ of payment. The case is referred against the court ("e-court") decision which is only served on the plaintiff.

After issuance of the order for payment, the court *ex officio* will revoke the order in the case the order may not be served because the place of the defendant's residence is not known or if it cannot be served within the territory of the country, as well as if after the issuance of the order for payment it will be found that the defendant at the moment of the claim submission did not have capacity to be a party in the civil case, and capacity to be a party in a given civil case, or any entity appointed for his/her representation (art. 505^{34} § 1 and 502^{1} of the Code of Civil Procedure). Prior revoking the order for payment of reasons of obstacles in serving the order, a deadline for the party to remove this obstacle must be defined, and the order for payment is revoked only after ineffective expiration of the said deadline.

Revoking the order for payment *ex officio* of the aforesaid reasons results in the case referred to by the "e-court" to the court according to general jurisdiction. The case is referred against the decision of the court ("e-court") which only is served on the plaintiff.

The form of appealing against the order for payment in electronic proceedings by writ of payment is more simplified than in ordinary civil proceedings (art. 505^{35} of the Code of Civil Procedure).[17] The defendant is only required to submit objection as to the improper specification by the plaintiff of the amount of the subject of the claim, objection as to the incompetence of the court, objection as to the existence of mediation agreement, objection as to the agreement on jurisdiction of foreign court, objection as to the arrangement on the arbitration court. However, the said does not

---

[16] W. Broniewicz, *Postępowanie cywilne w zarysie* (*Outline of Civil Law Procedure*) (Warsaw, 2008), 191–3.

[17] See: K. Weitz, Kodeks postępowania cywilnego. Komentarz pod red. T. Erecińskiego (*Code of Civil Procedure. Commentary*, ed. T.Ereciński), vol. 2, (Warsaw, 2009), 871.

constitute the requirement for the appeal. According to T. Zembrzuski, the appeal is limited to ordinary negation presented towards the plaintiff's claim.[18] Thus, the statement of the defendant in which he/she objects against the order for payment is sufficient. It is not possible to appeal against a part of the order for payment, even if the defendant agrees that the remaining part of the claim is grounded.[19]

The results of proper appeal against the order for payment are as follows. Firstly, the order for payment completely loses its force of law. Secondly, the "e-court" refers the case to the court according to general jurisdiction (art. 505^{36} of the Code of Civil Procedure), and the decision to refer the case shall be served both on the plaintiff (electrically), and on the defendant (electrically or traditionally). Then it will be required to adjust (supplement) previously filed statements of claims to requirements of common proceedings (art. 505^{37} of the Code of Civil Procedure).

Lack of appeal against the order for payment issues in electronic proceedings by writ of payment causes validation of the order.

At the end of this paper it must be noted that the use of electronic manner to deal with civil cases caused amendments in provisions regulating execution under the order for payment issued in electronic proceedings by writ of payment. (see art. 773, 781, 783, 795, 797, 805, 816 and 816^1 of the Code of Civil Procedure). However, the framework of this paper does not make it possible to discuss the aforesaid amendments. It shall be only noted that it is sufficient for the execution to use the printout from the system along with enforcement formula holding a special code which allows the judicial enforcement officer to verify the authenticity of the document.

10.6 Comment on the Future Use of Electronic Technologies in Polish Civil Procedure

Providing the citizens with electronic access to legal protection in civil cases is to basically increase efficiency and speed of civil procedure in Poland. Provisions of the Code of Civil Procedure in electronic proceedings by writ of payment, which are to limit acceptance of statements of claims with formal mistakes, and which are applicable in improvement of serving the documents as well as in creation of electronic court files, shall be positively evaluated within the said scope.

However, it must be emphasized that exclusion of the obligation to enclose evidence by the parties in electronic proceedings by writ of payment and lack of obligation to file the power of attorney to the files of the case creates real danger

[18] T. Zembrzuski, *op. cit.,* 78.

[19] Compare explanatory statement of a bill pp. 14–15, and *Legal Opinion of T. Zembrzuski to the draft of the Act on amendment of the Act – Civil Code and some other Acts (Sejm printout no. 859) developed against order of Office of Sejm Analysis* at: http://orka.sejm.gov.pl/rexdomk6.nsf/Opdodr?OpenPage&nr=859 [17.07.2009].

of conducting fake proceedings. There is also a fear of the parties being equal in the proceedings, automation of the judicial system and guaranty of safety for users of the online system, which is in force in electronic proceedings by writ of payment.[20]

Although there are some threats (dangers), referred to above, introduction of electronic proceedings by writ of payment to the Polish legal system must be completely accepted. It must be noted that achieving goals designed for electronic proceedings by writ of payment requires creation of an efficient online system that manages the proceedings, as well as the acceptance of the new form by potential users of the system. As for now these proceedings are popular which can be seen in the number of claims electronically filed to the District Court of Laws Lublin-Zachód in Lublin.[21] If electronic manner will be found useful in practice, it may be use in other proceedings, e.g. proceeding to issue enforcement formula or in mortgage and land register proceedings.

[20] Compare: D. Zawistowski, Kodeks postępowania cywilnego pod red. H. Doleckiego, T. Wiśniewskiego, Tom II (*Code of Civil Procedure*, ed. H. Dolecki and T. Wiśniewski), (Warsaw, 2010) 765-6, 768 and 769. It must be noted that online system shall not substitute human in issuance of decisions. In electronic admonition proceedings, modern technologies are only of important function in supporting the evidence.

[21] Within first year of electronic proceedings by writ of payment existence almost 700 thousand claims were filed to the District Court of Laws in Lublin. In 2010, above 613 thousand orders for payment were issued (data announced by the Ministry of Justice on 30 December 2010: www.ms.gov.pl/aktualności). In 2011, 1.825.078 claims were filed to the "e-court", and 1.467.843 claims were decided. Suability orders for payment is 3% (data announced by the Ministry of Justice on 21 December 2011: www.ms.gov.pl/aktualności).

Chapter 11
Enforcement of a Claim with the Support of the New Information Technology in the European Union, Romanian Case Study

Pantilimon Rikhárd-Árpád

11.1 Law Harmonization and Enforcement of Civil Claims in the EU

The success of the Common Market and the increase of economic activities within the European Union have generated a substantial growth in European cross-border transactions. The European Union works on the harmonization of Member States' legislation in order to facilitate the free movement of persons, capital and goods, thus making the economy of the Union as competitive as possible. The growth of economic activities has led to an imminent growth of cross-border conflicts. It has become a commonplace to say that developments in the European Market call for a harmonization of national legislations in Europe. The fast development and variety of civil and commercial relations within the EU generates an urgent need for efficient harmonization. In order to have a functional internal market we need an adequate civil procedure and an efficient enforcement of claims within the European Union. Despite the success of harmonization till now, the process is difficult and challenging in the same time. Harmonization takes into account the local factors yet applies general principles to make a consistent framework of law. Harmonization generally incorporates local factors under a relatively unified framework.[1] However harmonization had success in different fields of the European law system, harmonizing a part of the law of civil procedure has, however, only recently emerged. As a rule, the law of civil procedure was and still is considered part of the sovereign national law. In the absence of a general community competence for the harmonization

[1] M. W. Hesselink, "The Ideal of Codification and the Dynamics of Europeanisation: The Dutch Experience," in *European Law Journal* 12, no. 3 (2006): 279–305.

P. Rikhárd-Árpád (✉)
Institute of Legal Sciences, Sapientia University,
Department of Legal Sciences of the Sapientia University
e-mail: rikhard@pantilimon.com

M. Kengyel and Z. Nemessányi (eds.), *Electronic Technology and Civil Procedure*, Ius Gentium: Comparative Perspectives on Law and Justice 15,
DOI 10.1007/978-94-007-4072-3_11,

of national procedural rules, it is for the domestic legal system of each Member State to lay down the detailed procedural rules governing the actions intended to enforce the court decisions.[2] The variety of laws of civil procedure in Europe has created an already long-existing problem by which effective access to the courts is seriously hampered. The first successful step towards partial harmonization of the law of civil procedure within the European Union was made in 1968 by the acceptance of the Convention on Jurisdiction and the Enforcement of Judgments in Civil and Commercial Matters. Since then there was a substantial development in the field of civil procedure within the European Union. The Brussels I Regulation officially the Council Regulation (EC) No. 44/2001 of 22 December 2000 on jurisdiction and the recognition and enforcement of judgments in civil and commercial matters and other broadly similar instruments in content had a major impact on the development of civil procedure law in the European Union. However, we must state that the proper administration of justice implies that the enforcement of court decisions should itself be effective and efficient. There has been no shortage of efforts in the field of procedural harmonization: they have concerned all the "*linchpins of procedure*" (judges, lawyers, prosecutors, enforcement agents). While a common frame of reference has been constructed in relation to judges and lawyers, there has been far less success with regard to enforcement procedures, which still vary considerably. It is true that the progress made on European and international judicial cooperation is encouraging, but efforts aimed at harmonizing and radically improving national enforcement procedures are still in their infancy.[3] For the efficient functioning of the Common Market an effective structure of civil claim enforcement is a key element. The desirability of harmonized procedural rules, for the purposes of ensuring a smooth functioning of the internal market and a uniform application of Community law, has been advocated by some scholars.[4] As has been argued, even where the direct effect and supremacy of EC law are accepted and endorsed by the national courts, the uniformity of Community rules may still be jeopardized by the different procedural rules of the national legal systems.[5] A possible solution could be creating a more flexible and less invasive system, based upon voluntary compliance by the single Member States, the adoption of a soft law

[2]M. Eliantonio, "The Future of National Procedural Law in Europe: Harmonisation vs. Judge-made Standards in the Field of Administrative Justice" vol. 13.3 (September 2009), 1, see the document at: http://www.ejcl.org/133/art133-4.pdf [10.02.2011].

[3]M. Eliantonio, "The Future of National Procedural Law in Europe: Harmonisation vs. Judge-made Standards in the Field of Administrative Justice" vol. 13.3 (September 2009), 10, see the document at: http://www.ejcl.org/133/art133-4.pdf [10.02.2011]

[4]M. Eliantonio, "The Future of National Procedural Law in Europe: Harmonisation vs. Judge-made Standards in the Field of Administrative Justice" (Maastricht Faculty of Law Working Paper, 2008–8).

[5]M. Eliantonio, "The Future of National Procedural Law in Europe: Harmonisation vs. Judge-made Standards in the Field of Administrative Justice" vol. 13.3 (September 2009), 5, see the document at: http://www.ejcl.org/133/art133-4.pdf [10.02.2011].

instrument in the field.[6] It is evident that the development of such a system can contribute in an efficient and substantial way to the development and speeding up of the claim enforcement procedure on state level and on the EU level as well.

11.2 IT and Enforcement of Civil Claims in the EU

The development of IT has a major impact on our legal systems. In our modern society there is an urgent need for innovation and adaptation. The fast expansion and development of new technologies has challenged our legal system as well. IT can be used to substantially increase the speed of justice and of the claim enforcement. It is true that IT has a positive impact on our legal systems, but technology has also brought legal dilemmas, such as whether they violate the debtor's privacy and place restrictions on their personal freedom. The use of the new technology will help improving the efficiency of enforcement proceedings and liquidation of the debtor's assets. The member states of the European Union have agreed to take advantage of the opportunities offered by the communication technology and to intensify cooperation in this field. The modernization of the legal system and the integration of new theologies into our legal systems is a high priority. The modernization process aims to have a significant impact on the enforcement of claims as well. However we must sate that within the European Union the differences between the legal systems and information databases are obvious. The enforcement proceedings and the conditions to get access to information regarding the debtor's assets are very different in the EU, forming a legal gap which is hard to cross for the member states.

11.3 Enforcement of Claims in Romania

The provisions regarding enforcement in civil and commercial matters are contained in articles 372–580 of Romania's Code of Civil Procedure. In the Romanian legal system the enforcement represents the second stage in civil or commercial proceedings, and its main objective is to ensure effective and efficient recognition of rights conferred by a court decision or out-of-court settlement. As a result, the enforcement procedure enables the holder of a right established by a judiciary or non-judiciary decision to oblige the (natural or legal) person that infringed the right to comply with it and restore the rule of law with the support of the competent state authorities.[7] The Romanian Code of Civil Procedure provides a list of direct and indirect enforcement measures.

---

[6] M. Eliantonio, "The Future of National Procedural Law in Europe: Harmonisation vs. Judge-made Standards in the Field of Administrative Justice" (Maastricht Faculty of Law Working Paper, 2008–8).

[7] See the information provided by the European Commission at: http://ec.europa.eu/civiljustice/enforce_judgement/enforce_judgement_rom_en.htm [10.02.2011].

The following methods of enforcement qualify as direct if they concern the object of the obligation to be enforced:

- enforced consignment of movable assets, enforced consignment of fixed assets – articles 572–580
- the enforcement of an obligation to do something or refrain from doing something – articles 580^2–580^5 of the Code of Civil Procedure and articles 1073, 1075–1077 of the Civil Code.[8]

Among the main novelties of the new legislation proposed by the draft code of civil procedure relating to the debt recovery are:

- Consecration, for the first time in the Romanian legal system, of the principle of the respect for the rule of law and the obligation of the state to ensure the prompt and effective execution of any enforcement and liability of the state for damage suffered due to non-compliance with its obligations
- To establish the general competence of the executor body to carry out the enforcement in order to ensure a consistent and uniform regulations in relation to the execution
- Judicious delimitation of the categories of decisions – final and executor decisions: the law had given up to the category of "irrevocable" decisions, which is not known throughout Europe and was capable to cause mistakes in the law practice, creating difficulties in determining the status of decisions, likely to be challenged by extraordinary ways of attack

In the context of modernization, special procedures where implemented in order to clarify and to solve fast and in an efficient way the disputes generated during the claim enforcement procedure.

11.4 IT and Enforcement of Claims in Romania

Romania considered that the usage of intelligent technologies will accelerate the reform process of justice and will insure the standardization of the procedures, at the system level, as a whole. Romania became a member of the European Union on January 1, 2007. "*The IT Strategy for the Reform of the Judiciary* (2005–2009)" issued by the Romanian Ministry of Justice was part of a massive effort of the Romanian government to modernize the judicial system. The aim of the project was to create a "secured IT network and interconnection of all institutions of the judicial system, implementing a joint security policy within the entire judicial system, creating an electronic unitary and efficient system for registering the court sessions, hearing the withheld identity witnesses and distance hearing of the imprisoned detainees, interconnect the IT integrated judicial system with the other IT systems

[8] See the information provided by the European Commission at: http://ec.europa.eu/civiljustice/enforce_judgement/enforce_judgement_rom_en.htm [10.02.2011].

within the public service nationwide and at the European level."[9] Romania has made substantial progress in implementing the secured IT network and interconnection between institutions. The information technology sector is one of Romania's fastest-growing industries. Having targeted information technology as one of the country's top priorities, the government has created a Ministry of Communications and Information Technology as well as an Advanced Technology, Communications and Information Technology Commission, an information technology promotion group and, most significantly, a strong regulatory framework.

The ECRIS project marks a major milestone in the evolution of the Romanian justice system. The implementation of ECRIS in courts and prosecutors' offices shows that Romania is committed to reforming its legal system and enforcing EU regulations.

The main software application used in courts, including the High Court of Cassation and Justice is ECRIS, achieved as a result of Phare program 1997. Now the application – the version for the courts – is functional in more than 30 courts and is tested to be installed in more. The application has three main modules: management of the cases, random distribution of the cases on panel of judges and legislative and jurisprudence library.[10]

- Management of the cases modulus is useful for the registration of essential data about the cases and for printing subpoenas, meeting lists, official statements and other documents specific to court activities.
- Random distribution of the cases on panel of judges – this modulus insures the distribution of cases using objective criterions, without human intervention.
- Legislative and jurisprudence library modulus allows access to legislation and jurisprudence on electronic support for magistrates and auxiliary personnel. The modulus can be updated every day and has powerful search engines.[11]

In the recent years the http://portal.just.ro/ was launched in Romania, which is a website offering access to information referring to courts and cases around the country, with general information regarding every court and jurisdiction in Romania. People interested can search relevant information with the help of an interactive map of Romania.[12]

[9] IT Strategy for the Reform of the Judiciary – see the information provided by the Romanian Ministry of Justice at http://www.just.ro/Sections/SistemulJudiciar%C3%AEnRom%C3%A2nia/ITSTRATEGYFORTHEREFORMOFTHEJUDICIARY20052/tabid/249/Default.aspx [10.02.2011].

[10] IT Strategy for the Reform of the Judiciary – see the information provided by the Romanian Ministry of Justice at http://www.just.ro/Sections/SistemulJudiciar%C3%AEnRom%C3%A2nia/ITSTRATEGYFORTHEREFORMOFTHEJUDICIARY20052/tabid/249/Default.aspx [10.02.2011].

[11] IT Strategy for the Reform of the Judiciary – see the information provided by the Romanian Ministry of Justice at http://www.just.ro/Sections/SistemulJudiciar%C3%AEnRom%C3%A2nia/ITSTRATEGYFORTHEREFORMOFTHEJUDICIARY20052/tabid/249/Default.aspx [10.02.2011].

[12] See information provided by the Romanian Ministry of Justice at: http://portal.just.ro/ [10.02.2011].

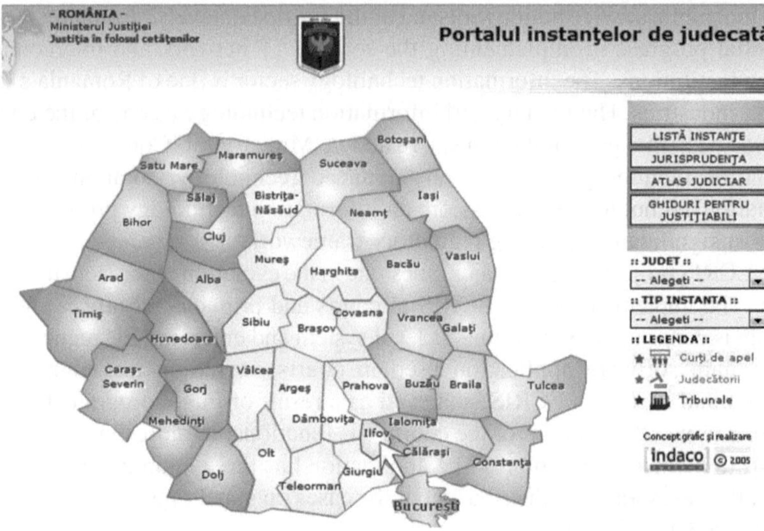

Another important characteristic of the web portal is that one can search after cases based on the name of the parties or based on the registration number of the case. The persons using the web portal can find information regarding the parties involved in each case, a short extract of the decisions taken and the timeframe for future terms as well.[13]

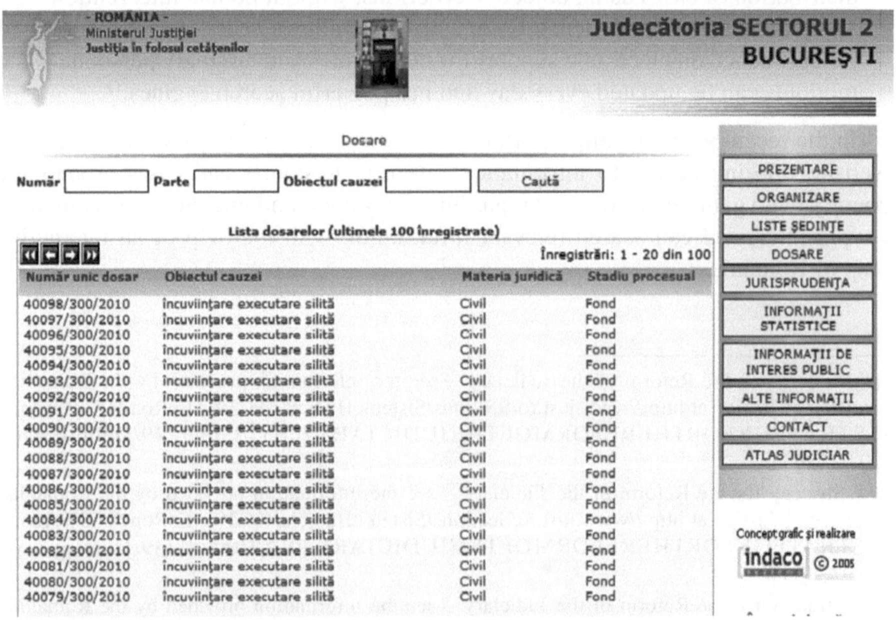

Număr unic dosar	Obiectul cauzei	Materia juridică	Stadiu procesual
40098/300/2010	încuviinţare executare silită	Civil	Fond
40097/300/2010	încuviinţare executare silită	Civil	Fond
40096/300/2010	încuviinţare executare silită	Civil	Fond
40095/300/2010	încuviinţare executare silită	Civil	Fond
40094/300/2010	încuviinţare executare silită	Civil	Fond
40093/300/2010	încuviinţare executare silită	Civil	Fond
40092/300/2010	încuviinţare executare silită	Civil	Fond
40091/300/2010	încuviinţare executare silită	Civil	Fond
40090/300/2010	încuviinţare executare silită	Civil	Fond
40089/300/2010	încuviinţare executare silită	Civil	Fond
40088/300/2010	încuviinţare executare silită	Civil	Fond
40087/300/2010	încuviinţare executare silită	Civil	Fond
40086/300/2010	încuviinţare executare silită	Civil	Fond
40085/300/2010	încuviinţare executare silită	Civil	Fond
40084/300/2010	încuviinţare executare silită	Civil	Fond
40083/300/2010	încuviinţare executare silită	Civil	Fond
40082/300/2010	încuviinţare executare silită	Civil	Fond
40081/300/2010	încuviinţare executare silită	Civil	Fond
40080/300/2010	încuviinţare executare silită	Civil	Fond
40079/300/2010	încuviinţare executare silită	Civil	Fond

[13]See information provided by the Romanian Ministry of Justice at: http://portal.just.ro/InstantaDosareLista.aspx?idInstitutie=300http://portal.just.ro/InstantaDosareLista.aspx?idInstitutie=300 [10.02.2011].

The Romanian Ministry of Justice is aiming to interconnect the IT system within the judiciary with other IT systems within the public service, nationwide and at the European level as well. The Ministry is also exploring the possibilities of implementing an e-filing system. E-filing can involve a number of functions, including case initiation, notification and summon of parties, exchange of documents and issuance of judgments and orders. The e-filing systems have been implemented in Finland, Great Britain and the federal courts of the United States. E-filing will be at first used only for commercial cases. In the implementation of a similar system there are risks as well including the reluctance of attorneys to embrace e-filing, the need for new procedures and rules, provision for security and system backups and handling exceptions to the process. Experience shows also that the introduction of sophisticated case management practices and integrated systems (including court recording, e-filing) requires a considerable investment of time and testing.[14] The ECRIS system can form the solid base for the civil claim enforcement IT system. The development of such a system is imminent. Romania should build the system based on the accumulated experience of other countries and learn from the "best-practices" of other countries which implemented such systems.

11.5 Progress in the Field of Bankruptcy, a Positive Example

In Romani a major development in the field of bankruptcy was achieved with the help of a project financed by the EU. The approved project had four parts: to improve the legal and institutional framework on bankruptcy in order to render the procedure more efficient; to create a Best Practices Manual; to develop a software application; and to deliver specialized training for judges, lawyers, liquidators, accountants and clerks.

A new document filing and management system, or an "electronic dossier" was created to implement standard forms based on the provisions of the new insolvency law.[15] This software application integrates the electronic standardized forms within the existing case management document system used by the Romanian judiciary.

An Insolvency Bulletin was created to provide online procedural notifications for the parties involved. The Insolvency Proceedings Bulletin is a publication edited by the National Trade Register Office, where the public service of summons, communication, convening notice, notification of the procedure documents issued by courts

[14]Improving the institutional framework of family and civil litigation in Romania – summary of final working sessions – September 2007, see the document at: http://pdf.usaid.gov/pdf_docs/PNADL213.pdf [10.02.2011].

[15]Speeding up the Bankruptcy Process Lessons from Romania – Bizclir, 3, see the document at: http://www.bizclir.com/galleries/bestpractices/01.128.08BP15_Romania.pdf [10.02.2011].

of law and judicial administrators is carried out, after the opening of insolvency proceedings, in compliance with the provisions of Law no. 85/2006 on insolvency proceedings.[16,17]

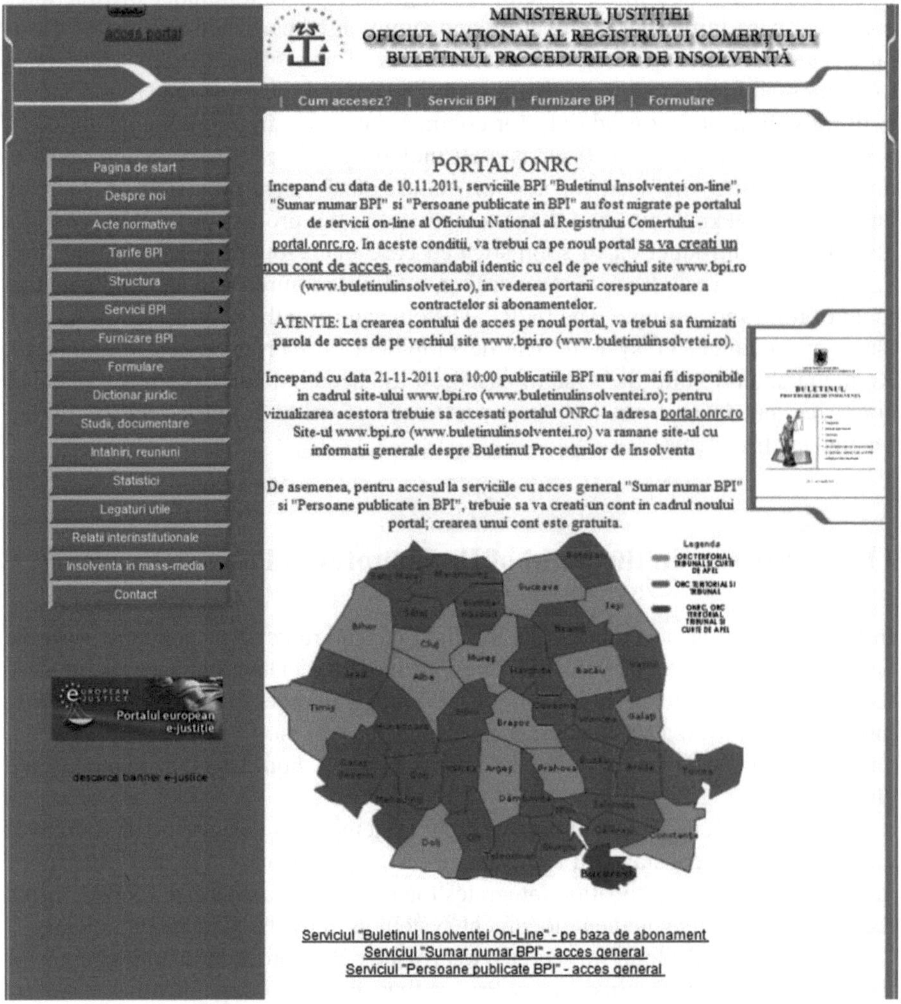

The software application which will be developed based on this project will provide the courts with an "*electronic dossier*". The software will generate progress regarding the:

- Length of the procedures;

[16] Speeding up the Bankruptcy Process Lessons from Romania – Bizclir, 3, see the document at: http://www.bizclir.com/galleries/bestpractices/01.128.08BP15_Romania.pdf [10.02.2011].

[17] See information provided by the Insolvency Proceedings Bulletin at: http://www.buletinulinsolventei.ro/ [10.02.2011].

- Performance of the electronic dossier; and
- Performance of the interaction between the subjects of the claim enforcement procedure;

The National Association of Practitioners in Insolvency (UNPIR) is currently analyzing the implementation of an online liquidation mechanism. The organization is seeking to create an online platform where the assets of the companies entered in bankruptcy are sold. Currently in Romania there are websites which are focusing on liquidation of assets, such as:

www.executari.com[18]

www.executarisilite.eu[19]

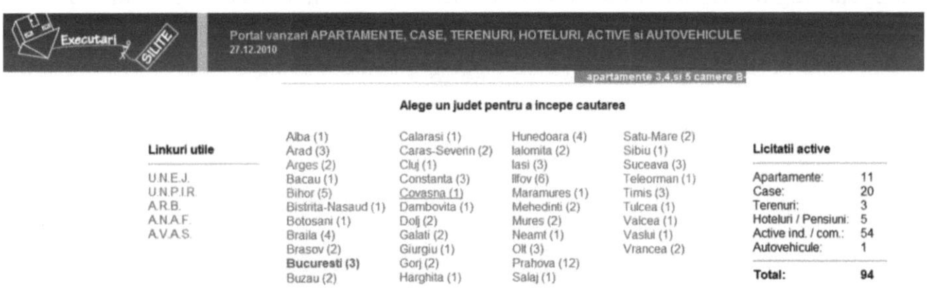

These websites are focusing only on the dissemination of information and they are not offering a platform for online liquidation.

I am confident that the substantial progress achieved in the field of bankruptcy will generate the development of new IT systems for the enforcement of civil claims.

---

[18] See information provided at: http://www.executari.com/ [10.02.2011].

[19] See information provided at: http://www.executarisilite.eu/ [10.02.2011].

11.6 The Future and the Challenges of Civil Claim Enforcement in the EU

Because of the increasing mobility and stronger relation of the entities and services in Europe, the enforcement of civil claims will increase significantly. The EU Member States aim to improve, simplify and expedite effective judicial cooperation between each other and to develop a functional claim enforcement system. The area of freedom, security and justice has continued to grow over the last few years. It is essential for transparency of enforcement expenses to go beyond the strictly domestic framework.[20] It is important for the EU to establish a strong network between the courts and the enforcing agents of all the Member States in order to assist the enforcement of European enforcement orders. This system must guarantee the transparency of the enforcement process to assure free and easy access to information. Such a system must attempt to draw a balance between the right of the creditors to collect the debt with the right of a debtors to be treated in a reasonable and respectable manner. The privacy of the debtor should be respected by both debt collectors and creditors. Every communication between the debtor, the creditors, debt collectors and courts must be protected by a secured network where every entity needs a certification. Access to data must be restricted strictly to the dept collection. In every communication it must be clearly disclosed that the purpose of the communication is to collect the debt and any information obtained in the process must be used for the purpose of the dept collection. The creation of an international enforcement procedure, by which foreign creditors could directly interact online with the courts and enforcing agencies of another State, is very important but challenging in the same time. This future system must be fast and safe, preserving and safeguarding the rights of the debtor and serving the interests of the creditors in the same time.

[20] European commission for the efficiency of justice (cepej) – enforcement of court decisions in Europe, Report prepared by the Research Team on enforcement of court decisions (University Nancy (France)/Swiss Institute of comparative law) and discussed by the CEPEJ-GT-EVAL at their 8th meeting, see the document provided by the Council of Europe at http://www.coe.int/t/dghl/cooperation/cepej/series/Etudes8Execution_en.pdf [10.02.2011].

Part III
Specific Applications of Electronic Technology Applications to Court Proceedings: Documentary Evidence and Other Applications of Information Technology

Chapter 12
Electronic Documents. Security and Authenticity*

Helmut Rüßmann

12.1 Scope and Identification of Issues

Looking at the themes covered by this colloquium as a whole, I should not be talking about the law of evidence in relation to electronic documents (topic 5),[1] or about courses of proceedings by electronic means (topic 1), or about the service of documents by electronic means (topic 2). What is left then, are questions regarding the security and authenticity of procedural documents and the results of proceedings when recorded in electronic form. We are talking about electronic case files and electronic registers.

An electronic case file documents the entire set of court proceedings – from when it begins with the initiating documents to when it ends with a judgment – in electronic form. An electronic file can be used alongside a paper file in order to make access and use of information relevant to the proceedings easier. In such cases, where the electronic file does not replace the paper file, the paper file remains the decisive version. Questions of security and authenticity can easily be answered by comparing the two files, and so no new problems are posed. This changes, however, if the electronic file replaces the paper one, and this is when questions regarding the security and authenticity of electronic documents arise. We need to find ways of transferring the safety and security standards of the old world of paper files into the new world of electronic files.

---

*This paper was presented at the conference in German. I am responsible for the English version, which is based on a translation furnished by Gabriele Bares.

[1] At least this is the way I saw it, when I presented my theses long before the congress took place asking for contributions and discussions of my theses. The only contribution to my theses from Prof. Dr. Dimitrios Maniotis of Greece, however, dealt explicitly with an evidentiary question relating to electronic documents: "On the evidentiary value of electronic documents". I have therefore supplemented my theses under Sect. 12.4 with a section on electronic documents as a form of evidence.

H. Rüßmann (✉)
Faculty of Law and Economics, Saarland University, Campus C3.1, 66123, Saarbrücken, Germany
e-mail: ruessmann@mx.uni-saarland.de

M. Kengyel and Z. Nemessányi (eds.), *Electronic Technology and Civil Procedure*, Ius Gentium: Comparative Perspectives on Law and Justice 15,
DOI 10.1007/978-94-007-4072-3_12, © Springer Science+Business Media B.V. 2012

When I talk about registers, I mean the land register, the commercial register, the register of associations, and other registers. The information in these registers is linked with public trust to varying degrees: Some rely on and protect the positive trust in the correctness of the documents (land registry, good faith, positive publicity) whilst others protect the negative trust in the absence of an undocumented fact where a duty to record such facts exists (commercial register, negative publicity). In both cases, the question of security and authenticity of the electronic documents is a central concern. But these questions are not simply about protection from forgeries, they also concern the sustainability and durability of electronic registers, which will still have to fulfil their documentary function for decades, if not centuries, to come.

12.2 Theses on Security and Authenticity

Electronic files and registers must offer (at least) the same standards of security against falsification as their paper equivalents. Documents of the paper world are certified as authentic by signature requirements. Signatures are required not only on the parties' statements of case, but also on any judicial decision (orders, determinations, judgments). Even third parties attest to the authenticity of a document with their signature (delivery certificates, receipts for payment). The integrity and correctness of the court file is ensured by way of pagination requirements, storage regulations, and access controls. There is, in practice, little real risk of changes to the file being made as a result of unauthorised or dishonest access. Should such changes occur, we trust in the forensic ability to detect them by way of the traces they leave behind on paper documentation.

12.2.1 Security and Authenticity of an Electronic Documentation

It is said that electronic documents can be changed without leaving behind these giveaway traces.[2] If this is the case, the electronic world needs special procedures in order to prevent changes to or forgeries of electronic documents, or at least to make them so difficult that a security level comparable to that of the paper world is achieved. This means in particular:

- The server, on which the electronic documents are stored, must be in a computer centre guarded by the strictest security protocols.
- Access to the information must be via a system, which differentiates reading and writing rights, so that only appropriate access – and where necessary completion

[2] See H. Rüßmann, "Moderne Elektroniktechnologie und Informationsbeschaffung im Zivilprozeß – Die Informationsbeschaffung für den Zivilprozess," in *Die verfahrensmäßige Behandlung von Nachlässen, ausländisches Recht und internationales Zivilprozessrecht*, ed. Schlosser, 137 (Bielefeld, 1997).

and alteration – rights are afforded (and even these only within the limits of each specific authorisation).
- All actions must be recorded.
- Entries should be tagged with a time stamp.
- Where the paper world would require a signature, the electronic world requires a qualified electronic signature.
- Where initials would suffice in the paper world, the electronic world requires an advanced electronic signature.
- For documents, which are to be kept even after the end of proceedings, and for registers, an archiving system which is secure with regard to the current technical situation is required.

12.2.2 The Electronic or Digital Signature

I have named advanced and qualified electronic signatures amongst the requirements for electronic security. These are signatures based on an asymmetric key system. Asymmetric keys work with mathematical formulae. They use random numbers to generate a pair of complex numbers which fit together, on the proviso that knowing one number must not allow you to work out the other.

One of these numbers is called the "public key" and the other the "private key". The public key is made accessible to all interested parties by way of specially equipped computers whilst the private key (ideally) remains with the owner of the key sequence, guarded by the strongest security measures.

12.2.2.1 Ensuring Confidentiality

If a lawyer wants to send the court a message, he searches the court server for their public key. He uses this to encode his confidential message (statement of claim) and sends it. Due to the mathematical attributes of the key, this information can now only be read with the court's private key. Not even re-applying the public key will render the confidential message readable. As long as one assumes the court has looked after its private key properly, one can also assume that only the court will now be able to decode the message. If the court wants to respond to the lawyer, they find his public key from the server and send him a message in exactly the same way as he used the court's public key to send his message to them.

12.2.2.2 Ensuring Authenticity

How, though, can the court be sure that the message really came from the lawyer? The court's public key is available to everyone, and therefore anyone could have sent this message. The solution lies with the private key of the lawyer, with which he seals the message. And then what happens? Simple: He writes his confidential message, and encodes it with the court's public key. The message is now safe from the prying eyes of others (and even from his own). The third step is to apply his own private key to the message. In the same way that the public key can only be decoded with the corresponding private key, a private key can only be decoded with the corresponding public key.

So the lawyer seals the encoded message with his private key. The court receives the message thus sealed and checks whether the message came from the lawyer. To do this, they apply the lawyer's public key to the message. As only his public key will successfully decode the message, the court can be sure – assuming that applying the lawyer's public key does in fact decode the message – that the message is from him. Now the court just needs to use its own private key to allow it to see the original text of the message.

12.2.2.3 Digital Signature

Where the lawyer's (primary) concern is not the confidentiality of the message, but rather simply to allow any manipulation of his data by third parties to be noticed, then he does not have to apply his private key to the whole message in the way I have just discussed. As the asymmetric key procedure is relatively slow, encoding the entire text would take some time. For this reason, amongst others, digital signatures – where authentication and the prevention of manipulation are more important than confidentiality – are usually generated by way of publicly available algorithms in a so called Hash procedure. This involves turning the message to be sent into a "Hash Value", which is no more than a check-sum. The lawyer then applies his private key only to this check-sum, resulting in the generation of a digital signature which is attached to the plain text. When the court receives a message from the lawyer signed with a digital signature, it applies his public key to the signature, which has two consequences: Firstly, the digital signature (the encoded check-sum) is decoded, and the court can see the Hash Value generated by the lawyer. Secondly, the plain text of the message is again used to generate a new Hash Value. If the new Hash Value generated by the court matches that contained in the digital signature, the court can be sure that the content of the message sent by the lawyer has not been altered. If even a single Bit of the message had been changed en route, the two Hash Values would no longer match. That all sounds like an awful lot of work, but in practice it is not a problem, because all the steps of the signature technique can be carried out automatically.

12.2.2.4 Assigning Keys to People

The problem now is making sure that the keys published on the servers really belong to the people they say they do. From a technical point of view, generating a key, which gives off the impression that it belongs to someone else is not difficult;

standard software generates the key sequences and then asks for the name of the user. If Frieda wants to fake Bob's key, she just puts in the name "Bob" and publishes the key on the server. Assuming that no other key already exists, anyone who searches for Bob's key will only find the "false" key from Frieda, which claims to belong to Bob. If there is already a key on the server which (genuinely or otherwise) claims to be Bob's, the person searching will simply be presented with a choice of multiple keys and may end up choosing the wrong one.

This being the case, we must ensure that the published keys really belong to the correct person. But how do we do this? There are several methods, some more effective than others.

Web of Trust

When it comes to it, the keys themselves are nothing more than digital information. As such, they themselves can, and even – due to the increased security it affords – should, be encoded and saved somewhere. But you can also sign them. If Bob and Alice meet up and happen to have their keys "with them", Alice can sign Bob's public key with her private key. This tells Peter, who later does business with Bob, that Alice is sure that Bob's public key really does belong to Bob. If Peter now comes across two keys on a server both claiming to belong to Bob, he will (hopefully) pick the one signed by Alice. This is particularly so if he knows Alice personally and trusts her. It is, of course, possible for Alice to sign Bob's key without ever having met him. This would, however, be reckless of her, especially when one considers that people (like Peter in the previous example) are relying on her "word".

Trust Centres

In the modern legal world it is unlikely that a company or public authorities who want to send Bob a message, and be sure that it reaches the correct person, would rely on the fact that someone unknown to them called Alice has signed Bob's key. They would far rather prefer that an institution conducted the signing of keys in a "sovereign" manner. This institution should have the highest possible security controls and work to as uniform a set of standards as possible.

In everyday life, such an institution would be the passport office, which generates an unambiguous method of proving identity – namely your passport or ID card. The digital life version of these institutions are the so called "Trust Centres". These guarantee that the key provided does in fact belong to the correct person (for example by sending an employee to check the identity documents of the owner and collect the key), or generate the keys themselves, having first checked the official identity documents of the party requesting the key. Of course this system demands that the employees of the Trust Centres themselves have no interest in falsifying the keys, but that's why they are called "Trust Centres". In order for a Trust Centre to be accredited as such, it must fulfil certain requirements which, in Germany, are laid down in the Signature Law.

Paragraph 2, numbers 1–3 of the Signature Law differentiate various types of electronic signatures. Following § 2, No. 1 of the Signature Law, electronic signatures are defined as data in electronic form which accompany or are logically attached to, and serves to authenticate, other electronic data. The next level up (No. 2) is "advanced electronic signatures". These are the electronic signatures which

- Are assigned exclusively to the signature key owner
- Allow for the identification of the signature key owner
- Are generated by a method which the signature key owner can keep within his sole control and
- Are linked to the data to which they relate in such a way that any change in the data can be detected.

Finally, the Law identifies "qualified electronic signatures", which must not only fulfil the criteria demanded of advanced electronic signatures but also

- Be based on a valid, qualified, certificate generated at the time of their creation and
- Be created by a secure signature-creation unit.

These qualified signatures, which are based on a qualified certificate, offer the highest level of security available. Only the so called Trust Centres are allowed to distribute qualified certificates. No approval is legally required to run a certification service (§ 4 Subs. 1, Sig. Law), but anyone who intends to start up such a service must notify the relevant authorities no later than the point at which they start trading (§ 4 Subs. 3, first sentence, Sig. Law). Additionally, the only people who will be permitted to run a certification service are those who can prove that they have the necessary reliability and expertise, as well as the insurance cover prescribed by § 12 Sig. Law. Even then, they will still have to satisfy the further requirements for the running of a certification service as set out in both the Signature Law and the regulations under § 24 No's. 1, 3 and 4 (§ 4 Subs. 2, first sentence, Sig. Law). A person is deemed sufficiently reliable under § 4 Subs. 2, second sentence Sig. Law if they, in their capacity as a certification service provider, guarantee to obey the regulations applicable to their operation. § 4 Subs. 2, third sentence, Sig. Law acknowledges the necessary expertise to be present if those persons responsible for operating the certification service have the knowledge, experience and skills which this operation requires. The further requirements for the operation of a certification service are fulfilled if the relevant authority is presented with a security plan whose measures ensure that the security demands placed on such organisations by § 24 No's. 1, 3 and 4 Sig. Law, and regulations there under, are met. The measures detailed in the plan must, of course, also be effectively implemented in practice (§ 4 Subs. 2, fourth sentence, Sig. Law). The allocation, content and blocking of qualified signatures is regulated in §§ 5 et seq. Sig. Law.

12.3 Electronic Files in Judicial Proceedings

In Germany, the Federal-State Commission (FSC) for data processing and development in the judiciary[3] formed a working subgroup especially for electronic court files. In 2008, this group was given the task of appraising the existing solutions, gathering approaches to improvements, and defining the requirements, from a judge's point of view, of a judiciary-specific solution for the substitution of paper files. Even today, the 23rd of February 2011, the group has not yet produced its final report. It is likely to be published on the internet at some point during 2011.[4] An interim report was presented at a meeting of the German Association for Computing in the Judiciary in September 2009.[5] On the situation in Germany it contained the following pronouncements regarding the legal position combined with the requirements for an electronic court file.[6]

12.3.1 Legal Parameters

12.3.1.1 No Comprehensive Set of Rules

There are no comprehensive rules for using an electronic file in judicial proceedings. Even the existence of paper files is more assumed than really regulated in the codes of procedure. The use of paper as a carrier of information is deemed so self evident by the procedural codes that there is no rule explicitly stating that a court file should be kept, what medium it should be kept on, or which configuration and form the individual files should take. That court files exist, however, is more or less explicitly assumed by countless individual procedural rules. The clearest instances of this can be found in the regulations concerning inspection of files (e.g.: § 299 ZPO; § 46 Abs. 2 ArbGG in combination with § 299 ZPO; § 147 StPO; § 474 f. OWiG; § 100 VwGO; § 78 FGG; § 120 SGG) according to which the parties may inspect the court file and any other files presented to the court. An electronic file would therefore run into specific legal issues besides data protection and privacy issues only if one stops merely complementing and begins to replace the paper file ("leading" electronic file, electronic file as the official file).

[3] Bund-Länder-Kommission (BLK) für Datenverarbeitung und Rationalisierung in der Justiz.

[4] See the information provided by the portal of the justice authorities of the federal and state governments at http://www.justiz.de/BLK/schlussberichte/index.php [31.01.2011]

[5] See the information provided by the German Association for Computing in the Judiciary at: https://www.edvgt.de/pages/startseite/18.-deutscher-edv-gerichtstag/arbeitskreise--mit-praesentationen-und-protokollen/blk-ii-elektronische-akte.php [31.01.2011]

[6] The author of this part of the study is a Judge at the Federal Administrative Court, Professor Dr. Dietmar Berlitz. The language of the report is, of course, German. The translation has been furnished by Gabriele Bares and me.

12.3.1.2 Rules of Procedure

Authorisation to Keep Electronic Files: With the exception of the Criminal Procedure Rules, which do not contain the necessary authorisation, the possibility of electronic court files is universally provided for – albeit with differences in the details – in the procedural rules (see § 298a ZPO; § 110b OWiG; § 46e ArbGG; § 55b Subs. 1,first sentence VwGO; § 52b FGO; § 65b SGG). According to these, the court files may be kept electronically. The precondition for this is the passing of a regulation permitting it, which not only sets the point in time as of which the court files should be kept electronically, but also lays down the organisational and technical basis on which the files are to be created, maintained, and stored. Thus far, no such regulation has been passed for the highest federal courts. Even at the level of regional courts, such a regulation is as yet unknown. A regulation on the admissibility of electronic files in proceedings in front of the federal patent court is currently being prepared, but no details are known yet. It must be assumed that there would be a strong link between the technical and organisational possibilities – such as using a DMS/EPS[7] system to complement the implementation of an already existing court software related to specific matters – and the regulation which sets out the detailed requirements for creating, keeping and storing electronic files. Indeed, it must be assumed that the regulation will, to a large extent, be dictated by the technical capabilities in these areas.

It would make no sense for the regulation to introduce an electronic court file, which was not technically feasible. The core function of the regulation is to create certainty and clarity of law by laying down the ground rules in these areas, within which alternative configurations are possible.

The Purpose of Files: The procedural codes themselves present no clear, binding, specifications for the form of an electronic court file. The principle that court proceedings should be documented is, however, an integral and intrinsic part of these proceedings. This flows from the very function of the judiciary, to resolve disputes through an orderly process, and it is therefore anchored in Art. 19 Subs. 4 GG. The requirement for the actions of those in positions of public authority to be recorded can also be derived from the general principle of the rule of law. The exact dogmatic heritage of the principle is rendered a secondary concern by the fact that the requirements for the form of an electronic court file which it has produced are both vague and in need of further definition and refinement. There are, however, (at least) three interrelated basic functions, which we can be sure an electronic court file must fulfil.

Documentary Function

The contents and structure of an electronic court file must allow for a safe and complete record to be made of whether:

[7] DMS Document Management System; EPS Event Processing System.

- The requirements laid down in the relevant applicable procedural code have, in fact, been met.
- Individual interim steps which are necessary in order for a case to progress (e.g. exchange of written statements of case, giving of notice, and service of documents) have been carried out in a procedurally correct manner (if not, such failures must be able to be identified)
- Other basic procedural rights and rules, including the right to be heard, have been observed.

The DMS must be able to guarantee that the status and progress of the proceedings are evident from the electronic file, that the electronic file acts as a central storage unit and contains all documents related to the particular set of proceedings with which it is concerned. These documents must be filed and made accessible in such a way that the judicial proceedings are comprehensively documented. In order to do so, the electronic file will have to be complete, authentic, true to reality and – within reasonable boundaries – safe from manipulation. This security must extend until a legally binding end to the proceedings is reached, no matter how long this may take. Particularly with regards to security, the electronic file cannot lag behind the paper ones, and must be immune to the specific risks (e.g. electronic manipulation of the contents of the file) associated with electronic data processing. They do not, however, have to offer "absolute" security, which even a paper file cannot provide.

Informative Function

The electronic file must present those charged with running the case (including those with quasi judicial functions in ancillary proceedings such as costs assessments) with a complete, reliable, and demonstrable evidentiary basis on which to make the correct decision. For this function, it is generally of little importance whether information is presented as an original or a copy, as long as the authenticity or evidentiary value of a particular document is not in question. When employed for its informative function, which builds upon the documentary function, it is rarely vital that one is able to comprehensively access the entire contents of a case file. Each process within the relevant interim stage of proceedings generally only requires a small section of the case file. As far as the judges are concerned, the informative function is the most important function of a case file. The DMS must be able to guarantee that an electronic case file will reliably provide the necessary information in order to allow those concerned with the case to quickly demonstrate that individual steps in the proceedings have been taken in a manner which conforms to the procedural regulations (e.g. the right to be heard has been respected as written statements of case have been exchanged, the parties have been properly notified of an oral hearing, questions of limitation affecting legal remedies can be answered by proof of service). Here too, the completeness, authenticity and integrity of electronic court files are of crucial importance in order to allow them to form a reliable basis on which the actual processing of a case can be done. These concerns are pushed into the background by the need for quick, easy, direct, structured and flexible access to the necessary documents.

- Working with electronic files must be facilitated and supported by the ability to search, filter, sort and structure the files.
- There must be overview functions, which allow one to electronically "flick through" the files.
- Also relevant, from a practical point of view, is the time taken to access specific documents and/or the file as a whole, as well as the ability to make working notes on it which are either only visible to the person making them, or at least are not permanently attached to the file (electronic "post-its").

Monitoring/Checking Function

The electronic file is the basis for external "monitoring" and "checking" of judicial proceedings. This can be done by various bodies in several situations including: On appeal to a higher court, in the Federal Constitutional Court if a breach of the constitution is alleged, in the ECHR (for example because proceedings are alleged to have taken an unreasonably long time), by the parties (through inspection of the file) and – within limits which will not be discussed in detail here – by parliament. This monitoring and checking function builds upon the documentary function, and again requires a guarantee that the electronic court file is complete, secure, authentic, and true to reality. Furthermore, all the additional information about the progress of the proceedings required to perform these checks must be demonstrably, accessibly, reliably, completely and securely stored in the files.

12.3.1.3 Location of Electronic Court Files

The independence of the judiciary as a special instrument of state power is an important consideration in relation to electronic court files. This is particularly so for the general and special administrative courts, whose duty it is to exercise checks and balances on the public authorities.

The content, security, and integrity of court files must be assured by the judiciary itself. The electronic court files must be saved in such a way that even the possibility of manipulation of their contents by persons outside the judiciary is, within the bounds of reasonableness, physically and technically impossible. Protection is not only needed against external private (non-state) third parties, but also against "internal attacks" from people or institutions in the service or under the supervision of other non-judicial people or institutions. The documentary function of an electronic court file demands a sufficiently separated, reliably functioning, data security concept.

12.3.1.4 Availability and Persistency of Electronic Court Files

As a central precondition to the transparency and comprehensibility of the judicial process, as well as the basis for satisfying the procedural requirements regarding the protection of legal rights, electronic court files, if they are used to the exclusion of paper files, must guarantee that proceedings can always be carried out smoothly, without any real technical hindrances or failures. This requires a high (which itself needs to be defined more precisely) level of accessibility to the files (protection from system failures); at the same time, the judicial prerogative to decide when a file should be worked on, which is closely linked to their independence, must be respected. Certainly, in order for them to be accepted, it is vital that electronic court files are "mobile" in that they must allow a judge to work on them from home, in a secure and data protection compliant manner. They must also be "portable" enough, for example, to be taken to site inspections and/or sittings at a different location. Long term accessibility necessarily entails durability. It has to be ensured that, at least for their "lifetime" (the time period until they are archived), files are accessible for use (reading, amending, viewing associated access data etc.) even on a long term basis (securing availability, completeness, integrity, confidentiality, authenticity and usability of files in the DMS/EPS).

In any event, it must be ensured that the documents contained and worked on within an electronic court file (including the metadata saved with the documents), and the information regarding the course of proceedings which is required to make the judicial process transparent and comprehensible, can be made available on subsequent systems without loss of information or evidence.

12.3.1.5 Electronic Court Files and Signatures

The procedural codes contain no global stand alone regulatory concept for the introduction of electronic signatures. For "court-internal" purposes and electronic court documents, any document which would normally require the signature of a judge or a court official can be properly completed by the responsible person writing their name at the end of the electronic document and sealing it with a qualified electronic signature as defined by § 2 No. 3 Sig. Law (see, for example: § 130b ZPO; § 110c OWiG; § 46d ArbGG; § 55a Abs. 3 VwGO; § 65a Abs. 3 SGG).

Qualified Signatures

A specific, exhaustive, list of documents, which require such a signature is not contained in the procedural codes. The use of a signature is explicitly provided for in the case of

- Judgments,
- Interim decisions and
- Transcripts.

Even where the law does not explicitly say so, a signature (and therefore a qualified electronic signature) is necessary in several situations including: Judgment forms, orders, letters rogatory, and – according to case law – for procedural decisions with an ongoing effect (e.g.: setting a deadline under § 87b VwGO [BVerwG

NJW 1994, 746] or setting a hearing date for a decision under § 130a VwGO [Buchholz 310 § 130a VwGO Nr 11]). Qualified signatures are also necessary on separate notes and endorsements which are to be attached to others (such as corrections to the facts and notices of publication attached to judgments); electronic "initialling" is insufficient in these cases. The consequences this has for the DMS are that:

- In so far as it is not achieved by specialist software, the DMS must allow for a qualified signature to be applied within any given document.
- It must also be possible to apply a signature to a document as a whole ("container signature").
- Particularly for judgments and orders, multiple signatures must be possible.
- The system must allow the users, without the need for action by its creators, to mark those documents and document types which require a hand written or qualified electronic signature, resulting in an automatic system check or reminder notice being generated when these documents are used.

Plain Text; Other Authentication

In court-internal matters, the orders, acknowledgments etc. do not explicitly require a qualified electronic signature if, and so long as, the person responsible for a specific order, measure, or action has been sufficiently securely identified to the system by some other method, and the processes undertaken are properly logged. This covers all the situations in which, on the current paper files, initialling is used. In these situations, it is a question of practicability (and a matter for future determination within the courts) whether a qualified electronic signature should be required, or whether a lesser form of identification will suffice.

Such cases also generally do not require a special (advanced) signature if, and in so far as, electronic actions can be traced with certainty to specific personnel through the system identification and log-on process. For this to be the case, it will be both necessary and sufficient for the system to record such information in a manner which makes it easy to reproduce. The conceptual guidelines envisage only introducing a more specific (advanced or qualified) form of signature where this is legally required, and otherwise to replace initialling with activity logging.

Incoming Electronic Documents

In regard to documents related to proceedings which can be transmitted to the court under the relevant procedural rules, a qualified signature is required for those documents which are deemed to be equivalent to a written document for which a signature would normally be required (§ 130a ZPO; § 46c ArbGG; § 41a StPO; § 110a OWiG; § 55a VwGO; § 65a SGG; § 52a FGO). Thus far, no use has been made of the option given to the courts to allow the use of an alternative secure procedure – other than qualified electronic signatures – to guarantee the authenticity and integrity of an electronically transmitted document. This being the case, a DMS would be required to:

- Be able to carry out an electronic signature check on any incoming electronic documents and/or automatically save the result of an external signature check with the incoming document so that this itself can be checked.
- Make it easy for the user/administrator to do this, either by linking the two documents (the incoming document and the document regarding the signature check) or by saving the result of the signature check as metadata.
- Make it easy for the user/administrator to set up within the system a list of documents which legally require signature checks. An exhaustive list of all electronic documents which are deemed equivalent to a written document requiring a signature would not be necessary to fulfil the aim of this concept.

12.3.1.6 Transfer of Medium

It must be assumed that, should electronic files be introduced, there is likely to be a lengthy transition period in which a large quantity of documents presented to the court during the course of proceedings will not be in the same format as that with which the file operates. § 55b, Subs. 2 VwGO provides that in this case – in the absence of a contrary regulation under the rules of admissibility – documents, which are not in the format used by the filing system should be converted to the appropriate format before being filed (a transfer of medium).

The specifics of such a transfer of medium are set out in § 298a Subs. 2, 3 ZPO, § 46e Subs. 2, 3 ArbGG; § 110b Subs. 2 OWiG, § 55b Subs. 4 VwGO, § 52a FGO and § 65b Subs. 2, 3 SGG. The core concepts of all these regulations are substantially similar, albeit with important differences in the details, in that if a paper document needs to be converted to an electronic one, a record must be made on the document of when and by whom this procedure is carried out. In the reverse scenario, where an electronic document is transferred to paper form, the printout should bear a record of the result of the integrity check carried out on the document, whom the signature check designated as the owner of the electronic signature on the document, and when the signature check records the electronic signature as having been applied. There is some dispute as to the form a transfer record should take when a paper document is transformed into an electronic one. One opinion is that it should be a specific mark either made on the document or linked to it, whilst others believe that including this information in the metadata – or logging it in some other – would be enough to create a lasting and secure link between the transfer mark and the document itself. Of these, the opinion that an inclusion of the transfer mark in the metadata will suffice is to be preferred. A clarification of the rules on admissibility of electronic documents is desirable. The possibility of making a visible "scan-transfer mark", whether automatically or by hand, on the document during the scanning process – similar to a receipt stamp – should also be considered. The law does not require the transfer mark to contain a signature or initials, nor does it prescribe a minimum size, so long as it is recognisable and legible.

12.3.2 Data and Privacy Protection Law

12.3.2.1 Generally

Data protection laws present a further constraint on electronic court files. Whilst judicial decisions at a regional level are excluded from the scope of the federal data protection law – "FDPL" (§ 1 Subs. 2 No. 2. b FDPL), the federal courts are, as instruments of the administration of justice, public federal bodies (§ 2 Subs. 1 FDPL), and therefore subject to the federal data protection law. Under the general subsidiarity clause (§ 1 Subs. 3, first sentence, FDPL), the provisions of the federal data protection law take precedence over other federal provisions relating to personal data, including those regarding its publication. This class of overriding provisions includes procedural regulations, which either implicitly require or explicitly demand the collection, processing, or use of personal data within judicial proceedings. They also regulate the filing of documents as such, and therefore the filing procedures, which make use of personal data within a document. If an electronic court file is kept, it must conform to the requirements of the federal data protection law, and must overcome the specific risks, which the increased accessibility of, and ease of access to, electronic files (can) present.

12.3.2.2 Technical and Organisational Protection Measures

The demands of the federal data protection law, in particular those contained in the Annex to § 9 FDPL setting out the technical and organisational protection measures required to comply with this law, apply fully to electronic court files (§ 9 FDPL). This law demands that systems controlling and limiting physical access (to the servers/computers), electronic access (with separate controls for the system as a whole and for individual documents within it), distribution of documents, data entry, data deletion and the availability of documents are put in place. In light of the standards already achieved by the judiciary with regards to processing personal (and other) data electronically, these requirements do not extend beyond the status quo set by the current measures. Even today, basic data are held electronically and can be accessed by a multitude of people within the courts. What, if any, new data protection and/or data security measures will have to be implemented depends on which judicial proceedings are going to be conducted with a "leading" electronic file in the future. It must be assumed that, due to the related security efforts they would require, proceedings which demand a very high level of protection, in which, for example, classified documents have to be processed, will not be conducted using an electronic file for a while yet.

12.3.2.3 Deletion/Storage

As well as the gathering and processing of personal data during the course of proceedings, the data protection regulations regarding the deletion of this information must also be considered.

Federal Courts

The federal courts are subject to the "Law on the Storage of Records of the Federal Courts and the Attorney General after the End of Proceedings" – the Records Law (v.22.3.2005, BGBl. I, 837). According to this law, records of the federal courts which are not required for the proceedings may only be stored after the proceedings have concluded for as long as is necessary to protect the interests of those involved or third parties, or is in the public interest. Records to which this applies include file indices, lists of names, index cards, certificates, files and lose-leaf documents, written pleadings, books, printed material, maps, plans, drawings, photos, films, disks, sound recordings and other objects which form part of or an attachment to the files. This would apply to all the equivalent data in the case of an electronic file. A regulation adding specific details to the general limit on storage (§ 2 Records Law) has not yet been passed. It is clear, however, that time starts to run for the purposes of the storage limitations at the end of the year in which the records were ordered to be filed, following the conclusion of proceedings.

The highest federal courts must take into account that any documents which are presented to the court during appeal proceedings will, following the conclusion of these proceedings, be joined with and returned to those used at first instance, thereby creating a single court file which covers all instances of that particular case. Storage and separation issues, and therefore deletion issues, present themselves particularly in relation to the index of proceedings (deletion of data sets within the separate specialist systems), but also in relation to a number of other documents including: the judge's notebooks, which are sometimes kept but do not officially form part of the court file, files for potential cases which have not yet been allocated at first instance, preparatory documents which never became part of or attachments to the actual files (unused material, drafts etc.), the "reference files" used by the judges in the decision making process (which contain working copies of file documents) and the attachments to any expert reports which the judges may order.

Regional Courts

Without closer inspection, there does not appear to be a regulation equivalent to the Records Law for the regional courts. Detailed provisions, which would have to be implemented in any DMS, are generally contained in filing and separation regulations.

12.3.3 Archiving Legislation

The federal courts are subject to the "Law on the Security and Use of Federal Archive Material" – the Federal Archive Law (FAL). Under § 2 Subs. 1 FAL, even the highest federal courts have to offer any documents which they no longer need to fulfil their public duties, including their duty to protect the security of the Federal Republic of Germany or one of her States, to either the federal or the appropriate state archive and, in the case of documents of lasting value as defined by § 3, transfer said documents as federal archive material. Any future DMS must therefore be able to offer, and where necessary transfer, electronic documents to the relevant archive in an appropriate format.

12.3.4 Conclusion Regarding Electronic Court Files

That is as far as the Federal-State Commission goes in discussing the legal parameters for introducing "leading" electronic court files in Germany.[8] It confirms the security and authentication requirements of an electronic court file, which I have set out at the start.

12.4 Electronic Documents as a Form of Evidence

Many legal transactions are carried out electronically, and there are many areas in which electronic documentation is replacing paper documentation. As soon as a dispute arises as to the content of an electronically processed transaction or the conduct of an electronically documented medical treatment, one is faced with the question of the status of electronic documents in the law of evidence. For the lawyers who are bound by the laws of the state, and for the courts who may have to decide such cases, electronic documents as a form of evidence present various complex problems:

1. Does the law allow for electronic documents to be used as evidence?
2. What category of evidence should electronic documents be assigned to in cases where there is a difference in the rules relating to different types of evidence?
3. Are questions of authenticity and content of electronic documents to be decided by an unfettered consideration of the evidence, or is the judge's consideration of the evidence restricted by the laws of evidence?

[8] The report essentially contains an inventory of the systems which can be found in practical use, a list of the technical requirements, and a description of scenarios in which they could be applied to civil, criminal, and administrative proceedings as well as to those in welfare and employment tribunals.

4. How can one present a reliable picture of the authenticity and truth of an electronic document?
5. How does one gain access to the electronic documents, which are not under the control of the party who bears the burden of proof?

12.4.1 *Admissibility of Electronic Documents as a Form of Evidence, and Classification of Electronic Documents Under the Laws of Evidence Contained in the German Code of Civil Procedure (ZPO)*

Under German law, electronic documents are undoubtedly admissible as a form of evidence. Unlike that of other countries, the German procedural law contains no limitations which would hinder the use of electronic documents as evidence in any way.[9]

The forms of evidence known to the ZPO are legal inspection (evidence the judge can see, feel, taste, hear, smell for himself), witness evidence, expert evidence, documentary evidence, and party evidence (given under examination). Of these, documentary evidence is the category, which immediately springs to mind, perhaps with real evidence as a safety net.[10] A document, as defined by the laws of evidence in the ZPO, is a written expression of thoughts, which is capable of proving disputed assertions made by the parties.[11] Electronic declarations of intent lack the necessary embodiment for this.[12] Even printouts of data saved on the computer cannot be considered documents capable of substantiating the declaration of intent, as they do not evidence any original human thought but merely the fact that that data was entered or programmed into the computer.[13] As such, electronic documents used as a form of evidence are not subject to the rules on documentary evidence, but rather those governing legal inspection.[14] In cases where specialist knowledge and

[9] On the difficulties presented by the law of evidence in other countries, see amongst others H. Rüßmann, "Moderne Elektroniktechnologie und Informationsbeschaffung im Zivilprozeß – Die Informationsbeschaffung für den Zivilprozess," in *Die verfahrensmäßige Behandlung von Nachlässen, ausländisches Recht und internationales Zivilprozessrecht*, ed. P. Schlosser, 137 (Bielefeld, 1997).

[10] H.-J. Ahrens, "Elektronische Dokumente und technische Aufzeichnungen als Beweismittel," in *Zum Urkunden- und Augenscheinsbeweis – Festschrift für Reinhold Geimer*, ed. R. Schütze, 1 (München, 2002).

[11] Greimer in: *Zöller*, ZPO, 28th Edn. 2010, Vor § 414, margin number 2.

[12] M. Köhler and H. W. Arndt, *Recht des Internet*, 6th ed., (Heidelberg, 2008), 289; J. Scherer and M. E. Butt, "Rechtsprobleme bei Vertragsschluss via Internet", in *Der Betrieb* (2000): 1009–16.

[13] Greimer in: *Zöller*, ZPO, 28th Edn. 2010, Vor § 414, margin number 2.

[14] Greimer in: *Zöller*, ZPO, 28th Edn. 2010, Vor § 414, margin number 2.

skills are required to complete the security and authenticity checks required, or to visualise the electronic data, the rules relating to expert evidence will be applied.[15]

12.4.2 The Evidential Value of Electronic Documents

Being able to classify electronic documents according to the laws of evidence in the ZPO does not, however, tell us anything about their evidential value. Whereas statements made in private documents are governed by the evidential rule in § 416 ZPO.[16]

> § 416 Evidential Weight of Private Documents
>
> Private documents, as long as they are signed by the author or certified by a notary, provide full proof that the statements they contain were made by the author.

the law in force up until 2005 subjected electronic documents to an unfettered consideration of evidence, as provided for by § 286 BGB:

> § 286 Free Consideration of Evidence
>
> (1) The court, taking into consideration every element of the proceedings and any evidence which has been presented, is free to decide whether or not it has been persuaded that a factual assertion made is to be regarded as being true. The judgment of the court must detail the reasons which the court found persuasive.
>
> (2) The court is only bound by the laws of evidence in the circumstances described by this law.

The judge therefore has to decide, in the individual case before him and taking into account any and all other evidence he has heard so far, whether he believes that the electronic document is "telling" the truth or not. The likelihood of him "believing" the electronic document will depend on the extent to which he can assume that whatever is documented in the electronic statement reflects the truth. Even when considering an embodied declaration in the form of a paper document, there are many different possible ways in which it could have been falsified. With electronic documents, there are even more, several of which will now be addressed.

12.4.2.1 The Electronic Lie

Threats to the reliability of a documentation can originate even from its creator. If it is created in such a way that, right from the start, it does not accurately document what actually happened, then the documentation will contain a lie, or at least

[15] H. Rüßmann, "Moderne Elektroniktechnologie und Informationsbeschaffung im Zivilprozeß – Die Informationsbeschaffung für den Zivilprozess," in *Die verfahrensmäßige Behandlung von Nachlässen, ausländisches Recht und internationales Zivilprozessrecht*, ed. Schlosser, 137 (Bielefeld, 1997).

[16] Regarding their content, cf. J. W. Britz, *Urkundenbeweisrecht und Elektroniktechnologie*, (München, 1996), 136.

an erroneous divergence from the truth. There is as little one can do about this in electronic documents as there is about written lies or errors in written documents. Whilst lasting changes to written documents can, in some circumstances or with specific analytical techniques, be spotted, changes made to an electronic document are essentially undetectable through inspection of the document itself. Where evidence is to be considered freely therefore, paper documents provide a greater level of security than electronic documents and their printouts. This can be demonstrated with a simple example:

If I write out a purchase order on my PC and this order is to be used as evidence in court, it is possible for me to change this order to suit my evidential needs without the court, or even the court specialist, being able to prove that I have done so. As far as the printout goes, this is self-evident: The printout tells you neither when the order was first written, nor whether any changes have been made to it since then. This is, however, not only impossible to see from the printout, but also impossible to tell from an analysis of the data saved on the disk, the electronic storage device. Although the computer operating system both registers and marks the data with the exact date and time it was saved – right down to the second – this information is not a reliable source of the actual point in time at which the data was saved: It takes little more than a few clicks to change the system date and time and thereby generate a time stamp for the data which has nothing in common with the actual time at which it was saved. This manipulation is not visible on the electronic storage device. Thus, anything simply saved in this way without further safeguards, will, if any doubt surrounds it, be rendered evidentially worthless. The situation changes, however, if the file is saved to an unchangeable data carrier, or if a changeable data carrier is secured by being placed in the care of a trustworthy third party in order to ensure that the party with a potential interest in manipulating the data has no opportunity to do so. The situation also changes if system, programme and data access are comprehensively logged in such a way that one can tell from the records which data was accessed by whom, when, and with which programme.[17] The operating systems and programmes we are all familiar with, however, do not offer such capability as standard.

12.4.2.2 Unauthorised Actions of Third Parties

There are also many ways in which third parties can manipulate declarations made electronically or even create declarations which those named as the makers never actually made.

It is easy, for example, to send an email which apparently comes from a different sender, simply by changing the settings of your own email account or programme to send emails under a false name and with a false email address. In this way, anyone could send emails from "Prof. Rüßmann". The recipient of the email has only limited ways of checking who the actual sender was. This is only really possible

[17] On data logging in general, cf. G. Runge, "Protokolldateien zwischen Sicherheit und Rechtsmäßigkeit", in *Computer und Recht* (1994): 710–4.

where the sender operates from a network where fixed IP addresses are assigned. If the sender has dialled in to the internet via a modem and has been assigned a so-called "dynamic" IP address (i.e. every call is assigned whichever address happens to be free at the time), tracing him or her becomes difficult: You would have to contact the provider via whom the connection was made and ask them to provide you with the telephone number of the customer who made it. Electronic signatures, which are discussed in more detail below, offer some protection against misuse.

Orders placed via online forms are also susceptible to abuse; there is nothing to stop someone entering someone else's data.

The special programmes used for online banking are, on the other hand, relatively secure from the unwanted attentions of third parties. They normally use a security system involving PINs[18] and TANs.[19] One of the dangers of this, however, is that the PINs and TANs fall into the wrong hands and are then used by unauthorised people. The danger of an unauthorised person making use of otherwise legitimate access details is one present in all systems, and presents the question whether and how such unauthorised access can be proved. Once again, data logging software which keeps exact records of where a call originated, the name of the computer making the transaction, and the name of the user who was logged on to that computer at the time, is able to do this, but is rarely offered as standard on everyday PC operating systems and programmes.

12.4.2.3 Interim Conclusion on the Evidentiary Value of Electronic Documents

In light of all this, electronic documents leave those who use them in judicial proceedings in a position of serious evidentiary uncertainty. In the opinion of the Federal Supreme Court,[20] even sender records do not constitute prima facie evidence of the delivery of a declaration, but rather are merely indicators to be taken into consideration when assessing the value of the evidence under § 286 ZPO.

Attempts have been made to circumvent these evidentiary problems by concluding "evidence agreements".[21] One of the main problems with these, however, is the question of whether such clauses – where they are contained in the General Terms and Conditions – satisfy the legal requirements for General Terms and Conditions, and therefore whether they are in fact valid. Additionally, it is not for the parties to prescribe the method by which a judge should assess the evidence.

Overall therefore, there is considerable uncertainty as to what evidentiary value can be assigned to individual electronic documents. This uncertainty is troubling to those involved in electronic trade who – to be on the safe side – are increasingly forced to revert to good old-fashioned writing.

---

[18] Personal Identification Number.

[19] Transaction Number.

[20] BGH, 7.December.1994, VIII ZR 153/93; NJW 1995, 665.

[21] M. Köhler and H.-W. Arndt, *Recht des Internet*, 6th ed., (Heidelberg, 2008), 289.

12.4.3 Digital Signatures, Documentary Evidence, and Prima Facie Proof

Digital signatures are seen as one possible solution to the problems I have identified. Digital signatures allow manipulation of the content of electronic messages to be detected, and the actual sender and recipient of the message to be identified. The digital signature was born of the need to guarantee secure electronic communication between multiple parties. This need for security applies to every form of indirect communication via media, meaning any communication not taking the form of two people who know each other speaking directly to each other.

The way, in which digital signatures work has been described at Sect. 12.2.2. What we are now concerned with is the evidentiary value of, and weight attached to, electronic documents sealed in this way.

12.4.3.1 The Evidentiary Value of Digitally Signed Documents

The evidentiary value of digitally signed documents is questionable. It has been established that a digital signature can provide us with a very high level of certainty that the person named in the digitally signed declaration is in fact the person who made the declaration, and that the message has not been forged. This does not, however, provide a conclusion as to what evidentiary weight is to be afforded to such a document under § 286 ZPO.[22] That electronic documents sealed with a qualified electronic signature are deemed equal to paper documents does not change anything (§ 371a Subs. 1, first sentence, ZPO). The law of evidence contained in § 416 ZPO, which this parity brings into play:

> Private documents, as long as they are signed by the author or certified by a notary, provide full proof that the statements they contain were made by the author.

is of the logical construction "if P, then P". This, however, is a logically true sentence, which is not capable of being false in any world. The cost of this guaranteed truth is the complete absence of any informative content. The prerequisite for this rule of evidence to be applicable is the fact that the document originates from its author, which is tantamount to the fact that the statements contained in the document were made by its author.

In 2005, the German legislature addressed the provision of parity within the context of the rules on legal inspection. They read:

> § 371a The Evidentiary Weight of Electronic Documents
>
> (1) Private electronic documents which are sealed with a qualified electronic signature are subject to the provisions regarding the evidentiary weight to be afforded to private documents. The appearance of authenticity of a declaration made in electronic

---

[22] J. Scherer and M. E. Butt, "Rechtsprobleme bei Vertragsschluss via Internet," in *Der Betrieb* (2000): 1009–16.

form, once established through tests compliant with the Signature Law, can only be dispelled by facts which cast grave doubt on the assertion that it was the owner of the signature key who made the declaration.
(2) Electronic documents in the proper format which were created by a public official within the authority of his office, or by a person placed in a position of public trust within their sphere of operation (public electronic documents), are subject to the provisions on the evidential weight to be afforded to public documents. If the document is sealed with a qualified electronic signature, § 437 applies.

The binding of judges regarding their evaluation of private documents as evidence, flows from the second sentence of § 371a Subs. 1 ZPO. The authenticity of a declaration of intent made electronically is, according to this law, fundamentally ascertained by way of the tests conducted under the Signature Law, which use the secret key stored on the signature chip card of the owner to confirm his identity. The owner of the key can only contest the authenticity if he can present and prove concrete facts which render a different version of events a realistic possibility. This affords more protection to the recipient of a declaration of intent that he would have under the rules of the ZPO if he were using handwritten documents as evidence. Under these rules, no prima facie situation applies, and the burden of proof remains upon the recipient of a signed document to prove the authenticity of a signature where his opponent disputes it (§ 439 Subs. 1 and 2, § 440 Subs. 1 ZPO):

§ 439 Declarations as to the authenticity of Private Documents

(1) A party who wishes to prevent another from relying on a private document must give a declaration as to its authenticity in compliance with the provisions of § 138.
(2) If the document contains a signature, the declaration must address the authenticity of this signature.
(3) If no declaration is made, the document is deemed to be accepted as authentic, so long as no intent to dispute its authenticity can be inferred from other declarations made by the party.

§ 440 Proof of Authenticity of Private Documents

(1) The authenticity of a private document must be proved.
(2) If the authenticity of a signature upon a document has been established, or the markings made have been certified by a notary, the text which appears above the signature or markings is presumed to be authentic.

The rules of § 371a ZPO are neither fish nor fowl. The creation of a presumption within the definition of, and therefore with the consequences prescribed by, § 292 ZPO

§ 292 Statutory Presumptions

Where a statutory presumption of facts occurs, evidence to the contrary is admissible in as far as this is not prohibited by law. Such evidence may take the form of an examination of the parties under § 445.

was clearly considered by legislators to be too strong. They therefore introduced a provision which not only differs very little in its practical effect from a presumption, but is also an anomaly within the system of civil procedure legislation in that it contains a statutory rule on the evaluation of evidence in a specific situation.

By using prima facie evidence – whereby the first impression is taken to be the truth – judicial practice has developed a flexible instrument with which to give help, in certain situations, to parties who bear what would otherwise be an insurmountable burden of proof. The judiciary would know how best to use this instrument in electronic business and judicial circumstances, even without instructions from the legislature. They have used it where the correct PIN has been used to fraudulently withdraw cash from an ATM,[23] and they would undoubtedly be able to master cases involving electronic signatures without the need for rules on the evaluation of evidence fixed by statute. If a statutory provision is required to calm the concerns and fears of those involved in electronic commerce, the legislators should make use of the existing provisions on presumptions, and oblige those who carry out their business electronically to demand not only electronic signatures, but also biometric access controls to their systems. PIN protected access to the chip card (on which the electronic signature is stored) is a serious weak point in the link between an electronic signature and a person.

Whether regulated on a prima facie basis or by presumptions, the situation, which presents the biggest problem is the interaction between user and key. There is always a possibility that the user and key become separated, and that the key is then abused by a third party. Even PIN security is only secure for as long as the PIN can remain uncompromised and non-transferrable.[24] This danger however, can never be completely eliminated, and true security can probably only be offered by an access procedure which uses biometric indicators.

A further significant weak point exists at the interface between screen and computer. Even though the former is increasingly seen as equivalent to the latter (and will, from some users, even receive the odd slap for the failings of the latter), this is of course not really the case. As such, the screen we see does not necessarily reproduce exactly what is happening within the computer. This makes it possible to give the user a false visual representation of the contents of the declaration he is signing, which does not reflect the real contents of that document at all.

In light of these weaknesses, the statutory acceptance of prima facie evidence has come in for some heavy criticism.[25]

It has been alleged that the preconditions necessary for the operation of the prima facie approach to evidence have not been fulfilled.[26] It is clear from the case law that the rules on prima facie evidence require there to be an established typical pattern of events. An assessment of evidence on a prima facie basis would therefore require typical life experience to dictate that a (digital) signature, so long as it is based on a

---

[23] See H. Rüßmann, "Haftungsfragen und Risikoverteilung bei ec-Kartenmißbrauch," in *Datenschutz und Datensicherheit* (1998): 395–400.

[24] In a pilot study on electronic communication in judicial proceedings, the lawyers simply handed their chip cards and PINs over to their secretaries in order to allow them to sign the written submissions. Cf. A. Roßnagel, *Die Simulationsstudie Rechtspflege* (Berlin, 1994).

[25] AGV, Opinion on the Law of 11.08.2000, 10 and Federal Association of Notaries, Opinion of 27.07.2000, 12.

[26] AGV, Opinion on the Law of 11.08.2000, 10 and Federal Association of Notaries, Opinion of 27.07.2000, 13.

certificate from an accredited certification office, has been intentionally applied by the key owner. Considering the fact that there is no established experience with digital signatures, this proposition is somewhat far-fetched.

On the other hand, one must remember that the requirements of prima facie evidence which are set out in case law, and to which the courts have bound themselves, only apply to those cases where this matter is not regulated by statute. There is nothing to stop the legislature from passing laws on the effect of prima facie evidence which pay no heed whatsoever to the requirements of case law which are currently binding in common law situations.

Whether the risk should lie with the key owner or not is a topic worthy of discussion. What speaks in favour of the new rules is that the greatest risk they present to the security of digital signatures originates from the key owners themselves, who must look after their keys properly. If they fail to fulfil this duty, it seems appropriate at first glance for them to carry the burden of disproving the prima facie evidence.[27] One must ask, however, whether an access system requiring a five digit PIN would not be too much for the key owner: No normal person is able to remember yet another PIN without writing it down somewhere! In which case, how can it be justified for him to bear the burden of a presumption or prima facie evidence, when there are other access control systems available to the key industry, which are able to guarantee the assignation of a particular signature key to a particular person without the need for a PIN. These other systems are access controls based on biometric indicators, which are immune to manipulation. The rules as they now stand actually hinder, rather than promote, the further development of such security procedures.[28] Under the new rules, one is obliged to rely entirely on market forces to get companies to develop new security procedures, which they will only do if offering these procedures to their customers will give them a competitive advantage.

12.4.4 Obligations to Co-operate in Relation to the Giving of Evidence by Way of Electronic Documents

One final question, which must be addressed separately to the admissibility and evaluation of electronic documents as evidence, is that of the duties placed upon parties to proceedings and external third parties to co-operate where evidence is being given by way of electronic documents.[29] This question can arise in a variety of situations. It is possible that a party bearing the burden of proof may wish to present evidence by way of either his own or someone else's electronic documents. These documents could be under the control of his opponent in court, who does not bear the burden of proof on this issue, or a third party.

---

[27] For an opposing view, see the Opinion of the Federal Association of Notaries of 27.07.2000, 13.

[28] AGV, Opinion on the Law of 11.08.00, 10 and Federal Association of Notaries, Opinion of 27.07.2000, 14.

[29] This question brings the procurement of information into play.

12.4.4.1 Your Own Electronic Documents

If a party bearing the burden of proof wishes to present evidence by way of one of their own electronic documents, they are obliged not only to give the court (and their opponent) access to the document itself – which is normally done by providing a printout – but also to all the information which is required to ascertain the reliability of the document. This will essentially entail the party giving the court, or the court expert, access to their system. If they do not do so, they run the risk of not being able to present the document in evidence, and consequently of losing the case for failing to discharge the burden of proof. In that sense, the burden of proof itself regulates the level of co-operation which is required whilst establishing the truth, and creates with it a de facto "burden of co-operation".

12.4.4.2 Electronic Documents Under the Control of Your Opponent

This "burden of co-operation" theory falls apart, however, as soon as the party bearing the burden of proof wants to present electronic documents in evidence which are under the control of his opponent. Following the quasi-natural law procedural principle of "Nemo contra se edere tenetur!", no one should be expected to damage his own flesh. The opponent in this case would therefore not only be perfectly entitled, whilst knowing that the documentary evidence is against him, to sit with his hands in his lap and do nothing, but because this would also deprive the other party of the evidence he needs to discharge the burden of proof, he would be rewarded for his failure to co-operate by winning the case. This result, albeit in a less extreme form, has found acceptance in some corners, with the basic principle being limited only where substantive rights to the information and its disclosure can be established, or where slight modifications have been made to the parameters of the model and/or to the model itself. These modifications include lowering the standard of proof required from the party bearing the burden whilst simultaneously raising the standard of defence required from the unburdened party and the inclusion of the legal concept of suppression of evidence.[30] Others have attempted to bring German civil procedure in line with "modern procedure"[31] by calling for a general procedural duty of clarification and co-operation to be established in relation to the party not bearing the burden of proof.[32]

---

[30] G. Lüke, "Der Informationsanspruch im Zivilrecht", in *Juristische Schulung* (1986): 2–7; Greger in: *Zöller*, ZPO, 28th ed. (2010), § 138, margin number 8a.

[31] P. Schlosser, "Die lange deutsche Reise in die prozessuale Moderne," in *Juristen Zeitung* (1991): 599–608.

[32] Fundamentally P. Stürner, "*Die Aufklärungspflicht der Parteien des Zivilprozesses* (Tübingen, 1976); *idem*, Parteipflichten bei der Sachverhaltsaufklärung im Zivilprozeß," *Zeitschrift für Zivilprozeß* 98 (1985): 237–56. The "Alternativ-Kommentar zur ZPO" has always recognised the better alternative: Schmidt in: AK-ZPO, § 138, margin number 17 et seq.

The traditional German academics on procedural law[33] and the case law of the Federal Supreme Court[34] have remained true to the principle of "Nemo contra se edere tenetur!" and, particularly in relation to electronic documents, have resorted to exhausting all the possible exceptions to this basic rule which had been developed over time. They have searched for rights under substantive law,[35] discovered special procedural rules,[36] considered the scope and extent of the standard of proof for claims and counter-claims,[37] and have even reached for the instrument of suppression of evidence and the procedural sanctions it carries.[38] The legislature have now finally put an end to these efforts and set both theory and practice on their way towards modern procedure (see below).

12.4.4.3 Electronic Documents Under the Control of Third Parties

There have been similar developments in relation to electronic documents under the control of third parties who are not participants in the proceedings. The German procedural law contained no answer to this problem. Third parties with no interest in the proceedings could only be forced to present either documentary or real evidence if the party wishing to present such evidence could prove that he had a right to that material under substantive law.

In terms of legal policy, this rule was unfortunate. It unnecessarily excluded opportunities for clarification and stood in unresolved contradiction of the almost unlimited duty of every third party, whether as a witness or an expert witness, to help resolve a disputed situation. No reason could be given for why one was not obliged to physically produce the very same material or objects about which the court was able to compel you to give oral testimony under

[33] P. Arens, "Zur Aufklärungspflicht der nicht beweisbelasteten Partei im Zivilprozeß," in *Zeitschrift für Zivilprozeß* 96 (1983): 1–24; G. Lüke, "Der Informationsanspruch im Zivilrecht," in *Juristische Schulung* (1986), 2–7.

[34] Cf. BGH, 11. June 1990, II ZR 159/89, ZZP 104 (1991), 203 with the official head note (Translation): The Code of Civil Procedure places no general duty of clarification upon the party who does not bear the evidentiary burdens of production and proof, or at least none which extends beyond the acknowledged duty to substantiate a denial. The BGH approves K. Schreiber, "Zur Frage, inwieweit die Parteien eines Zivilprozesses eine allgemeine Aufklärungspflicht trifft," in *Juristische Rundschau* (1991), 415–6. A critical opinion can be found in R. Stürner, "Zur allgemeinen Aufklärungspflicht der nicht beweisbelasteten Partei im Zivilprozeß," in *Zeitschrift für Zivilprozeß* 104 (1991): 208–17.

[35] Such as to the production of medical documents.

[36] Such as § 258 Subs. 1 HGB for trade books.

[37] This is the solution of the Federal Supreme Court in footnote 35.

[38] A skilled application of all of these instruments would almost certainly lead to the same results as have been achieved by recognising the duty of clarification placed upon the party not bearing the burden of proof.

§ 390 ZPO.[39] In 1977, therefore, the Commission for Civil Procedure Law published a suggested amendment to the law, which obliged third parties, in accordance with their duty to testify,[40] to present or hold ready physical evidence.[41] Twenty-five years later the German legislature followed the suggestion: Since the 1st of January 2002, third parties are obliged to produce and show any item about which it could be compelled to testify.

Even the party not bearing the burden of proof can be obliged to produce documentary and real evidence:

§ 142 Orders for the Production of Documents

(1) The court may order any party or third party to produce any documents or other records under their control to which any party has made reference. The court may set a deadline for production, and may also order that the documents, once produced, are to remain at the court for a period of time of its choosing.
(2) Third parties are not obliged to produce items if this would be unreasonable, or if they are entitled to refuse to give evidence by virtue of paragraphs 383 to 385. Paragraphs 386 to 390 apply.
(3) The court may order the production of a translation where the documents are written in a foreign language. Such a translation must be carried out by a translator authorised to do so under the regional provisions on the administration of justice. This order cannot be made against a third party.

§ 144 Legal Inspection; Expert Evidence

(1) The court may order the production or inspection of real evidence as well as an assessment by an expert. For these purposes, the court may order a party or third party to produce an object within their possession, and may set a deadline for such production. The court may also order that any inspection of real evidence ordered under the first sentence of this provision be tolerated, so long it does not affect a dwelling place.
(2) Third parties are not obliged to produce or allow inspection of items where this would be unreasonable, or where they are entitled to refuse to give evidence by virtue of paragraphs 383 to 385. Paragraphs 386 to 390 apply.
(3) The proceedings are regulated by all provisions which concern the taking of real evidence and assessments by experts ordered as a result of an application for such measures.

[39] § 390 Consequences of a Refusal to Give Evidence

Where a witness refuses to give evidence or take the oath and gives no, or no legally significant, reason for this failure, he will be liable for the costs caused by his refusal to give evidence, even in the absence of an application to this effect. He will also be subject to an administrative fine or, in cases where this is not practicable, a period of custody in default.

In cases of repeated refusal, an application may be made to compel the witness to testify. The period of custody set for the purposes of compelling testimony should not extend beyond the end of the proceedings at the instance to which it relates. The provisions regarding custody in enforcement proceedings apply

Any appeal of such an order is to be held immediately.

[40] The tie to the duty to testify gives access to privileges not to testify which may in a specific case allow not to disclose the information.

[41] See (Ed. Bundesministerium der Justiz) Bericht der Kommission für das Zivilprozeßrecht, 1977, S. 151 ff.

With this, the Federal Republic of Germany has joined the wave of international development. The right to information and evidence was a lot more pronounced in statutory regulations of other states than it was in the Federal Republic of Germany. Despite the fact that these states too started off with the basic principle of "Nemo contra se edere tenetur!", they have, sometimes with dramatic about-turns, freed themselves of it.[42]

12.4.5 *Summary on the Law of Evidence of Electronic Documents*

Admitting electronic documents as a form of real or expert evidence is unproblematic in Germany. Their evidentiary value, however, is much reduced by the many possible ways in which they can be falsified. In this respect, digital signatures might offer a higher degree of reliability.

The adoption of the prima facie evidence rule in § 371a Subs. 1, second sentence ZPO is a developmental error in the system of the law of evidence, and goes against the principal of free evaluation of evidence. It also does not give enough credence to the weakness inherent in allowing human access to electronic signatures saved on chip cards.

In 2002, Germany strengthened the right to information and evidence, and set both theory and practice on the road to modern procedure. This in turn has guaranteed that the information necessary to assess electronic documents will find its way into proceedings.

[42] See the report of H. Rüßmann, "Moderne Elektroniktechnologie und Informationsbeschaffung im Zivilprozeß – Die Informationsbeschaffung für den Zivilprozess," *Die verfahrensmäßige Behandlung von Nachlässen, ausländisches Recht und internationales Zivilprozessrecht*, ed. Schlosser, 137 (Bielefeld, 1997).

Chapter 13
Modern Communications and Information Technology and the Taking of Evidence

Georg E. Kodek

> "The student will not fail to observe the symmetry and beauty of this branch of the law, under whatever disadvantage it may labour from the manner of treatment".[1]

13.1 Introduction

While the role of information technology for the administration of courts, for filing, docket and calendar purposes is obvious, this paper is going to address the role of information technology in the taking of evidence. Here we find information technology not just in a supporting role. Rather, employing information technology for the taking of evidence goes go the "heart" of the trial and may affect the "cultural core" of civil litigation. In light of the many applications of information technology in this field, some already being implemented and some others contemplated for the foreseeable future, this is a vast topic. Hence the scope of this paper has to be limited to the gathering of evidence, the presentation of evidence and the preservation of the results of that process, i.e. court records and related aspect. E-discovery and electronic documents will be dealt with by separate papers in this volume.[2] Also, this paper will primarily focus on civil procedure since criminal proceedings pose problems of their own even in countries which have a law of evidence for both civil

[1] S. Greenleaf, *A Treatise on the Law of Evidence* (1899), 730.

[2] See the contributions in this volume, Chap. 2 by Richard Marcus and Chap. 12 by Bernhard Rüßmann.

G.E. Kodek (✉)
Civil and Commercial Law, Vienna University of Economics, Vienna, Austria
e-mail: georg.kodek@wu.ac.at

M. Kengyel and Z. Nemessányi (eds.), *Electronic Technology and Civil Procedure*, Ius Gentium: Comparative Perspectives on Law and Justice 15,
DOI 10.1007/978-94-007-4072-3_13,

and criminal cases, one obvious example being constitutional confrontation rights which may prevent remote testimony of prosecution witnesses.[3]

This paper is divided into two parts. The first part is designed to provide an overview of possible fields of application of information technology for the taking of evidence. This part will not aim to describe just the black letter law of one particular country (or a number of particular countries). Rather, the purpose of this part is to give the reader some idea of technologies that are already available in this area. The second part of this paper contains a discussion of "pros and cons" associated with these new methods, and will try to assess how they impact the "culture" of civil litigation.

13.2 Possible Fields of Application

13.2.1 Discovery

One area where information technology is already widely being used is discovery. In large proceedings, millions of documents have to be obtained, stored, indexed etc. This huge task is greatly assisted by commercially available software.[4] Special problems are posed by e-discovery, but these issues are the subject of a separate paper in this volume.[5] Information technology can also be employed for deposing witnesses. One field of application is Computer-Assisted Transcription (CAT).[6] The live transmission of text ("text only deposition") can be effected by way of software like DepoStream of realLegal.com.[7] While this technology has been ridiculed as a "baby steps version"[8] or "the poor man's remote real-time system",[9] this low-cost technology still is valuable for smaller cases or for cases on a tight budget.

Commercial court reporting firms already offer CD-ROMs which contain a sound and video recording of the deposition as well as a transcript.[10] In addition, several companies already offer the deposition of witnesses by way of video-conferencing[11]

[3] See United States v Yates, 391 F 3d 1182 (11th Cir. 2004); see also Amendments to Rule 26 (B) of the Federal Ruls of Criminal Procedure (April 29, 2002) (statement of Scalia, J.). On the other hand, in an often-cited case, the Florida Supreme Court has permitted satellite testimony of an eye-witness from Argentina. See Harrell v State, 709 So 2d 1364 (Fla 1998).

[4] F. I. Lederer, *An Introduction To Technologically Augmented Litigation*, http://wwww.courtroom21.net/About_Us/Articles/auglit.html [05.11.2011].

[5] See the Chap. 2 in this volume by Marcus.

[6] Lederer, op. cit.

[7] See www.DepoStream.com

[8] R. Porter, "The Next Step: Taking Depositions Online," in 37 *Trial* 12 (2001).

[9] Statement by Allen Williams, cited by Porter, op. cit.

[10] Lederer, op. cit.

[11] See www.i-dep.com or DepoCast of LegalSpan.com.

or live-streaming of the transcript. Using streaming technology and a multi-channel transmission, lawyers are transmitted sound and video as well as the transcript of the testimony.[12] Extra software enables several participants at the same time to relay notes, comments etc., put indices or highlight certain passages. While the secure transmission of text has been available for some time, more recent progress in technology permits secure transmission of sound and video as well.[13]

13.2.2 Preparation for Trial

In addition to legal databases which have been in use for at least about 25 years and the more recent possibility of electronic case evaluation,[14] commercial software is available designed to assist lawyers in preparing for trial.[15] Compared to the traditional method of hand-written notes on a legal pad, such modern methods make it easier to amend and rework the notes and also offer the possibility to include diagrams or other pictures which then can be shown at trial directly from the layer's notes. This takes us to technologically augmented litigation which will be discussed in the next chapter.

13.2.3 Technologically Augmented Litigation

In some countries, especially in the United States, the use of modern information technology at trial for opening and closing arguments, but primarily fort he presentation of evidence has already gained considerable importance.[16] The employment of information technology includes the use of video sequences, the presentation of electronically stored documents or graphics and the use of power point slides. In theory, also multi-media presentations are possible, although their use, even in the United States, seems till to be rare.[17]

---

[12] See R. Porter, "The Next Step: Taking Depositions Online," in 37 *Trial* 12 (2001).

[13] See R. Porter, op. cit, who discusses software solutions such as lexisONE and CertifiedMail.com. If only text is being transmitted, secure communication is already state oft he art. See again Porter, op. cit.

[14] See, e.g. http://www.e-jury.net oder www.virtualjury.com.

[15] Such virtual "trial notebooks" (e.g. Trial Director) are commercially available from a number of companies.

[16] See F. I. Lederer, "The Courtroom as a Stop on the Information Superhighway," in *Australian J.L. Reform* (1998); F. I. Lederer, *An Introduction to Technologically Augmented Litigation*. http://www.courtroom21.net/About_Us/Articles/auglit.html [05.11.2011].

[17] For a list of vendors of electronic evidence presentation technology, see http://www.ncsconline.org/d_tech/vendorlist/vendbyproduct.asp?id=19; for vendors of courtroom technology generally, see http://www.ncsconline.org/d_tech/vendorlist/vendbyproduct.asp?id=16 [05.11.2011].

The use of modern information technology can contribute to expedite the trial, because it is not necessary any more to search, retrieve and present documents manually. A more important aspect is the gain in clarity and intelligibility, thus ultimately contributing to the quality of the decision. However, in spite of the costs of litigation being much higher than in Europe, even in the US computer animation is still widely regarded as being too costly.[18] Also, the gain in clarity is arguably more important for jury trials. Thus, this aspect is of lesser importance for European countries where civil cases are usually tried by professional judges. Also another argument sometimes advanced, namely that jurors regard information brought to them by means of a computer screen as more reliable,[19] probably does not equally apply to Europe. For sake of completeness it has to be pointed out that the use of novel presentation techniques also brings about new demands for the court record which has to be able to adequately reflect what transpired during the trial.

An example for the possibilities this new technology brings about was provided some years ago by the Courtroom 21 Project. There, a holographic heart was presented. Wearing special goggles, jurors were able to be in a virtual operating theater, to view the location of any participant and to see what they could see.

Modern information technology has also entered appellate proceedings.[20] This is by no means limited to maintaining and monitoring the docket in electronic form, ideally by using data already generated by the trial court, or the electronic transmission of the contents of the court file. E-filing of submissions also belongs here which is already being used in the United States and a number of other countries.[21] In some countries, e.g. Austria, e-filing is even mandatory.

Such a use of information technology brings about advantages of the "handling" of files, i.e. on the clerical level. Even if this is by no means to be regarded lowly, more interesting is the possible gain in quality through the use of information technology. The remarkable speed of technical developments in this area is evidenced by a US case: In 1997 the California lawfirm Fish & Richardson filed a CD-ROM "paper" in an appellate proceeding. This included the submission itself, hyperlinks to all authorities cited, the record of the trial court proceedings, documents, diagrams and even videoclips und part of a video recording of a deposition.[22] The appellate court

---

[18] Lederer, op. cit. cites an example where in a lawsuit for damages of 1 million $ plaintiff had incurred costs of $2 million. The California Court of Appeals, *Science Applications International Corp. v. Superior Court of San Diego*, 46 Cal Cal. Rptr. 2d 332, decided in 1995 that "high power computer support" does not constitute recoverable litigation expenses. The court voiced concern that "[i]f costs are routinely awarded for high-power technology, most parties will be unable to litigate."

[19] See F. I. Lederer, op. cit. An Introduction to Technologically Augmented Litigation.

[20] See F. I. Lederer, op. cit. "The Effect of Courtroom Technologies on and in Appellate Proceedings and Courtrooms", in 2 *Journal of Appellate Practice & Process* 251.

[21] These include Indiana, Mississippi Montana, New Mexico, North Carolina und Texas (P. A. Talmadge, "New Technologies and Appellate Practice," in 2 *Journal of Appellate Practice & Process* 363 [370 FN 10]). For e-filing in federal courts, see J. B. Hillis, "A Review of Electronic Court Filing in the United States," in 2 *Journal of Appellate Practice & Process* 319 (321).

[22] See Lederer, op. cit.

granted the other party's motion to strike this submission. This decision was largely based on the fact that the lawyers had failed to obtain the court's and other party's consent to this (as it then was) unusual form of filing. The court, however, stressed the advantages of such form of filing and made it clear that it did not want to discourage the use of novel technologies.[23] One year later the same appellate court indeed permitted filing of a similar submission.[24] In the meantime several courts including the US Supreme Court have permitted such filings, with some courts actually encouraging them.[25] It is not surprising, therefore, that in the meantime appropriate software is readily available from a number of commercial suppliers, even if the use of this technology, not least because of the costs, is still somewhat limited.[26] Suitable cases for employment of this technology are unfair trade practices cases, cases with a technological background, copyright cases and cases with lots of maps, charts, diagrams, photos etc.[27]

Sometimes – in addition to the videoconference already mentioned – modern technology is also used during the appellate hearing. An example is an appellate hearing held in February 1999 before the U.S. Court of Appeals for the Armed Forces in Courtroom 21, in which an oral argument was made supported by a computer slide show with highlighted critical parts of the written filings, authorities and a photo.[28]

13.2.4 Modern Courtroom Technology

Modern courtroom technology was pioneered by the former "Court room 21" project, now named Center for Legal and Court Technology.[29] This is a joint project of William & Mary School of Law and the Center for State Courts, which is designed to test available technology in court proceedings. Originally the project focused on the test of recording techniques and the presentation of evidence including

[23] *Yukiyo Ltd v Watanabe*, 111 F 3d 883 (Fed Cir 1997): "[…] there is much to commend the filing of a CD-OM brief […]. By no means […] does the court intend to discourage the filing of CD-ROM briefs under appropriate rules and standards."

[24] *Rodime P.C. v Seagate Technology* 45 U.S.P.Q.2d 2023 (Fed. Cir. 1998).

[25] After several circuit courts the U.S. Supreme Court permitted such filings in *Harris v. Salomon Smith Barney*, 120 S.Ct. 2180 (2000). See also J. M. Snow, "CD-ROM Briefs: Must Today's High Tech Lawyers Wait Until the Playing Field Is Level?," in 17 *The John Marshall Journal of Computer & Information Law*, 615 (1999); F. X. Gindhart, "Documents, Transcripts, Exhibits are on Hand in Hypertext Briefs," in 217 *New York Law Journal* 5 (1997); M. D. Fibison, "CD-ROM Brief Foreshadows the Electronic Courtroom: The Visual Power of a Good Witness Can Sway a Judge's Decision," in *U.S. Bus. Litig.* 17 (1997).

[26] M. Devin, "CD-ROM Briefs: Are We There Yet?," in 2 *Journal of Appellate Practice & Process* 377, estimates that it had been used only in two dozen cases.

[27] M. Devin, op. cit. 2, *Journal of Appellate Practice & Process* 377 (391).

[28] *U.S. v. Rockwood.* Lederer, op. cit. "The Effects of Courtroom Technologies," in 2 *Journal of Appellate Practice & Process* 251 (266).

[29] See http://www.courtroom21.net/.

video conferences.[30] Subsequently the research was expanded to include areas such as e-filing, information management and questions of public access to information about court proceedings.[31] Similar projects exist in several states.[32]

CLCT conducts frequent technology demonstrations in the Courtroom. Presentations customarily include specific hardware and software demonstrations, as well as discussion of the legal and pragmatic implications of use of the given technologies. Presentations can be conducted live or via two-way videoconferencing. Using "six channel" 384K Tandberg videoconferencing, CLCT staff can conduct interactive presentations to any location in the world which has modern videoconferencing communications.

13.2.5 Video Conferencing

In the past years videoconferences have been used increasingly for the examination of witnesses. This development was pioneered in Australia and in the United States.[33] Obviously, such methods offer advantages particularly in cases where the witness and the court are separated by vast distances. However, this is by no means the only case where one might consider to resort to video conferencing. Other cases may include witnesses which are unable to travel or which are in custody.

Under the Federal Rules of Civil Procedure "[t]he court may, for good cause shown in compelling circumstances und upon appropriate safeguards, permit presentation of testimony in open court by contemporaneous transmission from a different location."[34] In the meantime, also several European legal systems[35] as well as the European Evidence Regulation[36] expressly provide for the taking of evidence by way of videoconferences. Pursuant to article 10 paragraph 4 of the Evidence Regulation, the requesting court may ask the requested court to use communications technology at the performance oft he taking of evidence, in particular by using video conference and tele conference. While admittedly a videoconference may not adequately convey the witness' body language and maybe even lead to a loss of

[30] F. I. Lederer, S. H. Solomon, "Courtroom Technology – An Introduction to an Onrushing Future," in *Proceedings of the Fifth Court Technology Conference* (Williamsburg: National Center for State Courts, 1997); F. I. Lederer, op. cit. The Courtroom As a Stop On the Information Superhighway.

[31] F. I. Lederer, op. cit. The Courtroom As a Stop On the Information Superhighway.

[32] An example is the University of Arizona Courtroom of the Future Project (http://www.law.arizona.edu/it/court/courtrm/html/).

[33] For the law in the United States see F. I. Lederer, "The Road To the Virtual Courtroom?" http://www.isrcl.org/Papers/Lederer.pdf [05.11.2011] FN 59 ff.

[34] Federal Rules of Civil Procedure P. 43 (a):

[35] See e.g. s. 128a German Code of Civil Procedure.

[36] Council Regulation (EC) No 1206/2001 of 28 May 2001 on cooperation between the courts of the Member States in the taking of evidence in civil or commercial matters, OJ L 174, 27.6.2001, pp. 1–24.

tonality, it offers a great step beyond traditional judicial assistance proceedings where the requesting court only gets a bare transcript.

From the examination of witnesses by way of videoconferencing it is but a small step to a videoconference with lawyers, parties and judges. This method has been used in the United States for a number of years.[37] Frequent examples include maintenance cases, sentencing hearings and appellate hearings.[38] In appellate proceedings some courts, by providing specially equipped courtrooms, in the meantime have institutionalized this procedure.[39] Not only lawyers may appear by way of videoconference, but also judges can take part in the hearing in this way.[40] From a technical point of view it is important that all participants get a chance to perceive also non-verbal reactions of the other participants, particularly of the judges. Thus, a simple voice-activated system is not sufficient. Instead, either several screens or a split screen image is required.[41]

A more economic alternative to videoconference is the use of commercial chatrooms, which of course only permit the transmission of text. This method has already been used in the United States for pre-trial-conferences.[42]

13.2.6 Court Records

Modern information technology also brings about significant improvements fort he court record. We all know, at least from American movies, stenograph machines. The traditional stenograph machine has often been replaced by computer-assisted

---

[37] An introduction into this topic is provided by F. I. Lederer, "Videoconferencing: Has the Future Arrived?" http://lawtechnews.com/december00/litigation_support_p133.html, F. I. Lederer, op. cit. The Road To the Virtual Courtroom?

[38] F. I. Lederer, op. cit. The Road To the Virtual Courtroom?

[39] F. I. Lederer, op. cit. The Road To the Virtual Courtroom?; F. I. Lederer, "The Effect of Courtroom Technologies," in 2 *Journal of Appellate Practice & Process* 251 (268); see also E. Touissaint, "Minnesota Court of Appeals Hears Oral Argument via Interactive Teleconferencing Technology," in 2 *Journal of Appellate Practice & Process* 395 (2000).

[40] In 1996 in *United States v Salazar* two of the five judges sat in different states and took part in the hearing by way of videoconferencing. In addition, an amicus curiae supplemented his brief by electronic submissions. See F. I. Lederer, loc. cit. The Courtroom As a Stop On the Information Superhighway. An additional example is provided by F. I. Lederer, "The Effect of Courtroom Technologies," in 2 *Journal of Appellate Practice & Process* 251 (269).

[41] F. I. Lederer, "The Effect of Courtroom Technologies," in 2 *Journal of Appellate Practice & Process* 251 (272 FN 6). In Italy six cameras are used for the video recording (not for a videoconference). See M. Fabri, "Introduction: State of the Art, Critical Issues, And Trends of ICT in European Systems," in *Justice and Technology in Europe*, ed. M. Fabri and F. Contini (2001) 13 FN 27; D. Carnevali and M. C. DiCocco, "An Innovation Process Embedded in a Strict Institutional Setting: ICT in the Italian Judicial System," in *Justice and Technology in Europe*, ed. M. Fabri and F. Contini (2001) 209.

[42] See the article "Judge holds Pretrial Hearing Online," http://www.apbonline.com/cjsystem/justicenews/2000/06/02/hearing0602_01.html/. The judge stressed that the case (which was about the installation of windows) did not pose particular data protection concerns.

transcription (CAT).[43] For several years conversion of the short-hand notes into full text is done by the use of computer programmes. This is essential for computer-assisted processing of the record. This aspect is particularly important for complex cases. In the meantime, such transcripts often are available in real-time (real-time record) and can be made available instantly to the lawyers (also to other lawyers working on the case somewhere else), but also to experts etc.[44] Voice recognition systems may be the next step, although it appears that these systems, while certainly having the potential to bring about a revolution in ordinary office work, are not yet sufficiently reliable to consider them as substitutes of more traditional forms of court reporting.[45]

The examples mentioned so far have only dealt with records which only contain text information. This traditional approach is already undergoing important changes in some countries. Many countries already use voice recording equipment. In some countries, such as Germany or Austria, the tape is only used by the judge who dictates a summary of the evidence.[46] This use of the tape makes it possible to use court personnel more efficiently, but does not bring about an increase in quality.

However, in the meantime technology has progressed much farther: In some countries the entire trial is taped. In recent years in some countries video recordings have been employed in lieu of the traditional court record. Examples include Italy,[47] Spain[48] and the United States where the use of video recordings in court proceedings was pioneered by Kentucky.[49] However, not until 1993 was there a right to videotape a deposition for use in a U.S. federal court. A video recording of testimony certainly preserves additional information, but on the other hand has the disadvantage that retrieval of certain parts of the record may be quite time-consuming.[50] Therefore, sometimes a transcript is provided in addition to the video recording,

[43] See B. Miller, "Court Reporting. From Stenography To Technology: Will court reporters evolve from stenographers to information managers, or be replaced by new technologies?" http: //www.govtechnet/magazine/gt/1996/mar/courts/courts.phtml [05.11.2011].

[44] F. I. Lederer, op. cit. The Road To the Virtual Courtroom? FN 31.

[45] See P. Bauer, "A Show Case for the Future: E-Justice in Austria," in *Justice and Technology in Europe*, ed. M. Fabri and F. Contini (2001) 55. See also http://www.speechmag.com. In the United States "voice writing" is used only by way of a court reporter who repeats everything that is being said, speaking into a special type of mask. F. I. Lederer, op. cit. The Effect of Courtroom Technologies, in 2 *Journal of Appellate Practice & Process* 251 (258).

[46] See section 212a, Austrian Code of Civil Procedure, section 160a, German Code of Civil Procedure.

[47] D. Carnevali and M. C. DiCocco, "An Innovation Process Embedded in a Strict Institutional Setting: ICT in the Italian Judicial System," in *Justice and Technology in Europe*, ed. M. Fabri and F. Contini (2001), 197 (209).

[48] I. Sanchez and P. A. González, "Towards the Integration of Case Management And Case Law: ICT in the Spanish Judicial System," in *Justice and Technology in Europe*, ed. M. Fabri and F. Contini (2001), 257 (271).

[49] For digital audio and video recordings, see F. I. Lederer, op. cit. The Road To the Virtual Courtroom, and F. I. Lederer, loc. cit. The Courtroom As a Stop On the Information Superhighway 4.

[50] See F. I. Lederer, "The Effects of Courtroom Technologies," in 2 *Journal of Appellate Practice & Process* 251 (257, FN 19).

either in all cases or in cases where an appeal is filed. This may soon belong to the past, however: There are already systems available which combine a full text transcript with synchronised digital audio and video recordings.[51] Similar systems are already used in the course of depositions during discovery on a routine basis. Such a system comprises multi-channel sound recordings on hard drive, and on DVD[52] or CD if required, and video recordings by voice-activated cameras from four or even more vantage points. The record can be read as a text, with the pertaining video sequence only a mouse-click away.

This may indeed turn out to be a quantum leap in trial technology, for it enables the court of appeals to obtain a direct impression of the witnesses' demeanor which is so important in evaluating the evidence.[53] This may have significant impact on appellate proceedings[54]: The reason why in many countries the findings of fact cannot be appealed at all or the trial courts are given great deference by appellate courts when it comes to fact-finding, is that here the impression created by the witnesses is so important which is not available to the appellate court.[55] How the availability of video recordings will affect the success rate of appeals is, of course, difficult to predict.[56] Experience gained in Kentucky seems to suggest that the success rate will be lower, i.e. that additional information available to the appellate courts lead to more decisions of trial courts being affirmed.[57]

Another question is, of course, whether a legal system views the additional costs associated with such an increase in information available to appellate courts worthwhile, particularly since the increase in available information is likely to be relevant only in a relatively small number of cases. Thus, from a practical point of view it

[51] An example is the TIMARO software which was developed in 1997. See F. I. Lederer, op. cit. The Courtroom As a Stop On the Information Superhighway 4.

[52] F. I. Lederer, "The Effect of Courtroom Technologies," in 2 *Journal of Appellate Practice & Process* 251 (256).

[53] Commercially available software combines this with an examination of the witness' credibility by way of a voice stress analyzer, a development which poses considerable concerns. See M. Devin, "CD-ROM Briefs: Are We There Yet?," in 2 *Journal of Appellate Practice & Process* 377 [385].

[54] See F. I. Lederer, "The Effect of Courtroom Technologies on and in Appellate Proceedings and Courtrooms," in 2 *Journal of Appellate Practice & Process* 251 (2000).

[55] See, e.g., Judge Learned Hand in *Petterson Lighterage & Towing Corp. v. New York C.R. Co.*, 126 F 2d 992, 994 f (2d Cir. 1942): "[D]ecisions [are] legion that when a judge ha[s] seen and heard the witnesses his conclusions [will] prevail unless clearly wrong", see also the graphic expression in *Gavin v. State*, 473 S 2d 952, 955 (Miss. 1985): "The trial judge who hears the witnesses live, observes their demeanor and in general smells the smoke of the battle is by his very position far better equipped to make findings of fact which will have the reliability that we need and desire."

[56] Some appellate courts have indicated that they might review the findings of facts more closely, although most courts appear to be reluctant to do that. See R. C. Owen, M. Mather, "Thawing Out the 'Cold Record': Some Thoughts on How Videotaped Records May Affect Traditional Standards of Deference on Direct and Collateral Review," in 2 *Journal of Appellate Practice & Process* 411 (416).

[57] J. A. Maher, "National Center for State Courts, Do Video Transcripts Affect the Scope of Appellate Review?," in *An Evaluation in the Kentucky Court of Appeals* (1990).

may well be questioned whether the improved review of judgments justifies this significant increase in costs. In case such a multi-media record is made available to the parties only if they pay for the extra costs, experience with verbatim transcripts (which are available in some countries, like Austria,[58] if the parties pay extra for them), suggests that parties may be reluctant to bear the additional costs this new technology entails.

13.2.7 Scanning of Documents

Another aspect of the use of modern information technology is the possibility to scan documents, either by way of making a digital "picture" (imaging)[59] or – in order to permit electronic processing – by way of optical character recognition (OCR).[60] This technology enables the lawyers to find documents easier, particularly in complex litigation,[61] but may also assist in the presentation of documents. By way of document cameras[62] or computers electronic images are displayed to the fact finder, judge or jury. These images appear either on large screens or on small computer monitor LCD displays. It is generally accepted that the use of technology to present evidence, which includes presentation of information in the form of opening statements and closing arguments, substantially shortens trials. Anecdotal evidence suggests that modern display technology may result in time savings of 10–50% compared to traditional methods.[63]

One advantage of this method is that it is possible to present a document on several screens simultaneously. While this may significantly improve the public's ability to observe the taking of evidence, where, particularly in civil proceedings, documents play an important part, one reason for the employment of this technology in the United States is that it may assist in compensating different reading speeds among jurors.[64] The new technology makes it possible that every juror can watch the document on his screen until he has read it completely. The fact that every

---

[58] See s. 280 Austrian Code of Civil Procedure.

[59] For imaging in complex litigation see National Center for State Courts, http://www.ncsc.dni.us/ncsc/briefing/image.htm/.

[60] See F. I. Lederer, "Courtroom Technology From the Judges' Perspective," http://www.courtroom21.net/About_Us/Articles/judicial.html [05.11.2011].

[61] Compare the impressive examples given by F. I. Lederer, "The Road To the Virtual Courtroom," who mentions cases with 30.000 and 1.5 million documents and 500 files on each side.

[62] Such cameras are often referred to as "Elmos" after the well-known manufacturer ELMO. However, similar technology is also offered by a wide range of other companies, including DOAR Communications, WolfVision, Samsung and Sony.

[63] See F. I. Lederer, "Technology-Augmented Courtrooms: Progress Amid a Few Complications, or the Problematic Interrelationship Between Court and Counsel," in 60 *New York University Annual Survey of American Law* 675 (2005) at 676.

[64] See F. I. Lederer, op. cit.

juror has a computer screen of his own may also contribute to the speed of the trial. Also, scanned documents may be used to compensate physical impairments such as blindness. Thus, in 2001 a blind witness testified with the assistance of scanned documents printed in Braille.

13.2.8 Selected Problems of Evidence

The increased use of modern information technology in business and private life brings about a number of additional problems in the area of evidence law. These problems range from the "inspection" of a computer pursuant to § 809 German Civil Code[65] to the use of experts to determine whether electronic documents have been manipulated or to find out the sender of an email. In the United States in the meantime a new category of experts has evolved, so-called software forensic consultants.[66] In November 2005, a group of European multidisciplinary experts formed, setting out to work on what is called the Admissibility of the Electronic Evidence Project.

For many years experts have been using computers for the preparation of their reports. One example is the use of computer simulation for the investigation of the cause of traffic accidents. These computer programs basically "try out" several versions of events until they find a version consistent with the evidence, e.g. the final position of the vehicles involved. Because of the apparent competence and authenticity of such computer simulations it is important to understand the underlying models and assumptions. These include the traction of tires, the forces stressing (and absorbed by) shock absorbers, wind conditions and driver performance. The most important problem is that "impact hypotheses" have to be used in order to estimate under what angles and with what speed the vehicles disengage after the collision. The traditional way of "working backwards", on the other hand, starts from the end position of the vehicles and uses, involving estimates of traction and effects of the brakes, the exact location of the impact and the speed of the vehicles involved. Special software for experts such as PC Crash™ is available. PC Crash™ is a collision and trajectory simulation tool that enables the analysis of a wide range of vehicle collisions and other incidents. Results can be viewed as 3D-animations and detailed reports, tables and graphs.

Another question whether and to what extent the internet can be used as a source of information in court proceedings. Clearly, the answer here cannot be a simple "yes" or "no". Indeed the various databases and other sources available differ

[65] Berlin Court of Appeals (Kammergericht), judgment of August 11, 2000, 5 U 3069/00, JurPC Web-Dok.249/2000. According to this judgment, § 809 BGB can provide a basis for inspection of computers in order to find out whether copyrighted software is being used.

[66] W. R. Leibowitz, "E-Evidence Demands New Experts," in *National Law Journal*, March 9, 1998; see also F. I. Lederer, op. cit. The Road To the Virtual Courtroom FN 55.

vastly in reliability and acceptance. One decision found no error in a case where the journalist responsible for an article was found out in this way.[67] On the other hand it would obviously go too far to consider all information available on the internet as public knowledge.[68]

13.2.9 *Virtual Trials and Cyber Courts*

In theory, the use of videoconferencing makes it possible to have a trial without a single participant being physically present. The technical resources for such an exercise are already available. In theory an increase in "authenticity" may be achieved for "virtual" court hearings by use of a CAVE, i.e. a cabin which, by use of three D-goggles, creates a three dimensional (3D) impression of the courtroom. While this possibility has at least once been contemplated in the United States,[69] it seems that it has not been actually used in practice. At any rate, the gain such technology could offer to civil litigation is not entirely clear.

On the other hand several states have been quite active in installing cybercourts.[70] The most far-reaching initiative probably is found in Michigan. There, in 2002 a cybercourt was set up which is part of the ordinary justice system,[71] This is done in order to provide an incentive to business, particularly in the IT business, to settle in Michigan.[72] The cybercourt is to solve commercial and business-related disputes and also offers alternative dispute resolution mechanisms. This development is part of the tendency in recent years in the United States to create special courts, particular for business related litigation.[73] The Michigan cybercourt differs from this project in that it focuses on the use of modern information technology for the entire proceeding. Not only are submissions to the court to be filed electronically, but there will be also virtual court hearings; all proceedings including hearings are to be held electronically (including audio and videoconferencing technology and internet con-

[67] Court of Appeals M. Frankfurt, January 1, 1999, 22 W 58/98, ZAP EN-Nr. 796/99.

[68] See also F. I. Lederer, op. cit. The Courtroom As a Stop On the Information Superhighway 7 ("slight expansion of the doctrine of judicial notice of facts which are readily verified").

[69] See F. I. Lederer, op. cit. The Road To the Virtual Courtroom?

[70] For a discussion of online dispute resolution in general, see G. Kaufmann-Kohler, T. Schulz, *Online dispute resolution: challenges for contemporary justice* (2004).

[71] State of Michigan, 91st Legislature, Regular session of 2001, Act No. 262 (enrolled house bill No. 4140), which entered into force on January 9, 2002. See also http://www.michigancybercourt.net

[72] Sec. 8001 Abs 2 leg. cit. "The purpose of the cyber court is to […] (a) Establish judicial structures that will help to strengthen and revitalize the economy of this state". Paragraph 2 (d) expressly mentions "technology-driven companies."

[73] Compare the Maryland Business and Technology Court Task Force Report, http://www.courts.state.md.us/finalbtreport.pdf/ [05.11.2011].

ferences).[74] The Cyber Court has concurrent jurisdiction over business or commercial disputes in which the amount in controversy exceeds $25,000.00. These disputes include disputes such as those involving information technology, software, or website development, maintenance, or hosting; as well as those involving the traditional type of business disputes such as the rights or obligations of shareholders, business contract disputes, commercial bank transactions, commercial insurance policies and commercial real property. Some particularly sensitive areas such as criminal cases, labour law cases, landlord-tenant disputes and tort cases are excluded.

The defendant can, within 14 days after expiration of the time limit for an answer, ask to have the case removed to the county court. If he fails to do so, this constitutes a waiver of his right to have a jury trial. Decisions of the cybercourt are subject to an appeal to the (non-virtual) court of appeals. Also lawyers not admitted in Michigan are allowed to appear. The decision is rendered by a single judge sitting without a jury. Whether the cybercourt will fulfil the expectations or whether it will become a footnote in legal history, time will tell.

13.3 The Impact of New Technologies

If one tries to assess the impact of modern information technology on court proceedings, particularly on trials, the resulting picture is quite ambivalent. Here one has to distinguish between the use of information technology by the court's clerical staff in a merely supporting role, and the possibility of a complete or partial "virtualization" of the trial by a reduction of personal contacts between the participants.

The use of video conferences enables persons to take part in the proceeding whose presence would not be possible or would be too expensive. This is an important progress compared to traditional judicial assistance, with video conferencing or at least a video record providing some form of "immediate" or "direct" impression of the witness and thus providing more information than could have been gained under traditional judicial assistance.

However, this technology should not to be employed on a routine basis without valid reasons. The lack of direct contact between the participants results in disadvantages which go beyond the social isolation described in connection with tele-working.[75] Rüßmann has already pointed out some specific legal aspects to be observed in this

[74] Sec. 8001 Abs 2 (h): "Establish virtual courtroom facilities, and allow the conducting of court proceedings electronically and the electronic filing of documents." Sec 8015: "All matters heard in the cyber court shall be heard by means of electronic communications, including, but not limited to, video and audio conferencing and internet conferencing among the judge and court personnel, parties, witnesses, and other persons necessary to the proceeding."

[75] H. Rüßmann, "Herausforderung Informationsgesellschaft," in *Procedural Law On the Threshold of a New Millennium*, ed. CLC/IAPL, (2001) 205 (206).

context.[76] The personal contact and interaction of the decisionmakers often makes a settlement easier; sometimes the personal contact may be essential for reaching a settlement.[77] This possibility would be lost in a completely "virtual" proceeding. More important are disadvantages relating to the court's evaluation of the evidence. A "virtualiziation" of the trial here brings about two shortcomings: On the one hand the judge is deprived of the possibility to obtain a direct impression of the parties or witnesses which is essential for assessing their credibility. It is therefore no coincidence that the appearance of lawyers by way of a video conference is encountered more often than the examination of witnesses in this way. On the other hand the witness, if testifying by way of video conference, is less likely to feel and understand the importance of the court proceeding and of his testimony,[78] even if this aspect may be impossible to capture in a precise analysis.[79] Finally the use of video conference may bring about the danger of manipulations, such as "assisting" the witness by giving signs, by prompting or even by use of a tele-prompter,[80] which are not possible if the witness is examined in the traditional form. It is for these reasons that a video conference is viewed with scepticism in difficult cases in the United States.[81]

Also it has to be pointed out that a video conference even out of court is always only used as a substitute for a personal meeting if this is not possible or too costly. In court proceedings, given their importance and their "exceptional" character, the need for inter-personal contact should be given even greater importance. At least from a European point of view it seems hardly necessary, nor indeed desirable, that parties, while strolling through a shopping mall, can "log in" into a court proceeding and thus take part "virtually".[82] That one has to take time underlines the importance of the proceeding. The physical presence of the judge also demonstrates the attention given to the parties; the parties thereby also get the feeling that they are actually being listened to. That the court is listening to the parties directly and visibly is an important aspect of humanity of proceedings which would be lost by a far-rcaching "virtualization" of trials.

[76] H. Rüßmann, op. cit.

[77] H. Rüßmann, "Herausforderung Informationsgesellschaft," in *Procedural Law on the Threshold of a New Millennium*, ed. CLC/IAPL, (2001) 206.

[78] An additional difficulty results from the fact that giving false testimony by way of videoconferencing may not be punishable as perjury in some states. See F. I. Lederer, op. cit. The Road To the Virtual Courtroom?

[79] In the United States a videoconference is felt not to be suited in cases where credibility of the witness is an issue. See Porter: op. cit. The next step: Taking depositions online, in 37 *Trial* 12.

[80] F. I. Lederer, op. cit. The Road To the Virtual Courtroom? Reports a medical malpractice case before the New Jersey Superior Court, in which the plaintiff who was completely paralysed, was observed by a retired judge during his testimony in order to prevent improper interference.

[81] See Porter: op. cit. The next step: Taking depositions online, in 37 *Trial* 12 mwN; F. I. Lederer, "The Effect of Courtroom Technologies," in 2 *Journal of Appellate Practice & Process* (260); A. Ramasastry, "Government-to-Citizen Online Dispute Resolution: A Preliminary Inquiry," http://writ.news.findlaw.com.comentry/20020206_ramasastry.html

[82] This is the scenario discussed by F. I. Lederer, op. cit. The Road To the Virtual Courtroom?

The possible influence of the atmosphere of the courtroom on the witness' willingness to testify truthfully leads us to another problem: A significant or even complete "virtualization" of proceedings could lead to a loss of symbolic functions of a court proceeding. This addresses an aspect of legal culture which may not be adequately captured rationally, but which nonetheless is important for the acceptance courts enjoy in society. If the state provides impressive buildings and rooms for courts, and requires judges and lawyers to wear robes, and if the court room is adorned with the state crest or flag or even, like in some European countries, a crucifix, all these factors are designed to demonstrate the importance of the proceeding.[83] Much of this symbolic contents would be lost in a virtual proceeding.

13.4 Modern Information Technology and Traditional Principles of Civil Procedure

From the practical advantages and shortcomings of the use of modern information technology, even if they may be in the center of public debate, one has to distinguish their influence on the structure of civil procedure. The facilitation of filings or their speedier processing only relates to the external management of the case; for the structure of the proceedings, its cultural nucleus[84] these aspects are of lesser importance. Assessing the importance of other aspects of the use of information technology in this context is more difficult. This chapter tries to analyze the effects of the use information technology against traditional principles of civil procedure such as oral and public proceedings, including the public tendering of the decision, and the principle that the decision has to be based on proceedings which transpired before the judge herself rather than before some detached official like an investigatory magistrate or the like.

Speedy disposition of proceedings per se, while of course desirable, probably is not a principle of civil procedure in its own right. Also, the contribution of modern technologies to the speed of the trial and the taking of evidence itself is not to be overestimated. More important may be their effect on procedural efficiency which, however, is not to be given precedence over other maxims. The principles of orality and immediacy, meaning that the entire proceeding has to take place orally in front of the judge who is to decide the case, are primarily important for the taking of evidence. In this respect most of the modern technologies discussed above do not bring about a significant change.

---

[83] Lederer, op. cit. 2 *Journal of Appellate Practice & Process* 251 (273), highlights that it is important "to give justice a pride of place and to enshrine it in physical form".

[84] See B. Heß, "Aktuelle Perspektiven der europäischen Prozessrechtsangleichung," in *JZ* 2001, 573 (582).

A notable exception could be video conferences. Here one has to distinguish: The video conference is always only a substitute for immediate personal contact. This applies equally to court proceedings and out of court. The extent to which one will accept to rely on this method depends on the importance and urgency of the matter, but also on the difficulties a "live" direct examination might pose. This balancing will lead to different results in the United States or Australia on the one hand and a small country like Austria on the other hand. In some cases, if the witness were not available for direct testimony, the video conference may actually bring an increase in information.

A more important aspect, at least potentially, of the use of modern technology at trials is its effects on the principle of publicity of proceedings. By publishing the record on the internet potentially a much larger audience is reached than by the public listening in the court room or by posting information on the court's notice board. Computerized court registers such as the land register or company register and, in some countries, the bankruptcy register provide an example. Their use is no doubt enhanced by electronic access. In civil litigation it is conceivable that either the trial is televised or that the record and the decision are published on the internet. A live stream of the hearing on the internet is already practiced by some appellate courts in the United States.[85] This gain in publicity, however, has to be balanced against disadvantages associated with this, such as the loss of natural and unimpeded communication and, thus, the danger of undue influence on the trial, aspects which cannot be dealt with in greater detail in the framework of this paper.

Sometimes court documents are also published by private parties, generally for business reasons. Thus, several years ago a Vienna lawfirm posted all written submissions, transcripts and decisions in a high-profile case involving an alleged violation of the duty of care on part of the Austrian banking supervisory authority on the internet.[86]

The question of factual access to information which ought to be publicly available takes us to a more general problem, i.e. access to justice. The use of some applications of modern technology may also raise concerns under article 6 ECHR, namely the principle of "equality of arms". In particular, the costs of technologically augmented litigation entails may turn out to be problematic in this respect, although at least in continental Europe, where juries are not used in civil cases, the practical significance of this technology has been limited so far. Ultimately, legal systems will have to strike a balance between means essential for the defense (which if necessary

---

[85] H. V. Samborn, "Plenty of Seats in Virtual Courtrooms," in 86 *ABA J* 68 (2000); S. McEwen, "TV or Not TV: The Telecast of Appellate Arguments in Pennsylvania," in 2 *Journal of Appellate Practice & Process* 405 (2000); for Florida see http://www.wfsu.org/gavel2gavel/, for Wisconsin www.courts.state.wi.us/WCS/scoa_search.html. See also Northwestern University's Oyez-Project, http://www.oyez.org. – A list of American courts publishing information on the internet is provided by the National Center for State Courts (http://www.ncsc.dni.us/NCSC/TIS/Tis99/Courtbus.htm)

[86] http://www.putz-rae.at/start.htm. The case was a test case the outcome of which was important for approximately 500 potential plaintiffs. The documents were shown in pdf format.

will have to be supplied by way of legal aid) on the one hand and permissible, yet not indispensible methods or "luxury" methods on the other hand.[87] Here we have to bear in mind that for reasons of procedural efficiency many legal systems already have limits for the costs to be refunded to the successful litigant or for the scope and kind of assistance provided under legal aid. Examples include representation by more than one lawyer, the examination of witnesses living far away in some instances or a verbatim transcript of the trial.[88]

13.5 Conclusion

Impressive as the new technologies available are, it is perhaps appropriate to warn from expecting too much from them in the context of trials themselves. Of course prognoses in a fast-moving field like information technology are extremely difficult. Yet a number of factors suggest that for the foreseeable future impact of these technologies on trials, in remarkable contrast to our lives in general, is likely to be limited. The reason for this is not so much the fact that the way a trial is to be held is governed by detailed provisions which at present in many countries make it difficult if not outright impossible to accommodate modern information technology since these provisions could be, and in some countries already have been, amended. More important are the fundamental differences between civil proceedings on the one hand and e-commerce or other forms of e-government, particularly in administrative law, on the other hand:

First of all – compared to the dimensions of e-commerce – court proceedings are much less frequent. Thus, any reduction in costs (and, therefore, also the pressure for such reduction in costs) which can be gained by employment of modern technology is arguably much smaller. Also, the complex structure of court proceedings imposes limits on the use of information technology. In addition, for most citizens, court proceedings – unlike e-commerce or maybe applications to administrative authorities – are not daily or frequent routine, but exceptional occurrences. Thus, the advantages to be gained from the use of modern information technology for the individual citizen are likely to be relatively small.[89] Also, court proceedings – unlike everyday dealings and business transactions – typically are less urgent, so that the need for speeding up proceedings by employing modern information technology is less pressing.

---

[87] See *Science Applications International Corp. v. Superior Court of San Diego*, 46 Ca. Rptr. 2d 332, 338 (Ct. App. 1995): "If a party litigant chooses unwisely to expend monies in trial presentation in excess of the value of the case, utilizing advanced methods of information storage, retrieval, and display, when more conventional if less impressive methods are available, the party must stand his own costs."

[88] For Austria see section 280, Code of Civil Procedure.

[89] This may be different for litigants such as institutional creditors who file complaints frequently.

The special function of court proceedings also result in special requirements as to data security, long term data storage and data protection which bring about in significant expenses.[90]

The special function of court proceedings probably is the most important factor militating against employing all methods that are technically feasible. Rüßmann has already pointed out quite correctly that the purpose of court proceedings must not be "immolated on the altar of modern technologies".[91] While no one is likely to dispute this statement in theory, practical experience shows that sometimes technical possibilities live a life of their own and, contrary to their purely assisting function, become the primary parameters for a reform. But if we consider the reform of trials, and thus the "cultural core" of civil litigation, not only, nor even primarily technical aspects have to be taken into account. Some factors which may be relevant here have already been pointed out. These include the reduction of the chances for a settlement if there is no direct contact between the parties, disadvantages in evaluating the evidence if otherwise available live testimony is replaced by video appearances, a loss of humanity of the proceedings. The employment of too much technology may also lead to a loss of the symbolic function of court proceedings and to a decline in their acceptance by the general public.

Because of these special features of civil proceedings it is submitted that technical developments will not bring about a fundamental change to the structure of civil proceedings. This is also supported by legal history. In the past, fundamental changes in procedure never resulted from progress in communication and information technology, not even from technical progress as such, but were always the result of social changes and a change of views. On a very general level, the oral and public trial we cherish today, is a result, at least in Europe, of the French Revolution and the subsequent nineteenth century debates, not of the invention of the typewriter, telephone or telegraph. This observation, however, of course only relates to the scope of changes technical developments are likely to bring about, but does not justify to refrain from using new technologies as they become available. No one will dispute that modern technology can contribute enormously to civil proceedings in a supportive role. Thus, the use of modern information technology is not to be seen as an alternative to more traditional methods, but as their supplement.[92] One always has to examine whether certain traditional procedures are to be explained by the limited technical possibilities of the nineteenth century, when most of our civil procedure codes were drafted, or whether they pursue independent goals which are still valid today. To find a meaningful integration of traditional principles of civil procedure and modern technology is a challenge for the future, for practitioners and academic lawyers alike.

---

[90] See F. I. Lederer, op. cit. "The Effect of Courtroom Technologies," in 2 *Journal of Appellate Practice & Process* 251 (265).

[91] H. Rüßmann, "Herausforderung Informationsgesellschaft," in *Procedural Law on the Threshold of a New Millennium*, ed. CLC/IAPL (2001) 206.

[92] See also *Wouter de Vos*, in *Procedural Law On the Threshold of a New Millennium,* ed. CLC/IAPL, 254: "The most appropriate solution for the future of civil litigation may, therefore, be to strive towards an integration of the best features of paper and electronic systems".

Chapter 14
Some Remarks Concerning the Probative Weight of Electronic Documents

Dimitris Maniotis

14.1 Introduction

The European policy for the promotion of electronic transactions is based mainly on Directive 1999/93/EC concerning the methods of certification of authentic electronic documents and on Directive 2000/31/EC, which regulates certain aspects of electronic commerce.

A type of electronic signature, the so-called "advanced electronic signature", is recognized by Art. 5 Para. 1 of Directive 1999/93/EC as having a legal validity equal to handwritten signature. Consequently, the ratification of Directive 1999/93/EC entails the treatment of electronic documents signed with the advanced electronic signature as the equivalent of private instruments signed by the issuer's handwritten signature, where the law requires this particular form.[1] According to Art. 9 of Directive 20001/EC, Member States must ensure that their legal system allows contracts to be concluded by electronic means of communication.

The UNCITRAL Commission elaborated a Model Law on Electronic Commerce and regards "durability" and "non-alterability" of electronic messages as harsh standards for their acceptance as writings.[2] The UNCITRAL Commission indicates that the promotion of electronic commerce requires the same degree of persuasiveness concerning the certification of authentic texts with both handwritten and electronic signatures.

[1] D. Maniotis, "Legal Status of Electronic Transactions: International Encyclopedia of Comparative Law," in *Cyber Law Hellas*, ed. R. Blanpain, 143 (The Hague, 2005).

[2] UNCITRAL Model Law on Electronic Commerce (Guide to Enactment) Nr 50, Art. 6 Para. 1 of the UNCITRAL Model Law qualifies data messages generally as writings.

D. Maniotis (✉)
Law Faculty of Democritus University of Thrace, Panepistimioupolis, 69 100, Komotini, Greece
e-mail: mzerdeli@law.duth.yr

M. Kengyel and Z. Nemessányi (eds.), *Electronic Technology and Civil Procedure*, Ius Gentium: Comparative Perspectives on Law and Justice 15,
DOI 10.1007/978-94-007-4072-3_14,

14.2 The Certification of the Genuine Advanced Electronic Signature (Directive 1999/93/EC)

During the evidentiary proceeding the court will come to a decision concerning the genuineness of a handwritten signature based on the inspection of the handwriting and on expert reports. In other words, the inspection by the judge or the experts will allow a direct conclusion concerning the subject matter of the evidentiary proceeding, i.e. the certification of the signatory's identity. Such a conclusion is based on the direct and visual examination of the handwriting.

The inspection itself cannot affirm persuasively the genuineness of the electronic signature. The person who tries to prove the authenticity of an electronic document signed with the method of the advanced electronic signature should work with indirectly inferred conclusions. This means that the genuineness of the electronic signature can be proved by reasoning based on the proof of facts other than the facts at issue.

The terminus technicus of advanced electronic signature has been introduced by Directive 1999/93/EC in order to indicate an electronic signature capable of the signatory's identification and the revelation of any subsequent change of the data message (Art. 2 Para. 2). The issuer of an electronic document has to relate his / her secret cryptographic key to the numerical value of the data message through a mathematical function, in order to produce an advanced electronic signature. The correspondence of this latter kind of electronic signature with the receiver's public key, which is mathematically related to the issuer's cryptographic key, allows the issuer's identification. Furthermore, the mathematical relation between the advanced electronic signature and the numerical value of the electronic document offers reasonable grounds for accepting its integrity.

The advanced electronic signature has probably been considered by Directive 1999/93/EC to be very reliable. It has to be examined whether such a consideration is to be regarded as a realistic one or not.

The authenticity of electronic documents should be accepted after the successful function of cryptographic keys. Such an acceptance based on direct observations does not constitute the subject matter of proof. For the establishment of the fact at issue the court should apply principles derived from experience. In other words, the successful function of cryptographic keys proves directly only the successful function of a certain method of cryptography which is used by the signatory or his/her authorized representative as a rule (in abstracto) for the production of his/her advanced electronic signature. The person who has worked in concreto on the signatory's cryptographic material in order to sign a certain electronic message with the advanced electronic signature can only be identified through the application of the following indirect syllogism.

Based on the ad hoc successful function of cryptographic keys, the court may arrive at a conclusion about the unknown relevant fact concerning the signatory due to the knowledge attained through common experience, according to which the owner of an advanced electronic signature data is legally interested in keeping it a secret, so that only the owner or his/her authorized representative can produce his/her advanced electronic signature.

The advanced electronic signature[3] can be relied on to detect an unauthorized alteration made to the concrete electronic document, if such an alteration has been caused by persons other than the contracting parties. In such a case, the received electronic text will not be related to the advanced electronic signature. The lack of such a relation can be easily understood by the receiver through the proper function of the public cryptographic key, which is in his/her possession. If one of the contracting parties chooses to change the data message without authorization, the court inspecting the two different data messages that constitute two different numerical values comes to no persuasive conclusion about the authentic text. In such a case there are two electronic documents before the court bearing two different advanced electronic signatures. The receiver's public cryptographic key will be related to both of them, because all three numerical values, i.e. two different advanced electronic signatures and the receiver's public cryptographic keys are related through a mathematical function to the signatory's secret cryptographic key, as explained above. The uncertainty described above constitutes one significant reason for dissuading contracting parties from using the electronic means of communication.

14.3 The Contribution of the Hosting Contract (Art. 14 of Directive 2000/31/EC) to the Authenticity of Electronic Documents

Security measures can be furnished by trusted third parties, so that the authentic data messages can be safely certified. For this purpose trusted third parties will receive, keep and submit for inspection the authentic data message, when it is asked by the court or any person having a legal interest.[4]

The legal relationship between trusted third parties, who will undertake the storage and custody of a data message, and the signatory is to be qualified as hosting according to Art. 14 of Directive 200/31/EC. The rule which introduces the new type of contract mentioned above does not provide specific regulations relating to the liabilities of the contracting parties. Consequently, its qualification is of great significance.

By the contract of hosting, the trusted third party takes delivery of a data message from the signatory in order to keep it in his/her custody, so that it can be presented in its authentic form when it is asked by the court or by any other authorized person. In the frame of the hosting contract, the contracting parties are bound by obligations which are to be regarded as analogous to the provisions governing the contract of deposit. However, such a qualification of the hosting contract depends on its correspondence to the legal prerequisites laid down for deposit. According to the law,

[3] For more detail about the method of producing it, see: T. Beth, "Confidential Communication on the Internet," in *Scientific American*, December, 1995, 70.

[4] M. Baum and H. Perrit, *Electronic Contracting, Publishing and EDI – Law* (New York: Wiley Law Publications, 1991), 350–1.

in case the depository receives a movable thing, which means a physical object or natural resources and energy,[5] there is no ground for analogous application of the rules concerning deposit. Besides, electronic messages incorporate original intellectual creations. Rules concerning intellectual property, at least in its continental meaning, provide a moral right the purpose of which is the protection of the personal connection between the author and his/her work just as in the case of economic rights.[6] The intellectual right is different from its "corpus mechanicus" and it is protected independently of the protection of its material substratum.[7]

Safe storage and custody of data messages are to be qualified as services within the framework of a hosting contract. Such a qualification is in accordance with the wording of Art. 14 of Directive 2000/31/EC. According to the rule, the person who undertakes such duties is called a service provider and the related activity is called an information society service.[8]

14.4 Difficulties in Proving the Facts Required for Proof of the Hosting Service Provider's Liability

Within the framework of a trial against trusted third parties who have undertaken the storage and custody of a data message, the plaintiff has serious difficulties in obtaining access to confidential information possessed by the opponent party. For example, when the service recipient (the issuer of an electronic document) has no access to the records of the service provider where data messages are stored, the burden of proving the receipt of a certain electronic document issued by the service recipient is not an easy task. Due to the fact that electronic messages are to be qualified as copyrightable works, the application of Art. 6 of Directive 2004/48/EC concerning the protection of intellectual property rights, which provides for the revelation of the means of evidence at the defendant's disposal, contributes to surmounting the evidentiary difficulties mentioned above.

Keeping certain measures of diligence makes the discussion concerning evidentiary difficulties unnecessary, because they prevent controversies concerning the authenticity of the data messages kept in custody. For this purpose "the acknowledgement of receipt of the data message" and "the confirmation of its content" are considered by the Guide to Enactment (§ 93) of the UNCITRAL Model Law of Electronic Commerce to be standard duties of the contracting parties during the

---

[5] I.e. Arts. 822 and 947 gr C.C.

[6] N. Deloukas and E. Perakis, *Commercial Law: Introduction to Greek Law*, ed. K. Kerameus and Ph. Kozyris, 237 (Alphen an den Rijin, 2008).

[7] M. Marinos, Copyright, Athens (2004), 62

[8] The German Court of Cassation (BGH Urt. v. 11.3.2004), (CR 2004, 765) defines hosting as a service contract aimed at the storage and custody of electronic messages belonging to another person.

development of an electronic transaction. Art. 14 of the same Model Law was formulated in accordance with Para. 93 of the Guide to Enactment. Art. 11 Para. 1 of Directive 2000/31/EC also provides that the person who offers services through electronic means of communication has to confirm the receipt of an electronic declaration. The rule goes further and regards the receipt of a data message as completed when the addressee can access it in a way that it is possible for him/her to read its content.

14.5 Conclusion

The reliable certification of an authentic electronic document is regarded as a very significant contribution to the development of electronic commerce. The advanced electronic signature, as provided by Art. 5 of Directive 1999/93/EC is a trustworthy method for the detection of any unauthorized alteration of the electronic document caused by a person other than the contracting parties. In case one of the contracting parties has changed the data message in an arbitrary manner, the advanced electronic signature cannot guarantee the certification of the authentic text. Such an uncertainty is to be regarded as one of the main reasons for dissuading contracting parties from using electronic means of communication. A hosting contract, as provided by Art. 14 of Directive 2000/31/EC, between the signatory and the trusted third party (hosting service provider), can offer a solution to the unacceptable situation described above. The hosting service provider undertakes the storage and custody of data messages. Proving the receipt of an electronic document with a specific content is important for the establishment of liability within the framework of the hosting contract. Bearing the burden of proving the service provider's liability is a very difficult task for the signatory (service recipient) because he/she has no access to the records of the service provider, who stores and keeps the data messages in custody. The disclosure of the means of evidence in the possession of the defendant (the service provider), as provided by Art. 6 of Directive 2004/48/EC, and the acknowledgment of receipt of the message together with the confirmation of its content, as described by Para. 93 of the Guide to Enactment of the UNCITRAL Model Law of Electronic Commerce, contribute to surmounting the evidentiary difficulties described above. Art. 11 Para. 1 of Directive 2000/31/EC adopted the aforementioned standards of negligence.

Chapter 15
Applied Technology in Litigation Proceedings (The Electronic Discovery Reference Model)

Zoltán Ambrus

15.1 Electronic Discovery (Also Known as eDiscovery or eDisclosure)

Electronic Discovery refers to any process used in civil litigation which addresses Electronically Stored Information [ESI] as evidence.[1]

ESI may have been avoided in the past but it cannot be ignored in today's discovery processes as 90% of company information is stored in electronic format. ESI could reside on different platforms, in multiple formats such as emails, spreadsheets, presentations, Word documents, messenger logs, calendars, etc. which are found on an ever increasing pool of personal computers, laptops, servers and mobile devices. eDiscovery uses advanced technology to provide quick and cost effective solutions to locate, process and present potentially relevant information in medium and large legal cases.

The electronic process should at every juncture maintain full chain of custody to ensure the output and processing steps taken are always defensible in legal proceedings. Therefore eDiscovery should use Computer Forensics[2] tools and procedures to retain original ESI properties.

[1] See the information provided by Wikipedia at: http://en.wikipedia.org/wiki/Electronic_discovery [23.03.2010].

[2] See more information provided by Wikipedia at: http://en.wikipedia.org/wiki/Computer_forensics [23.03.2010].

Z. Ambrus (✉)
Legal Services - eDiscovery and Digital Forensics, LDM Global, 11-21 Paul Street, London, EC2A 4JU, UK
e-mail: ambrusz@gmail.com

M. Kengyel and Z. Nemessányi (eds.), *Electronic Technology and Civil Procedure*, Ius Gentium: Comparative Perspectives on Law and Justice 15,
DOI 10.1007/978-94-007-4072-3_15, © Springer Science+Business Media B.V. 2012

15.2 EDRM (Electronic Discovery Reference Model)

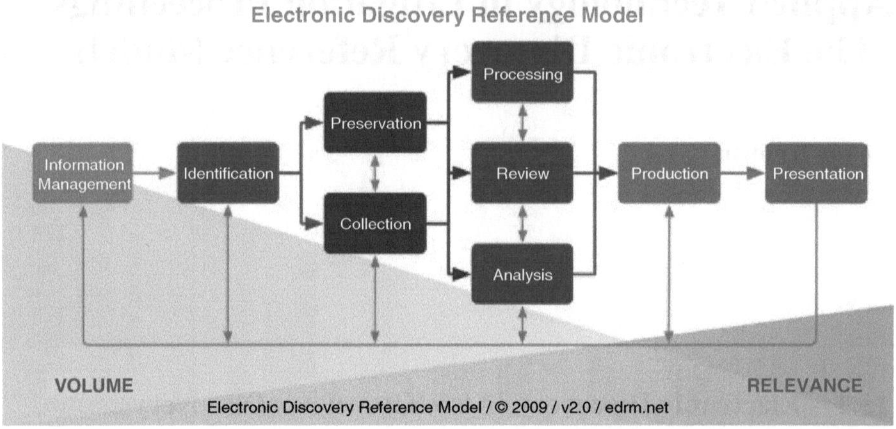

The EDRM model has been developing guidelines and standards for both eDiscovery consumers (law firms and corporates) and providers (litigation support departments and vendors) since 2005. The model has become an industry standard and most software companies and litigation vendors look to develop and market their products and services as EDRM compliant.[3]

The model defines stages and instructions from identification through trial presentation. Each box represents a major stage in the eDiscovery process while the colour arrows represent a common, general flow between these stages. The eDiscovery processes are often iterative and need to cycle back to previous stages. For example if a new custodian was identified at the review stage, the process can cycle back to identification by the grey arrows.

As you work through the discovery process you may find less data to review (yellow triangle) and higher percentage of relevant material (green triangle) in the remaining data set.

15.2.1 Information Management

This stage, which has been introduced to the EDRM recently, aims to define an embedded reference model called IMRM. It provides a common framework to help organisations to implement effective information management. As a first stage, it develops guidelines to bridge gaps between different groups within organisations (Legal, IT, Business). This step is not only referring to eDiscovery but also helps organisations prepare for a possible litigation.

[3] See the full EDRM specification at: http://edrm.net/ [23.03.2010].

15.2.2 *Identification*

At identification stage the legal team determines the key players in the case which can include the custodians and other people involved in the eDiscovery project. This level helps to initially scope the project and find all the relevant people and their relationship with the case. Once the scope of ESI is determined, the main data sources belonging to the key players within the organisation need to be identified. The in-house IT personnel may be able to assist with determining the structure and hosting platform of the data to collect and process.

Before submitting the relevant custodian data for ESI processing it needs to be captured. Whilst the method of data acquisition may depend on different requirements of the case, forensically sound data collection is however always recommended.

15.2.3 *Preservation and Collection*

To mitigate risks, data sources should be preserved in order to protect potentially relevant data in ways that are legally defensible, efficient and reasonable. Before doing the actual collection it is recommended to build a plan and select the collection method accordingly. It should be noted that many jurisdictions are moving towards a meet and confer regime between parties, so this collaborative approach should always be considered.

Forensic specialists usually perform this through a detailed landscape survey where the number of sites, custodians, source platforms, devices and the volume of data is finalised. Then the forensic analyst and the requesting parties will select the most appropriate method of data acquisition based on the survey information.

During a general forensic collection the source data is cloned twice to forensic image files in order to keep a pristine copy alongside a working copy. The pristine copy is kept in safe conditions in order to have a backup if the working copy becomes invalidated. The contents of a forensic image have to be verifiable at all times using hash values. The hash values are calculated during the data acquisition by applying mathematical algorithms on individual files and on the full data set.

There are different ways to collect the data on-site, as well as remotely. Targeted collection is usually applied on network shares or individual folders on the custodian's desktop computer. If it is determined or decided that further forensic analysis is required on the computer to restore deleted files or investigate file slack information the full hard drive will need to be imaged. Backup media is usually taken to processing centres as the restoring and acquisition process requires advanced equipment and software so this task cannot be carried out on-site.

15.2.4 Processing

The first step of ESI processing is the extraction when the data is processed at the lowest possible level whereby all items are extracted from their containers. Examples of containers include emails and attachments, archive items and embedded objects in documents. Document metadata is extracted and saved with all items into a database. Metadata is also called data about data[4] and includes several properties of an item such as format information, creation, modification, print dates, author information and email properties. These values can be used to search and filter documents to cull irrelevant material from the data set. The most common culling methods are:

File type filter: Key documents can be restricted to specific formats such as Office documents, Emails etc.

Deduplication: Eliminates unnecessary duplicates from further review. Duplicates can be found in email conversations between custodians or in documents, emails extracted from different archives.

Date restriction: Returns documents issued within specified date range(s) which is relevant to the case.

Full text searching: Keyword search terms can be applied to the data set in order to find relevant material responsive to targeted keywords.

NSRL filter: Removes known system and program files which are irrelevant to any litigation review. This is very useful on data collected from laptops and desktop computers and may reduce the size by 40–60%. The National Institute of Standards and Technology maintains the most comprehensive collection of digital file signatures called National Software Reference Library.[5]

The filtering methods can be combined in order to reach higher efficiency in removing irrelevant documents from further processing or review steps. This is a key step in the eDiscovery process as the most expensive step is the review, therefore the fewer documents sent to manual review the more cost effective resolution of the case.

The culling stage is also called pre-processing whilst the actual document preparation for litigation review is called processing. Processing usually involves file type conversion metadata manipulation and data formatting based on the review specifications.

Each step in processing has to be quality checked in order to find and handle exceptions. Due to variations in data sources and formats the range of possible

[4] See more definitions provided by EDRM at: http://edrm.net/resources/glossary/m/metadata [23.03.2010].

[5] See more information provided by NIST at: http://www.nsrl.nist.gov/ [23.03.2010].

exceptions are quite wide. For example there may be ineligible documents for keyword searching such as corrupted files, password protected files or not supported file types. The industry standard eDiscovery tools can handle most of the exceptions and support most of the common file formats.

15.2.5 Hard Copy Documents

Paper documents can come to the review at this point as they do not need advanced metadata extraction, culling or formatting. However there are several related steps which are recommended to take before submitting scanned images for litigation review.

Preparation can save significant time on scanning high volumes of paper documents. During the preparation process all physical boundaries (staples, paperclips, elastic bands, folders, etc.) get removed and replaced by barcode sheets. These sheets can mark document boundaries, host-attachment relationships and also provide information to reassemble the documents after scanning. The prepared batches will then be scanned by feeding into high volume scanners. Advanced scanning applications will scan and unitise the documents by the inserted control sheets.

In order to make documents searchable for reviewers the human readable text can be captured from the face of the scanned images with Optical Character Recognition (OCR). There is no relevant metadata to extract from scanned image files, however important document properties can be captured from the image such as Document Date, Author, Recipient, Document Type information which can be coded manually or by using auto coding applications.

15.2.6 Review

The first step at the review stage is scoping the project and determining what needs to be reviewed before the deadline. Based on the volumes and timelines, each member of the established review team needs to have his/her roles defined. Documents will be assigned to reviewers based on their position in the review level hierarchy or experience. The reviewer guide should help the review team to follow the workflow and determine relevant, responsive, privileged material. Redactions can be made to documents during the review process in order to hide personal or privileged information from other parties who are not allowed to see these details on the final disclosed documents. Documents can be also annotated to highlight or comment important sections of a document which may be relevant in further levels of the review process. The final responsive document set is usually produced for trial presentation or disclosed to other parties connected to the case. Before finally closing the matter, the review set can be archived for future records.

The modern litigation review systems provide numerous functions to support the review process and achieve workflow efficiencies. They form a centralised online repository accessible to anyone from anywhere. Since the same repository is shared between people, it provides a common platform to collaborate during the review process. Supporting team work on every level is essential as manual review is the most time consuming process in the eDiscovery project. With real time reporting the core team is always informed about the performance of the team and individuals as well as about the efficiency and partial results of the review. The user interface should be quick and easy to use in order to support a smooth review with searching, tagging, coding and annotating facilities. Advanced searching and data visualisation functions can assist to identify the review set and determine the workflow. Low level permission control may help to set up document level access rights to groups and people in the matter, even if it involves multiple/opposing parties or joint defence groups.

15.2.7 Production

At the end of the review process the relevant/privileged documents can be produced to parties for presentation and archiving purposes. The form of the production should be agreed between parties at early stages of the litigation process as it may impact on the ESI processing methods. A wide range of production formats are available to meet with various requirements. The most common format is still near-paper and paper, however the native and near-native productions are also becoming a standard output format.[6] Converting ESI from its native format to near-paper or paper is a time consuming and expensive process. Both review applications and presentation systems are being developed to support annotation on native documents, which could reduce project costs and timescales in the near future. Produced material should be quality checked very carefully in both technical and legal perspectives. The documents should be signed off by the legal team who need to make sure no information is given to other parties which may conflict with privilege and confidentiality rules. The technical team also makes sure the format of the production meets all the requirements stated in the production request.

15.2.8 Presentation

In the past, exhibits were presented in paper form and still are in many cases today. However more and more cases now require the legal team to present exhibits electronically or in native format due to the need to augment defensibility and also due to the increasing complexity of developing file formats. Trial presentation applications

[6] See EDRM's Production Guide for further clarification on the forms of production at: http://edrm.net/resources/standards/edrm-production-standards [27.03.2010].

can support both native and image formats. A quick and easy to use interface is an important requirement for these systems as pages and documents have to be displayed in conjunction with the presentation process. Sections of the selected page can be highlighted or magnified, annotating functions should also be available to make comments or hide privileged information if possible.

Technology used in court room processes is also developing in different areas. Transcription services can transform audio/video files and witness statements to searchable text which can be also uploaded into the review application.

E-bibles (or E-binders) are small, compact document databases stored in CD/DVD discs. All the trial documentation such as key documents, statements and exhibits can be stored in E-bibles for future presentation and archiving purposes.

15.3 eDiscovery, a Time and Cost Effective Solution

Company Data (GB)	1024
Custodians	5
Data/Custodian (GB)	3
Documents/GB	7500
Pages/GB	20000
Culling ratio	0.6
Reviewers	5
Review (Doc/Hr)	60
Review cost (GBP/Hr)	150
Collection (GB/Hr)	60
Collection cost (GBP/GB)	50
ED Processing (GB/Hr)	2
ED Cost (GBP/GB)	300

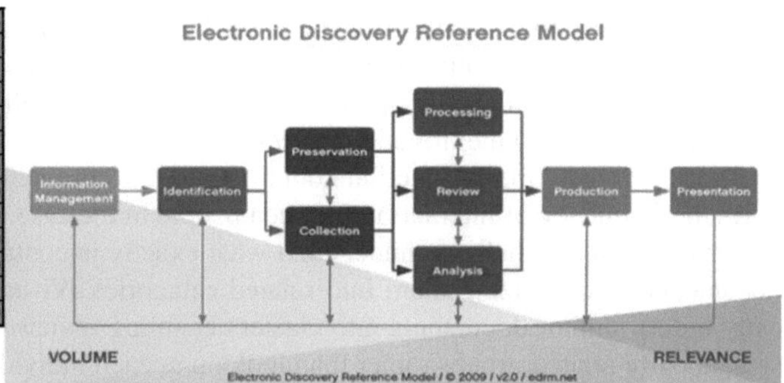

	Review without ESI processing	Processing	Review
Data Size (GB)	15	15	6
Documents	112500	112500	45000
Pages	300000	300000	120000
Reviewer Time	375	150	
Reviewer Cost (GBP)	56250	22500	

The main advantage of ESI processing is to filter out irrelevant material from litigation review. The most expensive and time consuming part of the process is the actual review, so the fewer documents submitted to the review team confers greater cost benefit. On the left hand side of the graph there are some processing metrics used by the industry to estimate project output.[7]

In this example the overall company data is 1024 GigaByte (1 TeraByte) which can be limited to an estimated 15 GB by identifying key custodians and their data sources within the organisation. Without any culling applied to the matter the estimated review time is about 375 hours for five reviewers. If we introduce ESI processing and apply keyword searching and deduplication to the data set, the total

[7] Based on metrics used by LDM Global (a premier international legal service provider) for project estimation purposes.

size to review can reduce by an estimated 60%. Since the actual processing requires mostly computer resources the time and cost saving can be quite significant with less documents to review manually.

15.4 Recent Developments and Future of eDiscovery

Recent developments in this technology focus on more cost and time effective solutions. Early case assessment systems are providing a new approach where a large amount of data can be filtered out before the acquisition stage. This technology is not only available when a litigation case comes but gets organisations prepared for a possible litigation. eDiscovery processing is still recommended as most of these solutions do not provide adequate low level item discovery but they are very effective to reduce the data size at the first stage.

Most of the market leading litigation review systems are implementing different scopes of visual representation of the case documents. With email threading the review team can easily follow email conversations between custodians. This is very useful to discover all the participants of an email topic as well as identifying possible new custodians to the litigation process.

The social networking function could help identifying key custodians as it displays the frequency of interaction between people in the case.

Visual clustering helps to understand what exactly is contained in a large set of documents by organizing them into related categories. Visual clustering dynamically orders similar documents into clusters enabling reviewers to browse data by semantically related groups rather than looking at each individual document.[8]

The conceptual search engines provide an advanced search platform. This technology looks into concepts rather than just keywords. Searching on the subject matter of the documents will find hits based on the meaning of an actual sentence and not only the direct match of a keyword.

Near duplicate detection is an efficient method to find non exact but nearly duplicate documents during the review. Identifying documents with these slight variances results in dramatic savings in time spent looking at the same document again and again.[8] Near deduplication can also be used efficiently on hard copy document sets as the possibility of exact duplication is very low in scanned documents.

15.5 eDiscovery Worldwide

Litigation support is an emerging industry and continuous growth is estimated in the legal/eDiscovery market over the coming years. While the United Sates still owns 90% of the market the technology is spreading across other countries as well.

[8] PureDiscovery is a market leading provider for visual/concept clustering and near deduplication technology: http://purediscovery.com/knowledgeGraph.html [27.03.2010].

UK, Australia, Canada and South Africa are expected to see large revenue growth over the next few years.[9]

Since the technology was originally developed around the US legal system the European migration of these technologies is quite slow. However a 2009 inquiry showed accelerating interest and awareness of discovery and disclosure in Europe.[9] Global companies involved in litigation may accelerate this development as in some international litigation cases the data protection rules do not allow ESI processing outside of the EU.

[9] Based on Gartner's research referenced at: http://www.cbronline.com/news/ediscovery_market_set_for_2010_boom_gartner_161209 [27.03.2010].

Chapter 16
DNA Databases: Methodological Approach, Family Searches and DNA Data Protection System in the EU

María José Cabezudo

16.1 Justification for the Work

We take as starting point the following genuine need: one of the main challenges facing the different countries, the EU and the International Community is to improve the fight against serious crime, particularly, organized crime and terrorism.

For this reason, in recent years, the search for new tools to fight serious crime has led to the implementation of DNA biometric technology in the field of computing. The result is a police DNA database that can fit into "Forensic Bioinformatics." The aim behind this database is to search automatically for a match between DNA profiles, in particular, between an identified profile, collected from a suspect and an unidentified profile or between two unidentified profiles, taken from the crime scene, of which one is already in the database while the other is the one we wish to identify. If the individual has re-offended, a match may be achieved linking one or more crime scenes with an unidentified or identified individual. In the second case, such a match may suggest, that at the very least, this person was present at the crime scene. These facts may be submitted at the trial as admissible evidence that can lead to a conviction or exoneration if no coincidence exists.

Therefore, in the three areas mentioned above, National, European and International, Institutions are adopting legislative measures in which the use of the appropriate technology has been taken into account, and in particular, the use of police DNA Databases. This plays a vital role in facilitating the automatic exchange of the relevant DNA profiles, in terms of biometric data, between the national DNA databases so as to achieve a "cross-border match" with the end result that such data may be introduced as admissible evidenced at the national court trying the matter.

M.J. Cabezudo (✉)
Departamento Derecho Procesal, Facultad de Derecho, Universidad Nacional de Educación a Distancia, Calle Obispo Trejo nº 2, 28040 Madrid, Spain
e-mail: mcabezudo@der.uned.es

M. Kengyel and Z. Nemessányi (eds.), *Electronic Technology and Civil Procedure*, Ius Gentium: Comparative Perspectives on Law and Justice 15,
DOI 10.1007/978-94-007-4072-3_16,

The aim with regard to criminal investigation and obtaining evidence is to achieve greater efficiency in the fight against serious crime. To this end, the automated processing of data kept in DNA files has been regulated at the national level in the United Kingdom, which was the pioneer in the field, in the US and other countries throughout the world.[1] Due to the increase in cross-border crime, especially organized crime and terrorism, most of these States have cooperated at an international and European level, through Interpol and Europol databases. Recently, in the EU, the Member States have strengthened their cooperation, as some States have approved the Prüm Treaty (2005), which was incorporated into EU Law with Decision 2008/615/JHA (Prüm Decision), implemented by Decision 2008/616/JHA. Both Decisions lay down rules with regard to speedier exchange of DNA profiles based on the principle of availability, through online searching and comparison of the DNA profile.

This is the intended purpose behind the legislative framework. But our analysis of the new regulations has identified significant legal problems which look likely to derail the effective exchange of DNA profiles and, ultimately, the provision of admissible expert evidence. Therefore it is crucial that an adequate solution be obtained that fully addresses these legal pitfalls.

16.2 The Methodological Approach

Due to these above mentioned regulations, new legal questions ought to be raised and analyzed in a comprehensively, using a common framework. Therefore, we have adopted the following methodological approach:

Up to now, we have considered DNA databases as a means of gaining efficiency in the fight against serious crime and, ultimately, obtaining an admissible expert evidence. This in turn, can lead to the maximum procedural efficiency, if two requirements are fulfilled (1) firstly, the DNA evidence has to be obtained lawfully, this is, the fundamental rights have to be fully respected; (2) secondly, the DNA evidence has to be obtained as reliably as possible. Therefore, the expert evidence has to be lawful and reliable. However, as the technology for DNA databases involves three phases, which are, (1) collection of the DNA sample, (2) analysis of the DNA profile and (3) processing of DNA in a databases, in which this work is framed, then such evidence will only be admissible if it is lawfully and reliably obtained in each of these three phases.

[1] INTERPOL, after consultation with its 188 Member States (172 respondents) indicated that 120 DNA profiles are used in police investigations and 54 have national databases of DNA. See the information provided by INTERPOL at: http://www.interpol.int/Public/ICPO/Publications/HandbookPublic2009.pdf

As far as my opinion is concerned, the European Institutions have adopted several rules that aim to fulfil both requirements. Below, a collection of rules associated with lawfulness and with reliability have been listed.

In order to obtain evidence lawfully in the processing phase of DNA databank data, which can affect fundamental right on data protection, Framework Decision 2008/977/JHA has been adopted. In the previous phase, regarding the collection of the DNA sample, which may impinge on fundamental rights to physical integrity, privacy, and the inviolability of the home, Decisions have been adopted concerning the collection of swabs from the crime scene[2] or from a specific person,[3] which may be transferred between Member States.

In order to ensure the greatest evidential reliability in the analysis of the DNA profile, two Decisions have been adopted: the first,[4] requires that laboratory activities carried out by forensic service providers are accredited at the national level as complying with EN ISO/IEC 17025 rules, the second,[5] invites the Member States, to use the 12 DNA markers or "loci", which form the current European Standard Set (ESS) and also to build up ESS analysis results in accordance with scientifically tested and approved DNA technology based on studies carried out within the framework of the ENFSI.

This legislation, which must still be developed among the Member States, raises critical issues regards to the procurement of lawful and reliable evidence. In particular we need to reconsider whether the DNA database is an effective tool, in a legal sense, in combating domestic and cross-border crime. In principle, the use of DNA technology databases is permissible. However, to achieve maximum procedural efficiency, the processing of such data will have to be carried out in accordance with a strict protection of fundamental rights, this is, evidence has to be obtained lawfully and also as reliably as possible.

In order to achieve this purpose, in other words, to answer this question, it is being necessary to identify and analyze new legal questions.[6] In this paper, an

[2] Framework Decision 2003/577/JHA and the Framework Decision 2008/978/JHA.

[3] Article 7 of the Framework Decision 2008/615/JHA.

[4] Framework Decision 2009/905/JAI.

[5] Council Resolution of 30 November 2009 on the exchange of DNA analysis results. Expanding the number of markers to 12, in order that the statistical value of DNA data corresponds to the contingent random probability and depends entirely on the number of DNA markers reliably analysed. These markers are: D3S1358, VWA, D8S1179, D21S11, D18S51, HUMTH01, FGA, D1S1656, D2S441, D10S1248, D12S391, D22S1045.

[6] Some of these issues have been raised by R. Williams, "DNA Databases and the Forensic Imaginary," in *Genetic Suspects. Global Governance of Forensic DNA Profiling and Databasing*, ed. Hindmarsh and Prainsack, 131–52 (New York/Cambridge: Cambridge University Press, 2010); B. Prainsack, "Key Issues in DNA Profiling and Databasing: Implications for Governance," in *Genetic Suspects. Global Governance of Forensic DNA Profiling and Databasing*, ed. Hindmarsh and Prainsack, 20–32 (New York/Cambridge: Cambridge University Press, 2010).

attempt will be made to address only two issues that are still being studied. The first one deals with the "family searches" in a DNA Databases, the second one, is focused on the limited application of the data protection system in the EU.

16.3 The "Family Searches" of DNA Databases

The purpose of police DNA databases is to identify the sample left at the crime scene, this occurs when a match is obtained between the profile sample and another profile already identified in the database. A match occurs if a specific number of "alleles" coincide in the two profiles being compared. Generally, there are two alleles at each "loci", so the number of matches would have to be 2 × the number of "loci" or markers to be determined. Each State has determined its markers. But, Decision 2008/616 requires that, at least, among the 24 "loci" which may contain a profile, seven constitute the ESS and Interpol Standard Set of Loci (ISSOL). Recently, because both national databases are increasing in size and number and because DNA data exchanged between the States, mainly in the EU, is increasing, it was necessary to improve the likelihood of matches between profiles. Therefore, as noted below,[7] this number has been expanded to 12 "loci." Although all Member States get to use these 12 ESS markers, problems arise in the exchange of DNA profiles between third countries, such as with the US,[8] because it uses 13 other markers.[9] Along with this want of reliability in "cross-border" matching, even at the national level, a perfect match is seldom achieved, therefore, some countries have authorized a new research method called "family search".

This technique consists in searching to identify not the specific individual who left the crime scene sample, but rather a relative of that individual, who will share a unique genetic similarity due to a shared ancestry. To this end and in the hope that the perpetrator has a relative in the database, a search is carried out that allows partial matches. In 2002, the United Kingdom was a pioneer in using this new form of search. In 2008 it was used for the first time in the US in California,[10] which explicitly authorized it to achieve a partial match of 15 alleles, and more States have followed, such as Nebraska. In particular, in the US, by means of less rigorous

[7] See below fn. 5.

[8] At least, Spain, Portugal and Germany have signed Agreements to exchange DNA profiles with US.

[9] See J. M. Butler, *Forensic DNA Typing*, 2nd ed., (Amsterdam/Boston: Academic Press, 2005), 94–5. These 13 markers were chosen because all 13 CODIS core loci are tested, the average random match probability is rarer than one in a trillion among unrelated individuals. These markers are: CSF1PO, FGA, TH01, TPOX, VWA, D3S1358, D5S818, D7S820, D8S1179, D13S317, D16S539, D18S51, D21S11 y Amelogenin.

[10] See E. Murphy, "Relative Doubt: Family Searches of DNA Database," This paper is available at: http://www.law.harvard.edu/faculty/faculty-workshops/erin.murphy.faculty.workshop.paper.spring.2010.pdf

searches carried out with CODIS, between 5 and 25 relatives of the person who left the crime scene sample can be identified, which will be considered "suspect" and therefore under investigation to determine if there is any link with the DNA sample left at the original crime scene. The importance of these controversial "family searches" lies in the difficulty of establishing appropriate safeguards to ensure lawful and reliable evidence as regards the identity of the individual who left the crime scene sample. However, despite these objections, appropriate regulation is being promoted by legislators.

When databases are larger, "family searches" have greater probability of success. Therefore, along with more database regulation and greater possibilities for exchanging DNA profiles between States, legislators are encouraging databases of increased size on four fronts namely: (1) by means of broadening the entry criteria profile: for instance the entry profile of a detainee has been authorized in several countries including the UK, some U.S. States and in Spain (article 3.1) of the LO 10/2007 as well as profiles for minor crimes, (2) through limiting profile removal criteria in the data base: for example there is no legal deadline for cancellation in the United Kingdom, which in turn has led to condemnation by the ECHR in the "Marper Case" of the December 4, 2008; (3) imprecise regulation with regard to the preservation of DNA samples: and lack of regulation with regard to the purpose and disposal of DNA samples, as is the case, for example, with the Spanish LO 10/2007 (article 9), (4) and the absence of specific purpose for samples stored in the "Biobanks" or genetic databases: whose use is allowed for other purposes, such as criminal investigation, for which they were initially collected, for instance as is the case with the Icelandic Biobanks Act 110/2000 (art. 5).[11] Thus, the tendency to increase the size of the databases has been authorized by legislation that violates fundamental rights to privacy and data protection, because they fail to respect the principle of proportionality and therefore fail to allow for the procurement of lawful evidence.

Objections to the use of "family searches"[12] can be organized into two groups. (1) as regards obtaining lawful evidence, namely: (A) the violation of the right to privacy and privacy surrounding family life that occurs because, for other reasons, family secrets on the paternal or maternal identity can be revealed or in turn because the lives of individuals and their families can be seriously damaged because of police suspicion and investigation; (2) as regards poor evidential reliability, it can be asserted that: (A) serious errors can be committed in the search, because matches can be obtained with innocent individuals who are not family members, (B) the difficulty involved in determining the minimum number of scientifically verified alleles dramatically decreases the likelihood of finding false matches; (C) the difficulty in finding a computer system essential for carrying this out in order to

---

[11] This Biobanks Act is available at: http://eng.heilbrigdisraduneyti.is/laws-and-regulations/laws//nr/3093

[12] Most of these arguments have been presented by E. Murphy, Relative Doubt: Familiar ..., 11–37. Also available, F. R. Bieber, C. H. Brenner and B. Lazer, *Finding Criminals Through DNA of Their Relatives*. See this paper at http://harvardextension.files.wordpress.com/2010/11/bieber-science.pdf

insure a minimal margin of error, granted the current repeated use of the ineffective CODIS software in "family searches" for which it was not designed[13]; (D) aggravation of the above errors, because the Police place over reliance on family matches at the expense of other more traditional investigation methods, such as eyewitness statements, which, in any case should be complementary to their investigation to avoid reaching false conclusions. In addition such searches could also lead to (non-racial) discrimination given the fact that suspicion inevitably falls on innocent relatives of the innocent and guilty alike. This, as for instance in the USA, may even exacerbate existing racial disparities in the criminal justice system, because of the high incarceration rate for Hispanics, who following on form blacks are three times more likely to be imprisoned as compared with whites, with the result that the families of these subpopulations are those primarily under suspicion.

To overcome these objections, States that opt to legally authorize "family searches" have to come to terms with the fact that it is an "especially" restrictive technique with serious repercussions for basic rights involving personal and family privacy and personal data protection. Consequently, regulation needs to strictly comply with the requirements of the proportionality principle. Some of these requirements will need to include : (1) the regulation of specific research purposes, (2) in compliance with the principle of necessity, that such methods can only be used if there are no other less burdensome option available to achieve the specific research purpose, (3) limitation to particularly serious crimes, (4) the search should be limited to those convicted, (5) legal inclusion of the minimum number of scientifically determined alleles, dramatically decreases the likelihood of finding false matches, and (5) the use of an effective computer system capable of carrying this out with a minimal margin of error.[14]

16.4 The Protection of Data System's Limited Application to "DNA Data Exchange" Within the EU

According to Decision 2008/615/JHA or the Prüm Decision (derived from the Prüm Treaty), implemented by Decision 2008/616/JHA, the Member States may make automated search and comparison of DNA profiles in order to achieve a match.[15]

[13] In this sense, CODIS alleles are looking for shared patterns based on the level of requirements that is specified in the search, it does not take into account the wide variations in the population of a certain combination of alleles in comparison with others, E. Murphy, "Relative Doubt: Family ....".

[14] Also the use of Monte Carlo simulations to investigate the likelihood of success in identifying biological relatives has been raised by F. R Bieber, C. H Brenner and B. Lazer, Finding Criminals Through…

[15] According to the ENFSI document on DNA-database management 2010, at the time of publishing of this work, Austria, Germany, Spain, Luxembourg, France, Holland (some of the States that are parties to the Prüm Treaty), Finland Bulgaria and Romania have exchanged DNA profiles. It is available at: http://www.enfsi.eu/page.php?uid=98

In such cases, the national contact point of the Member State initiating the search will automatically receive the reference data for which a match has been found. But European legislation has conditioned such exchange so as to raise and harmonize the protection level for DNA data. Thus it requires development through the domestic law of the Member States on the specific rules for the protection of DNA data as laid down in Decision 2008/615/JHA (articles 24–32) and alternatively applicable general rules pursuant to Framework Decision 2008/977/JHA. The problem is that the European protection system applies only to data which are or have been transferred under the Decision (art. 24). This limited scope of application means that there will be a dual system of data protection: one applicable to cross-border DNA data and another embracing national DNA data.

The cross-border protection regimen, applicable to DNA data exchange, will be one that incorporates into the national law of the Member States, the corresponding provisions of both European legislation, which must be complied with before the beginning of the exchange (art. 25.2 Decision 2008/615/JHA). This requirement shall not apply to Member States who exchange DNA data under the Prüm Treatry. In Spain, the cross-border protection system will impact on the Data Protection Act because this legislation excludes it from its data protection regime: "the files set for the investigation of terrorism and other serious forms of organized crime "(art. 2.2 c). Therefore, Spain will have to put in place a general system of data exchange protection within the EU context applicable to criminal proceedings.

The national DNA data protection system and, in particular its regulation in Spain, has been specifically developed in LO 10/2007 (articles 8–9), although, as noted above, the Act establishes the direct application of the Data Protection Act, it does not apply to files for the investigation of terrorism and organized crime. For this reason, Spain will have to identify the data protection regime it intends using in criminal proceedings.

This dual system may lead to disastrous consequences in terms of the measure's effectiveness[16]; in fact some of these adverse effects were debated both in the European Parliament[17] and the EDPS[18] during the processing of the original Framework Decision 2008/977/JHA. Such objections also apply to data protection rules laid down in Decision 2008/615/JHA. In particular: (1) the difficulty to determine, at a specified time, the system applicable to national or exchanged data, owing to the fact that when the data is collected or processed it is not known whether it

[16] See M. J. Cabezudo Bajo, "La protección de los datos personales tratados en el marco de la cooperación policial y judicial en materia penal," in *La Carta de Derechos Fundamentales de la Unión Europea,* ed. De la Oliva, 335–6 (Madrid, 2008).

[17] See the Explanatory Memorandum to the Report on the Proposal for a Council Framework Decision on the protection of personal data processed in the framework of police and judicial cooperation in criminal matters [COM (2005) 0475 – C6-0436/2005-2005/0202 (CNS)] of the Committee on Civil Liberties, Justice and Home Affairs of 05/18/2006. The reference is PE 370.250v02-00.

[18] In this sense, see the Second (2007/C 91/02), paragraphs 10–17, as well as the Third Opinion (2007/C 139/01), paragraphs 16–19 issued by the EDPS on this Framework Decision.

will or will not be the object of further exchange between Member States; (2) the increasingly frequent occurrence of different data protection levels contained in many criminal proceedings in the Member States, some originating from other Member State authorities while others are domestic; (3) the difficulty of strengthening mutual trust, given the lack of common rules for national and cross-border data, will make the acceptance of data exchanged between Member States all the more difficult; (4) the weakening of the EU position in its negotiations with third countries, such as the USA, because is will be unable to make the transmission of personal data to an appropriate level of internal protection.[19]

To avoid these negative effects the following two complementary solutions deserve consideration:(1) Member States have to develop a national protection system similar to DNA data exchange protection system; (2) In the future, European institutions may extend the cross-border regime, already approved in the EU legislation, to national data. Perhaps, this could happen if advantage is taken of the possibilities offered by the Lisbon Treaty. Indeed, the rules concerning the Area of Freedom, Security and Justice set forth in Title V on the Functioning of the European Union Treaty (TFU), mean that issues regarding that Area will be addressed from a more inclusive and supranational point of view. Article 87 of the TFU, which opens the Chapter on police cooperation, establishes that the European Parliament and the Council may adopt measures relating to "the collection, storage, processing, analysis and exchange of relevant information", under ordinary legislative procedures. Thus ordinary legislative procedure is placed on an equal decision making basis with that of the Parliament and the Council and, consequently, interests other than those represented in the Council can find a more appropriate channel to influence forthcoming regulations on the processing and transmission of data matters in the framework of police cooperation. In this sense, it is to be hoped that the European Parliament, as co-legislator, will play a different and complementary role than the Council has hitherto, with a trend towards more political integration and guarantees for the protection and development of individual fundamental rights.

16.5 Conclusions

The methodological approach to carry out this work has been established: Up to now, we have considered DNA databases as a means of gaining efficiency in the fight against serious crime and, ultimately, obtaining an admissible expert evidence. This in turn, can lead to the maximum procedural efficiency, if two requirements are fulfilled (1) firstly, the DNA evidence has to be obtained lawfully, this is, the fundamental rights have to be fully respected; (2) secondly, the DNA evidence has

[19] The EDPS, in its First Opinion (OJ C 47/39, 25.2.2006), paragraph 101, noted that if it were easier to transfer personal data to third countries than others Member States, this would enable the "information laundering".

to be obtained as reliably as possible. Therefore, the expert evidence has to be lawful and reliable. However, as the technology for DNA databases involves three phases, which are, (1) collection of the DNA sample, (2) analysis of the DNA profile and (3) processing of DNA in a databases, in which this work is framed, then such evidence will only be admissible if it is lawfully and reliably obtained in each of these three phases.

I propose the following preliminary conclusion in relation to the two problems addressed: With regard to "family searches" we can confidently affirm that it is very difficult to approve a regulation that takes into account the serious objections raised concerning "family searches" of DNA databases. Therefore, States that decide to authorize them will have to take into account that they are a technique involving "especially" restrictive repercussions for fundamental rights. In the light of which this will require the strict fulfilment of several requirements: (1) from a legal point of view, the regulation must absolutely respect the fundamental rights concerned and, in particular, the principle of proportionality, (2) from a technical perspective, the regulation has to include the minimum number of alleles, that in a scientifically determined way dramatically decrease the likelihood of finding false matches, and an effective computer system has to be in place that is fully capable of carrying this out with a minimal margin of error.

Given the DNA data protection system currently in place in the EU and in the Member States, we can affirm with some confidence that the restricted scope provided under the European Acts with regard to the protection of personal DNA data – limited as it is to "transmitted DNA data" – will end up creating a dual system of data protection. Thereby hindering the exchange of DNA profiles and their procedural efficiency. However, this negative scenario can be avoided, if States approve a similar regulation to cover both cases or if the forthcoming European Acts also extend the system of "cross-border data" protection to include "national data." The Lisbon Treaty may open up this possibility.

I have addressed these two issues in order to highlight some preliminary conclusions that remain subject to further study.

Acknowledgements The present work has been carried out in the frame of the Research Project supported by Plan Nacional de I+D+I (2009–2012), DER 2009–08071, MICINN, Spain.

Chapter 17
Applications and Trends of Digital/Electronic Evidence in Chinese Litigation

Baosheng Zhang and Huangxun Chen

17.1 Introduction

The development of information technology has greatly changed our lives and had important impacts on judicial practice in both civil and criminal cases. Digital evidence has become one of many new types of physical evidence. Since the first civil law case involving digital evidence was tried by the Beijing Haidian District Court in 1996, there has been a tremendous increase in the use of digital evidence in the litigation process within China.[1]

Digital evidence has two unique properties, making it different from traditional written evidence: (1) The method of storage and type of information rely on computer hardware or software as electronic carriers subject to alteration or recovery[2]; (2) The collection and the presentation of digital evidence require computer knowledge and skills and therefore display characteristics of scientific evidence.

Although digital evidence has its own unique features, it still shares some of the general properties of evidence. Therefore, the general rules of evidence should still

[1] See Chu Zhang and Likai Liu, "An Email with a Fake Name: Yange Xue v. Nan Zhang for a Fraudulent Email," in *Case Studies on Telecommunication Law,* 136–41 (Beijing: Law Press, 2005).

[2] Chunyu Zhao and Yunquan Zhang, "Characteristics of Digital Evidence and Their Impacts on Evidence Collection," in *Journal of Heilong Jiang Political Science & Law Management College* 1 (2006).

B. Zhang (✉)
Key Laboratory of Evidence Science, China University of Political Science and Law (CUPL), Beijing, China

Ministry of Education, Beijing, China
e-mail: bensenzh@cupl.edu.cn

H. Chen
China University of Political Science and Law (CUPL), Beijing, China

M. Kengyel and Z. Nemessányi (eds.), *Electronic Technology and Civil Procedure*, Ius Gentium: Comparative Perspectives on Law and Justice 15,
DOI 10.1007/978-94-007-4072-3_17, © Springer Science+Business Media B.V. 2012

apply to the standard use of digital evidence. This paper analyzes several recent cases and recommends general regulations governing the collection and the presentation of digital evidence in China.

17.2 Case Examples

17.2.1 Futaihong Precision Industrial Company v. Xiangjun Liu and Shaoqing Si

In June of 2006, Futaihong Precision Industrial Company Ltd (Plaintiff) (a part of Foxconn Group in Shenzhen Municipality) filed a civil lawsuit in Shenzhen Municipal Intermediate People's Court (hereafter, Shenzhen Court) against Xiangjun Liu and Shaoqing Si (the Accused) from Biyadi Company for invasion of commercial secrets and demanded RMB 70 million in compensation for damages.

On 10 August 2006, the Shenzhen Court preserved the evidence by copying all of the documents from the computers belonging to both of the accused and transferred them onto a portable external drive (hereafter, Disk A).

On 20 August 2007, the Shenzhen Court requested Jiuzhou Forensic Examination Centre of Intellectual Property in Beijing (hereafter, Examination Centre) to perform a forensic examination. The Examination Centre transferred the documents from Disk A onto four smaller disks (hereafter, Disk B), conducted its examination, and submitted its Forensic Examination Reports (Reports No. 117 and No. 118) on December 24, 2007.

On February 25, 2008, the Shenzhen Court opened the trial and allowed a discussion of the reports in court. During the debate, the defence examined both sets of disks and argued that Disk B contained an extra 17 documents (an extra 20 megabytes of additional information) that Disk A did not have when examined during a preliminary hearing on 28August 2006. The defence argued that Disk B was altered and the examiner from the Examination Centre could not provide a reasonable explanation. On 27 February 2008, the plaintiff withdrew the lawsuit from the Shenzhen Court and received permission from the court to abandon the suit on March 6, 2008.

17.2.2 Foxconn Group v. Xiangjun Liu

In June 2006, Foxconn Group filed a separate complaint at the police station in Baoan District in the Shenzhen Municipality, accusing Xiangjun Liu, an employee of Biyadi Company, of invasion of commercial secrets. The police subsequently conducted a criminal investigation.

In November 2007, the police copied the documents from Disk A from the Shenzhen Court onto an external drive (hereafter, Disk C) and asked the Jiuzhou Forensic Examination Centre of Intellectual Property in Beijing for its own forensic examination (the same Examination Centre that had previously been appointed by the Shenzhen Court). The Examination Centre transferred the documents from Disk C onto Disk D for its examination and submitted its Forensic Examination Reports (Reports No. 124 and No. 125). The Examination Centre alleged that the computer data contained a total of 100 documents belonging to Foxconn Group and, amongst them, 55 were considered to contain commercial secrets with an estimated value of RMB 2.28 million.

On 26 September 2007, the police station forwarded a formal charge to the Baoan District Procuratorate Office. On January 24, 2008, the prosecution brought an official charge against Xiangjun Liu for invasion of commercial secrets based upon digital evidence from the Forensic Examination Reports (Reports No. 124 and No. 125).

On March 31 of 2008, the Baoan District People's Court opened a trial, reached a guilty verdict for Xiangjun Liu and sentenced him to 4 years imprisonment. The Defendant Xiangjun Liu immediately entered an appeal against his conviction.

17.3 Rules of Digital Evidence Applied in Chinese Litigation

17.3.1 Collection Rules of Digital Evidence

Due to its unique nature, the collection and examination of digital evidence requires strict compliance with technical protocols.

One of the unique characteristics of digital evidence is the fact that it can be duplicated—the evidence can be reproduced if the command of "copy" is executed on a computer. In general, under normal working conditions, there should be no discrepancy between a copied version and the original version. The *U.S. Federal Rules of Evidence* (FRE) 1001 defines "writings" and "recordings" as evidence consisting of letters, words, or numbers, or their equivalent, set down by handwriting, typewriting, printing, photostatting, photographing, magnetic impulse, mechanical or electronic recording, or other form of data compilation. According to FRE 1003, "('Admissibility of duplicates') the preference for original writings or recordings can be excused and other 'secondary' evidence or copies of the contents can be admitted if the absence of the original can be explained or justified."[3]

Duplication of digital evidence, however, presents certain differences from traditional written evidence, suffering from any modification or being subject to fraud, alteration, deletion and damage during its storage, delivery or use in a much more accessible manner. The key element in collecting digital evidence is to ensure that the evidence has not been subject to any alteration or damage. In the ligation process, the use of duplicates must comply with strict procedures and standards,

[3] See Ronald J. Allen, Richard B. Kuhns and Eleanor Swift, *Evidence, Text, Problems, and Cases* (Austin: Aspen Law & Business, 2002), 693. [hereinafter Allen et al., *Evidence*].

ensuring that the duplicated version is exactly the same as the original version. According to FRE 1001 (4): "A duplicate is a counterpart produced …by mechanical or electronic re-recording,… or by other equivalent techniques which accurately reproduces the original." The phrase "Accurately reproduces the original" contains the fundamental requirement in reproducing any digital evidence.

In 2000, the experts from the G8 (Canada, Britain, France, Germany, Italy, Japan, Russia and the U.S.) proposed six principles in governing the collection of digital evidence[4]:

1. All general forensic procedures must be applied;
2. Upon seizing digital evidence, action must be taken to ensure no changes have been made to the evidence;
3. If a person requires access to the evidence, the person must be trained to handle the evidence;
4. Any seizure, access, storage, or transfer of digital evidence must be fully documented;
5. Any person possessing digital evidence must be responsible for all the actions relating to the handling of it; and
6. Any agency responsible for seizure, access, storage or transfer of digital evidence must fully comply with these principles.

In 2005, China's Ministry of Public Security issued its *Regulations on Examination of Computer Crime Scenes and Digital Evidence* (hereafter, *Examining Regulations*). The *Examining Regulations* specify similar principles as operational protocols governing the collection and examination of digital evidence.

The *Examining Regulations* provide: "the purpose of preserving and sealing digital evidence is to ensure its integrity, authenticity and originality. As evidence, all the storage means, electronic devices and digital data shall be preserved and sealed at the scene for future use."[5] It specifies the method of preserving digital devices and storage means: "(1) The sealing method in use shall ensure that no one shall have an access to the sealed storage means and digital devices unless an unsealing procedure takes place; (2) Photos shall be taken before and after the sealing procedure of the sealed digital and storage devices, and a 'Sealing Log of Digital Evidence' shall be prepared. Photos shall be taken from different angles, displaying the condition before and after the sealing, especially the sealing location and the sealing tag."[6]

Article 14 then states: "Methods of preserving storage devices and electronic/digital data shall include the following: (1) The Integrity Value Method shall measure the integrity value of electronic/digital data and storage devices, and shall record the information in the 'Preservation Log of Digital Evidence.' (2) When storage

[4] Scientific Working Group on Digital Evidence and International Organization on Digital Evidence: Standards and Principles, in Forensic Science Communication, 2000.

[5] Regulations on Examination of Computer Crime Scenes and Digital Evidence, Art. 12.

[6] Ibid. Art. 13.

devices and duplicates are immeasurable by the integrity value or unduplicated, the Sealing Method shall be used and the reasons be explained on the field note or the examination report according to Article 13 mentioned above."[7]

Article 29 also provides: "Duplication or copying of the original storage means shall comply with the following principles: (1) After duplication, reseal the original storage devices. (2) Camera recording shall be used if an examiner performs any critical action, such as unsealing, beginning of duplication, end of duplication, and resealing. (3) After the duplication is complete, a new sealing shall be conducted and a 'Preservation Log of Digital Evidence' should be prepared and all the actions taken shall be recorded according to Article 13."[8]

During the investigation process of two example cases, however, the examiners did not comply with the regulations discussed above, making the following serious mistakes in retrieving and handling the digital evidence:

1. What the police station in the Baoan District sealed and retrieved was not the original data, nor did the duplicated data reflect the original. According to Article 12 in the *Examining Regulations* all the storage, electronic devices and digital evidence must be preserved and sealed at the scene for future use as evidence. In the case under discussion, the Shenzhen Court copied the original documents and produced only the duplicated Disk A. However, the Shenzhen Court did not preserve or seal the duplicate and therefore failed to comply with the requirement set by Article 13 of the *Examining Regulations*, that states that when storage devices and duplicates are immeasurable by the integrity value, the Sealing Method shall be used, indicating the reasons on the field note or examination report. In fact, the police station in the Baoan District only copied the digital data from Disk A onto an external drive (Disk C), without any measures of ensuring "integrity, authenticity, and originality". From a stricter technical approach, before duplicating a computer hard-drive, technical measures should have been taken to remove any information left on the storage device to ensure it is in zero-contaminating condition. Along the same line, instead of some common copying methods, a RAID mirroring technology should have been used (from the Disk A to Disk C). Without this mirroring technology, a duplicate via a Windows system may produce discrepancies, such as a change of data sector and an increase or decrease of supplementary information.
2. The police station in the Baoan District did not comply with the requirements set by the *Examining Regulations*. Firstly, Article 13 (2) provides that before and after sealing Disk C, "photos shall be taken and a Sealing Log of Digital Evidence prepared". In addition, "photos shall be taken from different angles, displaying the condition before and after the sealing, especially the sealing location and the sealing tag." Secondly, Article 29 provides in particular that camera recording shall be used if an examiner performs any unsealing, beginning of duplication, end of duplication, and resealing.

[7] Ibid. Art. 14.

[8] Ibid. Art. 29.

3. When sealing Disk C, the Baoan District police did not comply with Article 14 on the requirement of the "Integrity Value." The investigation team simply copied the digital information (allegedly involved in the commission of a crime) from Disk A (an external drive) onto Disk C (an external drive) without measuring the integrity value. The police should have verified whether the duplicated data was identical to the original data using special verification software and recorded the results in the Preservation Log of Digital Evidence. In fact, the police did not measure "the integrity value of the digital data and storage device", nor provided "any reasons on the field note or the examination report". Thus, there was no measure taken by the police in this case to ensure that the digital evidence was not altered.

In summary, during the retrieving and examination process, the investigators in the case did not strictly follow the requirements of preserving and sealing digital data set forth by the *Examining Regulations*. As a consequence, the integrity, authenticity, and originality of the digital evidence obtained in the case were not guaranteed and its capacity to act as proof as legal evidence was fundamentally jeopardized, giving rise to an appeal.

17.3.2 Examination Rules for Digital Evidence

In litigation, an examination requires a scientific approach to evidence investigation. A forensic examination refers to the process where an examiner employs scientific technology or special knowledge, examines evidence through analysis and judgment, and provides forensic expertise on technical issues related to the litigation.[9] Forensic expert evidence is recognized as scientific evidence and classified as one of the types of evidence in China. An examination develops into forensic evidence when an examiner or expert performs an examination, analysis, makes a judgment, and draws inferences. "Experts can generate evidentiary facts themselves. Such individuals provide basic facts for the fact finder. Experts may present inferences and conclusions to which fact finders may defer."[10]An examination of digital evidence can provide important guidance to the fact-finding process and therefore should be standardized through the use of strict regulations.

In 2005 the Ministry of Public Security in China issued its *Examination Regulations* stating: "[e]xamination of digital evidence in the *Examination Regulations* refers to an examination process of designated sample(s) in which an examiner from a digital evidence examination unit within a public security agency employs special knowledge, equipment and devices, and technology, performs

[9] Decisions on the Issues from Management of Forensic Examination, Article 1, Standing Committee of National People's Congress, 2005.

[10] Allen et al., *Evidence*, supra, note 3, 732–33.

examination, analysis, verification, and judgment, and provides an examination conclusion."[11]

During the examination process of the related cases mentioned above, however, the examiners did not comply with the 2005 *Examination Regulations*. As a consequence, there were many serious mistakes in the Examination Reports issued by the Examination Centre.

Examination Reports No. 124 and No. 125 issued by the Examination Centre were based on an examination of Disk D which was a copy from Disk C which in turn was a copy from Disk A. The duplication process did not comply with the related regulations on preserving and sealing storage means on the scene. Therefore, the integrity, authenticity, and originality of the digital evidence were not ensured at all.

From the transcript of the preliminary hearing on 28 August 2006 at Shenzhen Court, the information belonging to "Xiangjun Liu" in Disk A amounted to 9.62 gigabytes in 7,440 documents (contained in 99 folders); the information belonging to "Shaoqing Si" in Disk A amounted to an additional 3.41 gigabytes in 2,757 documents (contained in 292 folders).

The Baoan District police made Disk C from Disk A at the Shenzhen Court and submitted Disk C to the Examination Centre in Beijing. The Examination Centre in turn made a further copy (Disk D) from Disk C and performed an examination on Disk D. The Examination Report (No. 124), however, indicates that the information belonging to "Xiangjun Liu" is equal to 9.64 gigabytes in 7,455 documents (contained in 99 folders), which contains an extra 0.02 gigabytes and extra 15 documents. According to examination report (No. 125) the information belonging to "Shaoqing Si" is contained in 2,759 documents with an extra two documents. In conclusion, the examination process failed to comply with Article 41 in the *Examination Regulations* that "the examination unit of digital data from a public security agency shall take technical measures and ensure no alteration to any original storage means and electronic devices during a forensic examination."[12]

In addition, Examination Reports No. 117 and No. 118 issued on December 24, 2007 by the Examination Centre dealing with Disk B (a supposed duplicate of Disk A) indicated that Disk B had been altered from the original. According to the trial transcripts on October 25, 2006 by the Shenzhen Court, the court printed out 13 files belonging to "Shaoqing Si" on Disk A. Among the 13 documents with the Foxconn labels and personal signatures, ten files have the name of "Foxconn Material Management Regulations for Product Information and Environmental Management." However, the Examination Report (No. 117) from the Centre did not record these files in Disk B.

The names of the first file (Operational System) and the fourth file (Foxconn Product Environment Quality) were not identical to the names of the first file (Operational System) and the fourth file (Environmental Quality of Foxconn

[11] Regulations on Examination of Computer Crime Scenes and Digital Evidence. Art. 2.

[12] Regulations on Examination of Computer Crime Scenes and Digital Evidence. Art. 41.

Products) in Disk B.[13] According to the trial transcript on Disk A, on a second review on October 25, 2006 by the Shenzhen Court, the court located 24 documents that revealed suspected invasion of commercial secrets, printed these files, and sealed these documents including the 2nd file named "Operational Procedure for Benchmarking Products" and the 17th file called "Operational Procedure for Recognizing External Purchase of New Products." However, the Examination Report (No. 118) did not contain any reference to these two documents.

In summary, the examinations of Disk B and Disk D by the Examination Centre lost credibility due to the suspicion that it had been tampered with and the Examination Centre has lost its probative force in its reporting, since it has reported unreliable digital evidence.

17.3.3 Rules of Proof of Digital Evidence

A fact-finding process consists of three main components: proof, cross-examination, and ratification.

The proof stage refers to the activities by which a party attempts to prove or disprove *factum probandum* with evidence. The proof includes production and cross-examination.

Production of digital evidence refers to those activities that testify to alleged statements recorded therein. An electronic file is one type of document that may not represent a typical document in a strict sense, yet the *Regulations* on writing documents still apply to it.[14] In several countries, the best evidence rule has been applied to written documents and digital evidence. The best evidence rule creates a preference for the production of originals.[15] In FRE 1001 (3) "an 'original' of a writing or recording is the writing or recording itself or any counterpart intended to have the same effect by a person executing or issuing it.... If data are stored in a computer or similar device, any printout or other output readable by sight, shown to reflect the data accurately, is considered an 'original'." The original does not exclude any duplication. However, it must fulfil the requirement that it "reflect[s] the data accurately." In Canada, the *Uniform Electronic Evidence Act 4* states "where the best evidence rule is applicable in respect of an electronic record, it is satisfied on proof of the integrity of the electronic records system in or by which the data was recorded or stored."[16]

[13] See Table 3 "Examination Report" (No. 117), 158: "Nonpublic Examination Results of 116 Files from the Plaintiff in the Computer Hard Drive."

[14] Electronic Evidence: Computer-Produced Records in Court Proceedings, Introduction [2], Ken Chasse, Toronto, Ontario, June, 1994.

[15] Allen et al., *Evidence*, supra, note 3, 693.

[16] Canada Uniform Electronic Evidence, Act 4.

The current laws and regulations in China also specify the best evidence rule regarding written documents and digital evidence. For example, the *Judicial Interpretation for the Execution of Criminal Procedural Law* of the Supreme People's Court ("*SPC*") provides: "Documentary evidence by investigation shall be the original document. If it is difficult to provide the original carrier, copies or reproductions may only be provided."[17] The *Specific Provisions on Evidence in Civil Actions of the SPC* ("PECA") provide: "the investigators who investigate upon and collect computer data or audio-visual materials such as sound recordings and visual recordings, etc. shall request the person investigated to provide the original carrier of the relevant data. If it is difficult to provide the original carrier, a reproduction may be provided. In the case of a reproduction, the investigators shall specify the source of the evidence and the process of its making in the investigation notes."[18]

The *Specific Provisions on Evidence in Administrative Actions of SPC* ("PEAA") provide: "Any computer data or audio-visual materials such as sound recordings and visual recordings, etc. provided by a party concerned to the People's Court should meet the following criteria: (1) Submit related original carrier. If it is difficult to submit, a reproduction may be provided; (2) Indicate clearly production method and time, the person in charge, and purpose of reproduction."[19]

17.3.4 Rules of Cross-Examination Relating to Digital Evidence

Cross-examination refers to "The questioning of a witness at a trial or hearing by the party opposed to the parry who called the witness to testify."[20] Because digital evidence can be easily falsified and altered, it is important to have valid cross-examination during the fact-finding process. For instance, the Canadian *Uniform Electronic Evidence Act* provides: "(1) A deponent of an affidavit referred to in Section 7 that has been introduced in evidence may be cross-examined as of right by a party to the proceedings who is adverse in interest to the party who has introduced the affidavit or has caused the affidavit to be introduced. (2) Any party to the proceedings may, with permission of the court, cross-examine a person referred to in paragraph 5(c)."[21] In China, cross-examination refers to any party concerned or the legal representative designated or appointee who can challenge the evidence

---

[17] The SPC Judicial Interpretation for the Execution of Criminal Procedural Law (1998) 3 Gazette of the Supreme People's Court of the People's Republic of China 101. Art. 53.

[18] Specific Provisions on Evidence in Civil Actions of the SPC (2002) 1 Gazette of the Supreme People's Court of the People's Republic of China 22 [hereinafter PECA]. Art. 22.

[19] Specific Provisions on Evidence in Administrative Actions of SPC (2002) 4 Gazette of the Supreme People's Court of the People's Republic of China 132 [hereinafter PEAA].Art. 12.

[20] *Black's Law Dictionary,* 7th ed., (West Group, 1999), 383.

[21] Canada Uniform Electronic Evidence Act 8.

presented by the opposing party through inquiry and questioning. The *Criminal Procedural Law* provides: "the testimony of a witness may be used as a basis in deciding a case only after the witness has been questioned and cross-examined in the courtroom by both sides, that is, the public prosecutor and victim as well as the defendant and defenders, and after the testimonies of the witnesses on all sides have been heard and verified."[22] The *Civil Procedural Law* also provides that: "evidence shall be presented in court and cross-examined by the parties concerned."[23] In summary, both criminal and civil trials in China require that digital evidence shall be presented and cross-examined.

17.3.5 Ratification Rules of Digital Evidence

Ratification refers to the process by which judges evaluate evidence in court in terms of its relevance, competence and probative value based on certain regulations and experience.[24] No evidence which includes digital evidence has any predesigned abiding force. Judges must apply logical reasoning and rules of experience to evaluate and verify all the evidence of the case comprehensively, objectively, and justly, weigh and balance the relevance, admissibility and probative value of evidence, and provide explanations for the reasons behind their decisions.[25]

17.3.5.1 Relevancy and Admissibility of Digital Evidence

Both PECA and PEAA clearly exclude the admissibility of the following digital evidence: "evidence that cannot exclude alteration," "evidence that is associated with doubts" or "evidence that cannot match the original file."[26] When one party challenges the probative value of evidence, the other party must asks its producer, witness, and person in charge of the custody of it to identify or authenticate it in court. The Civil Procedural Law provides:

> The People's Court shall verify audio-visual materials and determine after their examination in the light of other evidence in the case whether they can be taken a basis for determining a fact.[27]

---

[22] The Chinese Criminal Procedure Law (1996) 2 Gazette of the Supreme People's Court of the People's Republic of China 39. Art. 47.

[23] Ibid. Art. 66.

[24] See Jiahong He, *Brief Evidential Law* (Beijing: China Renmin University Press, 2007), 185.

[25] See PECA, supra note 18, Art. 64. Also see PEAA, supra note 19, Art. 54.

[26] See PECA, Ibid., Art. 69. Also see PEAA, Ibid. Art. 71.

[27] The Chinese Civil Procedural Law (1991) 2 Gazette of the Supreme People's Court of the People's Republic of China 3. Art. 69.

The relevance of digital evidence includes the requirement of evidence sufficient to support a finding. When presenting evidence, the producer can be identified based on the content of the digital evidence. However, digital evidence does not necessarily share this characteristic. It is relatively difficult to identify the producer of digital evidence without other relevant evidence or computer techniques.

For example, in once instance police in China received a report from a person named Li who alleged that his email account has been stolen and somebody was using his email account to distribute pornography. The police examined the content of several pictures and searched the server's log from the City ISP. The police examination uncovered the IP address, the telephone number and the internet account details. All three accounts were under the name of "Liang." The police also knew that the server of the City ISP, the computer name from which the emails were sent, matched the computer name under "Liang." In addition, Liang did not have an alibi: he was at home online at the time of the alleged crime. He had the opportunity to commit the crime. As a result, the police questioned him, confiscated his computer, issued a warning and fined him 1,500 RMB. Liang contested the police's decision, arguing it may have been done by a hacker and demanding a withdraw of the administrative lawsuit.[28]

This was the first case in China using IP information as evidence at trial. During the trial, the key point was whether the act of distributing the illegal pornography could be verified by the existing digital evidence, which included the IP address, emails, and server's log, together with the statements from both parties and the testimony from witnesses. The Court of first instance concluded that the pornographic emails at the recipient's end were sent from Liang's computer. The IP address, online telephone number, and email account all belonged to Liang. The server's log from the ISP indicated that Liang was using his telephone line for an internet connection. The police later confirmed from examination of Liang's computer that there was no hacking during that period of time of the offence. Based on the IP address of the internet, online account, and password used, the police concluded that the emails were sent by Liang.

However, the case was later challenged by experts. In 2005, Mr. Zeming Yang from the China Academy of High Energy Physics explained the case in detail at an academic conference. He argued that it was not sufficient to just rely on the IP address or "Internet behaviors." The trial court should obtain evidence from the internet, recovery of disks, or disk sectors, and conduct thorough analysis and examination.[29]

[28] Jun Deng, Nanjiang Zhao, Qichun Zeng, Jin Yang and Zeming Mao, "A Lost Lawsuit for Internet Users with IP Address as Evidence," *Southern Daily*, May 27, 2003.

[29] See Zeming Yang: *Computer As Evidence and Daily Journal Analysis*, the Annual Reports from China Internet Society and Computer Network and Information Security at the Annual Urgent Meeting of China's Computer and Network Security, 2005.

17.3.5.2 Probative Force of Digital Evidence

"Probative force" denotes the strength to support or negate *factum probandum*.[30] Whether or not digital evidence has probative force relies on the reliability and integrity of the digital records system. Therefore, the *United Nations Commission on International Trade Law Model Law on Electronic Commerce* in 1996 provided useful guidance "as regards the assessment of the evidential weight of a data message, and how the evidential value of data messages should be assessed (e.g. depending on whether they were generated, stored or communicated in a reliable manner, and whether such reliability method of information integrity was maintained, so that the methods or factors to identify the sender can be established)".[31] The probative force of digital evidence depends on the reliability of its generation, storage, delivery, and chain of custody. *Digital Signature Law of PRC* provided a similar regulation, requiring that "to verify the authenticity of digital documents as evidence, the following elements shall be included (1) reliability methods of maintaining generation, storage, and delivery of digital documents; (2) reliability methods of maintaining content integrity, (3) reliability methods of identifying the sender; and (4) other related factors."[32]

The *Uniform Electronic Evidence Act of Canada* clearly provides: "In the absence of evidence to the contrary, the integrity of the electronic records system in which an electronic record is recorded or stored is presumed (a) by evidence that supports a finding that at all material times the computer system or other similar device was operating properly or, if it was not, the fact of its not operating properly did not affect the integrity of the electronic record, and there are no other reasonable grounds to doubt the integrity of the electronic records system; (b) if it is established that the electronic record was recorded or stored by a party to the proceedings who is adverse in interest to the party seeking to introduce it; or (c) if it is established that the electronic record was recorded or stored in the usual and ordinary course of business by a person who is not a party to the proceedings and who did not record or store it under the control of the party seeking to introduce the record."[33]

Chinese scholars have proposed three rules for determining the probative force of digital evidence: (1) The probative force of digital evidence by a public notary service is greater than that of digital evidence without any public notary; (2) The probative force of digital evidence made during regular business activities is greater than that of digital evidence made for a litigious action; (3) The probative force of digital evidence by the adversary party is greatest, followed by digital evidence by a neutral party, and the weakest one made by own party.[34] Mr. Xing Lu added the

---

[30] Terence Anderson, David Schum and William Twining *Analysis of Evidence*, 2nd ed., (Cambridge/New York: Cambridge University Press, 2005), 44–5.

[31] The United Nations Commission on International Trade Law Model Law on Electronic Commerce: Section 2 of Article 9.

[32] Digital Signature Law of PRC. Art. 8.

[33] Canada Uniform Electronic Evidence. Act 5.

[34] Jiahong He, *Research on Digital Evidence Law* (Beijing: Law Press, 2002), 158.

following three additional rules: (4) Digital evidence with forensic expertise has greater probative force; (5) Digital evidence that has been verified has greater probative force; (6) Digital evidence that has been recognized by a certification agency has greater probative force.[35] Finally, Mr. Ye Li supplemented: "The digital evidence provided by a certification agency has greater probative force than that provided by any party concerned."[36] These rules collectively provide important referential values to judges for evaluating the probative force of digital evidence.

17.4 Future Trends

As the pace of the application of digital evidence in China's litigation increases, more issues with the current legislation governing its application have become exposed to the public. Currently all specific rules and regulations of digital evidence are limited to those found in judicial interpretation by the Supreme People's Court or governmental provisions. In the future, we can expect Chinese uniform provisions of digital evidence.

17.4.1 Trends for Further Standardization of Digital Evidence Collection

While there have been no clear regulations set for collecting digital evidence in China, academics are engaged in active discussion on how to introduce stricter provisions for gathering digital evidence, necessary on account of its unique characteristics. Several American scholars represented by Kevin Mardia propose detailed labels for each carrier during evidence collection, chain of custody for evidence integrity and prevention of potential alteration during collection and maintenance.[37] Chinese scholars suggest the following procedures: (1) Investigation and collection of evidence should consist of two stages: delegation and acceptance of the case, and; examination and identification.

During the first stage, specific requirements should be made known for the delegation of responsibilities to relevant professional units, as well as setting out time frames for acceptance and processing of the case.

During the second stage, examination and identification involves collection and sealing of digital evidence as well as requirements and flow charts for examination

[35] Xin Lu, *Research on Digital Evidence in Civil Proceeding* (Beijing: China University of Political Science and Law Press, 2006), 46.

[36] Hua Li, *Research on Probative Force of Digital Evidence* (Guangzhou: Jinan University Press, 2007), 32.

[37] Kevin Mardia, Chris Prosise and Matt Pepe, *Emergent Responses and Forensic Examination*, trans. Qingqing Wang (Beijing: Qinghua Press, 2004), 167–8.

and analysis. (2) Search warrants shall be presented when searching for evidence and the search procedure shall be prescribed by the scope set out in the warrant. Specific items such as computers, internet servers, terminal storage, and IP addresses shall be clearly identified. (3) After collecting the evidence, the following shall be indicated clearly: source, time, personnel involved, detailed records, the whole process of chain of custody, and information under custody and signatures on the records from two different witnesses. (4) Special attention shall be paid to the individual privacy of the party concerned during the collection and use of digital evidence.[38]

It is absolutely necessary that forensic evidence should be made from the evidential sample according to the appropriate technical procedure and methods. Similar regulations on digital evidence have also been proposed by local municipalities. For instance, the *Collection of Digital Evidence and Examination Protocols by People's Procuratorate of Huangpu District in Shanghai* established a basic system for forensic examination of digital evidence. Because digital evidence has certain characteristics, such as being high-tech, secret, easy to alter, and coming in a variety in forms, the following principles were established: (1) Standardization Principle, involving a flow chart of steps and requirements for a strict procedure; (2) Monitoring Principle, to ensure reliable evidence examination involving video tape recording of the main procedures for collecting digital evidence and having a third party as an independent witness; (3) Safety Principle, using stable and reliable equipment for non-destructive examination ensuring no damage to the original source file (it is necessary to use the Hash value to ensure the identification between an original file and a duplicated file) which must be sealed, with only a duplicated file being used for analysis and recovery, and; (4) Confidentiality Principle, whereby it is necessary for any operating personnel involved to maintain the privacy of personal information and case details, following related confidentiality requirements.[39]

17.4.2 Trends for a System on Provisions of Digital Evidence

To date there is no uniform law of evidence in China. Regulations on digital evidence come from five uncoordinated sources: related laws, judicial interpretations, agency regulations, related international regulations, and other standard official documents.[40] Justice Deyong Shen of the China People's Supreme Court once stated that: "The articles of many laws, regulations and judicial interpretations are inconsistent, and do not operate in a harmonized manner. To some extent, that situation creates

[38] See Feng Gao: *Prosecutorial Organization's Forensic Examination of Electronic Evidence* at: http://www.procedurallaw.cn/zjfx/zdwz/200905/t20090508_216810.html [08.05.2009].

[39] Ibid.

[40] See Jiahong He and Weiping Zhang, *Brief Evidential Law* (Beijing: Renmin University of China Press, 2007), 8–12.

confusion in the application of evidentiary rules in the trial. For instance, some insufficient laws are being used to guide the application of electronic evidence, etc. The reformation and improvement of the evidence system has become an important and urgent task for current Chinese judicial reform."[41]

Since 2001, China's academic community has engaged in some lively discussions on improving evidential legislation and scholars have proposed several different "Expert Proposals on Evidence".[42] In August 2006, with the support of the China Supreme Court's research office, the Institute of Evidence Law and Forensic Science of the China University of Political Science and Law undertook the task of drafting the *Uniform Provisions of Evidence of the People's Court: A Proposal for Judicial Interpretations*[43] (hereafter, the *Proposal*). The draft was completed in September 2007 and currently is being tested as part of a pilot program in seven local courts.

The *Proposal* provides systematic regulation of digital evidence based on case experiences from Chinese judges and foreign regulations on digital evidence.

First, as to Forms of Electronic Evidence, the *Proposal* provides: "Audio-video and electronic evidence shall have explanatory labels indicating the name of the author or collector and the time, place and process of the creation or collection… Electronic evidence shall have written explanations describing the handling and reproduction process, noting the time and place of the handling and reproduction, the form, category, file type as well as the handler, possessor and custodian who handles and reproduces the electronic data."[44]

Second, in regard to exhibition of electronic evidence, the *Proposal* provides: "Electronic evidence shall be exhibited in a recognizable means such as monitor, printout or word descriptions with a clear subject matter to be tested. When audio/video or electronic evidence involves business secrets and personal privacy, exhibition shall be conducted in a setting not open to the public."[45]

Third, in terms of the content of digital evidence in a cross-examination, the *Chinese Procedural Law* does provide certain regulations that can be applied to digital evidence, yet the regulations are vague on what content shall be debated. Cross-examination on digital evidence shall focus on the congruence between the reproduction and the original document. The *Uniform Digital Evidence Law of*

[41] See Chief Justice Xiao Yang, China Intensifies Its Efforts for the Reformation of Evidential System (May 30, 2006) in Xiao Yang, *The Central People's Government of the People's Republic of China*, at http://big5.gov.cn/gate/big5/www.gov.cn/jrzg/2006-05/30/content_295901.htm [28.05.2009].

[42] See e.g. Yuqiang Bi, etc.: *Draft for China Evidential Law with Suggestions and Argumentations*, 2004; Guangzhong Chen: *Expert Draft for PRC Evidential Law (Articles, Interpretations, & Argumentations)*, 2004; Wei Jiang: *China Evidential Law Draft (Proposal) & Legislative Reasoning*, 2004.

[43] See Baosheng Zhang, *Uniform Provisions of Evidence of the People's Court: Proposal for Judicial Interpretations and Drafting Commentary* (Beijing: China University of Political Science and Law Press, 2008). [hereinafter Zhang: Uniform Provisions of Evidence].

[44] Zhang: *Uniform Provisions of Evidence*, supra note 43. Art. 20.

[45] Ibid., Art. 92(2).

Canada provides: "This Act does not modify any common law or statutory rule relating to the admissibility of records, except the rules relating to authentication and best evidence."[46] This Canadian regulation points out that authentication and best evidence are the main focus of a cross examination on digital evidence. Therefore, the *Proposal* also provides clear regulations on authentication and best evidence for digital evidence, stating that "When electronic evidence and demonstrative evidence is exhibited but before being admitted into the court, if the opposing party objects, the proponent of the evidence shall introduce the producer, the collector and/or the custodian of the evidence to testify in court in order to confirm the identification and authentication of the evidence."[47] The precondition here lies in the objection of the opposing party which constitutes an important measure for a cross-examination requiring identification and authentication. Through identifying the sources and the chain of custody of the digital evidence, producers, handlers and custodians are able to ensure if a document is an original one, if the identification between the duplicated and original documents can be proved, if the new file still possesses original features, and if any alteration and changes have been attempted.

The *Proposal* states the authenticity of electronic evidence: "when the opposing party has raised an objection, it shall be identified and authenticated by its producers, witnesses, custodians and other persons who have the knowledge of the process of producing and custody of such evidence. The contents of identification and authentication of electronic evidence include, but are not limited to, the following elements: (1) Reliability of the methods for producing, saving, delivering and storage; (2) Environmental elements and agreements related to its producing, saving, delivering and storage; (3) Properties and characteristics of the electronic files; (4) Persons that may enter the information exchange system and their level of familiarity with the system; (5) For electronic evidence with a password, digital signature and account name and number, the password, digital signature, person who sets up the account, user and owner of the account and the use of the name or account; (6) De-encryption in the transmission; (7) Whether the system hardware is sound, whether the software is reliable, whether the system operates normally, whether it has been infected by a virus, whether it is possible that the stored data has been changed or altered; and (8) Whether the method of reproduction reflects the contents of the original accurately and completely."[48]

Fourth, with regard to ratification and admission of duplicates, Article 169 in the *Proposal* states: "If the other party in the lawsuit does not object, or although an objection has been raised, the identification and authentication of the evidence can be determined and proved, or when there is no sufficient opposing evidence to disprove such evidence, adjudicators may admit the following types of evidence presented by one party: (1) Photocopies, photographs, duplicates or excerpts that have

[46] Canada Uniform Electronic Evidence, Section One of Art. 2.

[47] Zhang: *Uniform Provisions of Evidence*, supra note 43, Art. 94.

[48] Ibid. Art. 100.

been verified to be identical to the original document evidence; (2) Duplicates, photographs, records or demonstrative evidence that have been verified to be identical to the original document; (3) Duplicates that have been verified to be identical to the audio/video and electronic evidence."[49] Chinese scholars further emphasize the identification and authentication of digital evidence. Great attention shall be paid to the accurate results of the duplication and the systems of digital evidence for identification and authentication. "When one party in the lawsuit does have objections to the authentication of a fact or its related document, the proponent of the evidence shall present to court the time logs, online records, recipient's phone number, IP cards or the information of the computer in use, its operational information, disks, decoded files, and related CDs/DVDs. Related witnesses shall also be summoned to court."[50] The new development of electronic information technology, especially in the area of e-commerce has raised some unique issues to identification and authentication of digital evidence. For instance, "for digital evidence with encrypted files, accounts with digital signatures, or regular accounts, the party concerned or public prosecution agencies shall present and prove the identity of the person who sets up the account, who uses it, or any related persons in relation to the encryption, digital signature, and account, as well as any information related to the account or the name associated with the account. If necessary, the court shall consider the consequence of revealing the code to the public during a trial as digital evidence and may rely upon indirect evidence to prove the identity of the person who sets up the account, who uses it, or any related persons in relation to the encryption, digital signature, and account, as well as any information related to the account or the name associated with the account."[51]

Fifth, in terms of the greater role of the expert witness for digital evidence during a cross-examination, the *Proposal* specifies that: "Parties to a lawsuit may invite persons with specialized knowledge as an expert witness who—with permission of the People's Court—will appear in court to express opinions on specialized issues."[52] In any litigation related to digital evidence, if the attorney does not have relevant knowledge such as a scientist or an engineer should have, he can invite experts in the area to assist his duties, for example, to provide explanations of the digital evidence or forensic expertise presented to court, to question the examiner of digital evidence from the opposing party and cross-examine the forensic expertise, or to rebut the expert evidence from the opposing party, to answer questions from the judge or the party concerned with the purpose of advising any questions related to unique issues or specialized knowledge from digital evidence.

---

[49] Ibid. Art. 169.

[50] Wei Tang, *Evidential Law in Civil Proceeding*, Article 210, Expert Proposal.

[51] Wei Jiang, *China Evidence Law Draft (Proposal) & Legislative Reasoning* (Beijing: Renmin University of China Press, 2004), 542.

[52] Zhang: *Uniform Provisions of Evidence*, supra note 43, Art. 107.

17.4.3 Our Thinking on the Admissibility of Digital Evidence

Digital evidence is one type of scientific evidence available to litigators. Questions remain whether the admissibility of digital evidence should be founded upon the reliability of scientific principles and methodology, or the reliability from inference based on scientific principles and methods. In the *Daubert* case, the Court said that in determining admissibility of expert testimony the "focus, of course, must be solely on principles and methodology, not on the conclusions that they generate".[53] In other words, even if an expert applies a reliable methodology, the expert can still draw an inference that may ultimately fail in a test. If this is true, what is the significance that can be placed on the reliability of scientific principles and methodology? In judging the reliability of digital evidence, in addition to the reliability of scientific principles, should we put more emphasis on the reliability of scientific inference?[54] While the admissibility standard of scientific evidence shall be applied to admission of digital evidence, how does the admissibility standard of general evidence relate to the admission of digital evidence?

Although the admissibility of scientific evidence can be improved on a continuous basis, judges are laypersons to science and less experienced to evaluate the suitability of scientific principles and methodology. In essence, judges, as legal experts, can skilfully apply the general admissibility standards giving weight to the admission principles of scientific evidence.

There are two limitations to the admissibility of duplicates in the *U.S. FRE* 1003 states: "A duplicate is admissible to the same extent as an original unless (1) a genuine question is raised as to the authenticity of the original or (2) in the circumstances it would be unfair to admit the duplicate in lieu of the original."[55] In the *Bryant v. State* case, the defendant was accused of child abuse and the only evidence in the case was a piece of digital image that had been edited and enlarged by the prosecution for the trail. The appellant court held that "only if the proponent can ensure a fair and accurate duplication of the fact to be proved and reflect the truth of the incident, can the duplicates be admitted as evidence."[56]

Chinese scholars generally agree that the general rules of traditional evidence in the law shall also be applied to the collection and proof of scientific evidence. It is fair to delegate discretion to judges in determining if the admission of scientific evidence exceeds its proved value in substance, or if the admission is consistent with the requirement of a fair trial. For example, it is a legal issue to obtain delegation and permission from agencies or parties concerned in collecting digital evidence. In criminal proceedings, the police must follow the law in conducting a

[53] Allen et al., *Evidence*, supra, note 3, 753–4.

[54] See Baosheng Zhang, *Evidence* (Beijing: China University of Political Science and Law Press, 2009), 228–9.

[55] FRE 1003. ADMISSIBILITY OF DUPLICATES.

[56] Bryant v. State, 810 So. 2d 532 (Fla. Dist. Ct. App). [2002].

search, placing the evidence in custody, or making online interception with approvals from agencies of authority and without any invasion of citizen's rights.[57] In civil proceedings, any party shall obtain permission from the owner or the custodian of evidence and then collect related digital evidence.

In conclusion, in order to provide for greater application of digital evidence in fact-finding, the following considerations are important: stricter technical protocols and standards should be written into the law in China; general rules in the law of evidence should not be ignored but applied to digital evidence with the emphasis upon uniqueness and technicality of digital evidence; and the judge's discretion carries very great weight.

[57] Yinghui Song, "Issues Related to Legislative Improvement of Search and Taking Custody of Digital Evidence," in *Evidential Forum* 7 (2004).

About the Contributors

Zoltán Ambrus is an experienced litigation support professional who has worked on many litigation cases for global law firms, corporations and public organizations. Zoltan is specialized in E-discovery, Digital Forensics and Litigation Hosting technologies and processing practices with extensive knowledge in software development, process automation and database design for the legal market.

Current employer:
LDM Global
11–21 Paul Street, London
EC2A 4JU, United Kingdom

Position:
Software and Innovation Developer

Postal Address (Personal):
H-6800, Hódmezővásárhely
Szent László utca 59.
Hungary

María José Cabezudo is an Associate Professor of Procedural Law at the Universidad Nacional de Educación a Distancia (UNED), Madrid, Spain. She is currently leading a research project supported by the Science and Innovation Spanish Ministry (DER 2009–08071) entitled "Personal data exchange and protection in the European Union: DNA evidence and its legal effect in Spanish criminal proceedings".

Profession: Associate Professor of Procedural Law at the Universidad Nacional de Educación a Distancia (UNED), Madrid, Spain.

Current job: Associate Professor of Procedural Law at the Universidad Nacional de Educación a Distancia (UNED), Madrid, Spain.

M. Kengyel and Z. Nemessányi (eds.), *Electronic Technology and Civil Procedure*,
Ius Gentium: Comparative Perspectives on Law and Justice 15,
DOI 10.1007/978-94-007-4072-3, © Springer Science+Business Media B.V. 2012

Postal address:
Departamento Derecho Procesal,
Facultad de Derecho
Universidad Nacional de Educación a Distancia
Calle Obispo Trejo nº 2
28040 Madrid
Spain

Petronio Calmon is a Doctor (Ph.D.) in Procedural Law from the University of Sao Paulo. He is Vice-President of the Brazilian Procedural Law Institute, and prosecutor on federal districts, Brazil. He is the author of the book *Comments on the Brazilian Law of the Electronic Process.*

Paul D. Carrington is a native of Dallas. A graduate of the University of Texas (B.A. 1952) and Harvard University (LL.B. 1955), Professor Carrington taught in 15 American law schools as well as in other law schools in Germany and Japan. He had served as the Dean of the Duke school of law, from 1978 to 1988, and is currently a Professor. From 1886 to 1993, he served as Reporter to the committee initiating amendments to the Federal Rules of Civil Procedure. His fields of interest are the history of the legal profession, civil procedure, and international civil litigation, and his current book *Lawyers in American History: Serving the Public Good?* will be published in 2012.

Huangxun Chen is a doctoral candidate at China University of Political Science and Law (CUPL), Beijing.

José Eduardo de Resende Chaves Júnior is a Judge-Assistant of the Presidency of the National Judicial Council, Brazil and Principal Judge of the 21st Federal Labour Court Circuit of Belo Horizonte, Brazil. He is the Chairman of the Board of the Judicial School of Latin America, EJAL. He holds a Ph.D. in Fundamental Rights. He is also the Coordinator of GEDEL – *Justice and Electronic Law Study Group* of the Judicial School of Federal Labour Court of Minas Gerais State, Brazil and a Member of the Brazilian Institute of Electronic Law, IBDE.
Postal address: Rua Antônio de Albuquerque, n. 1159/101 – Bairro de Lourdes – Belo Horizonte – Minas Gerais – Brazil
Zip Code: 30112–011

Nikolaj Fischer (born in 1971, Dr. Iur. 2002, Dr. Iabil. 2004) teaches as Privatdozent (Lecturer) at the Institute for Comparative Law at the Johann Wolfgang Goethe University of Frankfurt/M. Venia legendi for German and European Civil Law, German and European Civil Procedure Law and Judicial Law. For further details see (online):
http://www.jura.uni-frankfurt.de/l_Personal/priv_doz/fischer/index.html

Viktória Harsági is an Associate Professor, Head of Department at the Pázmány Péter Catholic University (Budapest), Faculty of Law, Department of Civil Procedure Law and lecturer at the Andrássy Gyula German Speaking University (Budapest).
E-mail: harsagi@jak.ppke.hu

Fernando Gascón Inchausti is a Professor for Civil and Criminal Procedural Law at the Universidad Complutense of Madrid. He has published widely on civil and criminal procedure, with a special interest on International and European procedural law. He has made research in Germany (Hamburg, Freiburg, Berlin), in Italy (Florence) and the USA (Berkeley).
Current Data:
Fernando Gaṣcón Inchausti
Departamento de Derecho Procesal, Facultad de Derecho
Ciudad Universitaria s/n
28040 Madrid
Spain

Fernando Gascón
Departamento de Derecho Procesal, Facultad de Derecho
Universidad Complutense de Madrid
Ciudad Universitaria s/n
28040 Madrid
Spain

Masanori Kawano has been teacher of civil procedural law in Japan since 1973 and holds a Doctor of Law from Kyushu University 1995. He joined Nagoya University Graduate School of Law in 1999 where he was Dean for the period 2002–2004. His main fields of research are civil procedure, bankruptcy, and arbitration. He is a member of the International Association of Procedural Law. He was also a member of the joint American Law Institute/UNIDROIT Study Group and contributed to the preparation of the "Principals of Transnational Civil Procedure" from 2000 to 2003.

Georg E. Kodek is a judge at the Austrian Supreme Court and a professor of civil and commercial law at the Vienna University of Economics (Wirtschaftsuniversität Wien). He has published widely in the fields of civil and commercial law, bankruptcy law and conflicts of law and international civil procedure. In addition, he has repeatedly worked as an expert in civil procedure for the Council of Europe.

Dr. Robert Kulski graduated in law from Lodz University in 1998, and in 2006 received his Ph.D. in law from Lodz University. He holds the post of the Assistant Professor at the Chair of Civil Procedure II of the Faculty of Law and Administration of Lodz University. Currently his research is focused especially on complex litigation. He has conducted research in the USA and the Great Britain.

Address:
Work
Lodz University
Faculty of Law and Administration
Chair of Civil Procedure II
Kopciński Street 8/12
90-232 Łódź

Home
Neptunowa Street 32
42–218 Częstochowa

Michele Angelo Lupoi is an Associate Professor of the Faculty of Law of the University of Bologna, where he teaches Civil Procedural Law and Comparative Civil Procedure (in English). He has written and researched extensively in matters on international and European civil procedure, matrimonial proceedings, arbitration and trust law. He is also a practising lawyer in Bologna. He may be contacted at miclupoi@tin.it.

Professional data:
Associate Professor of the University of Bologna. Practising lawyer.
Address: via Cesare Battisti 33, 40123, Bologna, Italy
miclupoi@tin.it

Prof. Dr. Dimitris Maniotis holds a J.D. Law Faculty of Aristoteles University of Thessaloniki (1976). and a Ph. D. from Law Faculty of Albert Ludwigs Universität of Freiburg i. Br., Germany, and since 1998 is a Professor of Law at the Law Faculty of Democritus University of Thrace/Komotini, Greece.

Subjects: Civil Procedural Law and Administration of Justice.

Address
Law Faculty of Democritus University of Thrace, Panepistimioupolis, 69 100 Komotini – Greece.
Private Address: Aristotelous Nr. 22 Thessaloniki 54 623 Greece.

Richard Marcus holds the Horace O. Coil (1957) Chair in Litigation at the University of California, Hastings College of the Law in San Francisco. He is the author of leading American casebooks on Civil Procedure and Complex Litigation and of several volumes of the leading American civil procedure treatise, *Federal Practice and Procedure*. He has served since 1996 as Associate Reporter of the Advisory Committee on Civil Rules of the US Judicial Conference.

Pantilimon Rikhárd-Árpád is an attorney at law, member of the National Association of Romanian Bars – Mures Bar and visiting lecturer at the Law Faculty of Sapientia University, Cluj Napoca. Rikhard is a Ph.D. candidate in law at the University of Pécs – Faculty of Law. He has a joint master degree in "Private European Union law" by the University Babes-Bolyai, Cluj Napoca and University of Pecs. He is an international mediator with significant international experience in conciliation and mediation.

Helmut Rüßmann (1943) holds the chair for civil law, civil procedural law, and jurisprudence at Saarland University

Address
Saarland University
Campus C3.1
66123 Saarbrücken

Baosheng Zhang is a Professor of Law and Vice President at China University of Political Science and Law (CUPL), Beijing, since 2004; Dean of Key Laboratory of Evidence Science(CUPL), Ministry of Education of China, since 2005. He got Ph.D. in Law at Renmin University of China, Beijing. He is a nationally recognized expert in the fields of evidence, procedure, and philosophy of law.

Office Address
China University of Political Science and Law
25, Western Tucheng Road
Haidian District
Beijing 100088, China
Tel: (+86 10) 5890 8030 (Office)
Fax: (+86 10) 5890 8029
E-mail: bensenzh@cupl.edu.cn

Home Address
1104, Building 18, Mingguangbeili Street
Haidian District, Beijing 100088, China

FSC
www.fsc.org
MIX
Papier aus verantwortungsvollen Quellen
Paper from responsible sources
FSC® C105338

If you have any concerns about our products,
you can contact us on
ProductSafety@springernature.com

In case Publisher is established outside the EU,
the EU authorized representative is:
Springer Nature Customer Service Center GmbH
Europaplatz 3, 69115 Heidelberg, Germany

Printed by Libri Plureos GmbH
in Hamburg, Germany